The Equation of Knowledge

From Bayes' Rule to a Unified Philosophy of Science

CHAPTER 19 ▪ The Bayesian Brain   355

SECTION IV Beyond Bayesianism

CHAPTER 20 ▪ It's All Fiction   375

Foreword

Arriving in a small town, with a heavy suitcase, you head for the taxi station, where only one car is parked. Unfortunately, by the time you get close to it, a faster traveller has already taken it and disappeared before your eyes. What conclusion can you draw from this misadventure? That there seem to be taxis in this city - given its size, it was far from certain? And that thus, if you wait patiently, another taxi will eventually show up? Or should you conclude that one of the few taxis in the city has just driven away from you and that, given the size of the city, such a chance will not come back soon? These two interpretations are correct, but both depend on what you knew - or believed - before you got off the train.

The traveller who arrived in an unknown city, made hypotheses about the number of taxis and revised his hypotheses according to his observations is not very different from a baby who arrives in an unknown world, or from a researcher who, surprised by what others have been taking for granted, wonders why the sun rises every morning. Both explore the world, make assumptions, and revise them based on their observations.

What can we learn from our experiences? What can we know about the world? These are the questions that the magnificent book of Lê Nguyên Hoang invites us to examine.

On these questions, one point has been crystallizing the controversies for more than a century: Is it possible to associate a numerical value to a hypothesis that measures its likelihood? For some, such as Hans Reichenbach, this is the very purpose of probability theory. In particular, any observation that confirms a hypothesis increases its probability of being true: each observation of a black raven increases the probability that the hypothesis that all ravens are black is true. For others, such as Karl Popper, the assignment of a numerical value to such a hypothesis is an illusion. By observing a black raven, we can only conclude that our hypothesis that all ravens are black remains consistent with our observations.

At the heart of this controversy is a disconcertingly simple formula, Bayes' rule, "the equation of knowledge", which gives its title to this book, and which allows computing the probability that we must attribute to a hypothesis after having made an observation - and thus makes Reichenbach right - but only on condition that we knew how to attribute a probability to this hypothesis, before making this observation - and thus makes Popper right.

If this question seemed clear-cut - in Popper's favour - in the 20th century, the evolution of data collection techniques is renewing it today. When we believed, in the 20th century, that there were white crows, we would interpret the fact that three observed crows are black as a coincidence. If we observe, today, a thousand, a million, or a billion crows, and if they are all black, it takes a certain courage - even a certain obstinacy - to claim that no, not all crows are black, and the agreement of our observations is only coincidence. At least we are forced to concede that there must be a large proportion of black crows among all crows, and probably even that white crows are the exception. This objection to Reichenbach's thesis, which constituted the problem of the hypotheses a priori highlighted by Bayes' rule, is now put into perspective by the flood of data. Other problems, however, are emerging: How were these data collected? Doesn't the collection method introduce bias or even discrimination against white crows? Once again, we see how technological developments, particularly in scientific instrumentation techniques, are changing the way questions are asked in the philosophy of science.

This is what makes Lê Nguyên Hoang's book so exciting. It has been written at the time of an upheaval, when technological developments are changing the way we look at Bayes' rule and its place in the edifice of knowledge.

It has also have been written at a time when communication techniques are changing the way we talk about science. Trained in the hard school of online videos, Lê Nguyên Hoang has found a new tone to talk about science, a tone that is both rigorous and narrative, where examples illuminate the most abstract questions.

Gilles Dowek
Research at Inria
Professor of the École Normale Supérieure de Paris Saclay

Acknowledgment

Writing this book was a formidable journey, with many ups and downs. I would not have been able to achieve it without the help, kindness, and wit of the numerous people I am highly indebted to.

I would like to thank Thibaut Giraud, Julien Fageot, Maxime Maillot, David Loureiro, and Marie Maury, whose detailed reviewing and feedback were extremely precious. More generally, I would like to thank all those who accompanied my Bayesian reflections or proofread my writing, in particular Peva Blanchard, Rachid Guerraoui, El Mahdi El Mhamdi, Alexandre Maurer, Julien Stainer, Hadrien Hendrikx, Sébastien Rouault, Matej Pavlovic, Clément Hongler, Christophe Michel, Sébastien Carassou, Mithuna Yoganathan, and Zachary Stargensky, among many others. I am also greatly thankful to the École Polytechnique Fédérale of Lausanne, the publisher CRC Press, and Gilles Dowek for being kind enough to take the time to write the foreword of this book. I would also like to thank all of those who follow me or whom I follow on social media, who all greatly contributed to my everyday desire to explore the topic of this book. I am particularly thankful towards fellow science communicator, especially on YouTube.

But above all, I would like to thank you, dear reader. The thought of sharing my Bayesian adventures with you is a wonderful source of joy and motivation. Thank you.

Preface

Logic has long been regarded as the primary foundation of knowledge. It is often said that if logic proves some fact, then this fact necessarily holds. If logic says so, then we humans should believe so.

Yet resting knowledge upon the foundations of logic is arguably flawed. Indeed, logic only handles one kind of reasoning, called *deductive reasoning*. Deductive reasoning derives conclusions from hypotheses. But much of science is actually about figuring out the adequate hypotheses, given collected data. This is called *inductive reasoning*. Unfortunately, logic does not seem fit to address this equally fundamental type of reasoning. In particular, it is not clear how logic should exploit (messy) empirical data to infer knowledge.

There is another more fundamental flaw of logic. Logic restricts itself to true and false. While this dichotomy allowed brilliant advancements in mathematics and some fields like computer science and fundamental physics, it seems very limited to make predictions in more complex settings, such as those of biology, social sciences, and everyday decision-making.

To fix these flaws, a small but growing number of mathematicians, philosophers, and computer scientists have proposed to replace logic by some other foundation of knowledge. Namely, they proposed to rest knowledge upon the laws of probability. This dramatic epistemological revolution was eventually named *Bayesianism*, after Thomas Bayes, one of the key figures in the history of probability theory.

Amazingly, Bayesianism seems to generalize all the desirable features of logic, while avoiding the pitfalls caused by its dichotomic view on knowledge and proposing a compelling framework for reasoning. To this end, it argues that knowledge should be phrased in *probabilistic* terms. In Bayesianism, nothing is known for sure. Instead, everything is a matter of *credences*, that is, confidence levels measured by probabilities with values between 0 and 1.

As data pours in, Bayesianism imposes us to *update* our credences, depending on whether the empirical data fit our theories or contradict them. Crucially, this Bayesian update is rigorously determined by a fundamental equation known as *Bayes' rule*. It is this fascinating equation that this book is about. It is this equation that we shall refer to as the *equation of knowledge.*

Disturbingly, Bayesianism is far from being consensual within the science community. In fact, most scientists probably ignore the existence of this philosophy of knowledge. Worse, some even argue against Bayesianism. Their arguments, mostly based on the *subjectivity* of Bayesian approaches, have long seemed very compelling.

In fact, for much of the 20th century, Bayesianism was very much frowned upon. It was even considered unscientific by many leading statisticians. Nevertheless, over the last few decades, an impressive variety of fields, from social sciences to biology, meteorology, and astrophysics, have been relying more and more on so-called *Bayesian methods*, to construct more precise and more predictive models.

Perhaps more impressively, the rise of artificial intelligence through machine learning and massive data has led to a formidable gain of interest in Bayesian computations. Bayes' rule can be found at the heart of numerous state-of-the-art algorithms, such as the one that reconstructed the first image of a black hole[1]. According to Stanford's philosophy encyclopedia, the empirical successes and compelling theoretical foundations of Bayesianism have recently even made it consensual among philosophers as the right philosophy of confirmation[2].

Yet Bayesianism is not perfect. In fact, there is however, one extremely compelling argument against Bayesianism. Namely, Bayesianism requires unreasonable amounts of computation. In fact, computer scientist Ray Solomonoff proved that, in arguably its purest form, Bayesianism is in fact *incomputable*. While Solomonoff also showed that any computable philosophy of knowledge was necessarily flawed (or, more precisely, *incomplete*), the incomputability of Bayesianism seems like a definite reason to give up on pure Bayesianism[3].

---

[1] *First M87 Event Horizon Telescope Results. IV. Imaging the Central Supermassive Black Hole*. The Event Horizon Telescope Collaboration (2019).

[2] The article on "abduction" says: "In the past decade, Bayesian confirmation theory has firmly established itself as the dominant view on confirmation; currently one cannot very well discuss a confirmation-theoretic issue without making clear whether, and if so why, one's position on that issue deviates from standard Bayesian thinking."

[3] In fact, as Turing showed in 1936 through the famous *halting problem*, much of mathematical knowledge is also out of our reach because of incomputability.

This is in fact what much of this book is about. After explaining the building blocks of pure Bayesianism and defending the epistemological superiority of pure Bayesianism through numerous theoretical and historical arguments, I shall argue that today's big challenge in epistemology is the design and implementation of what may be called *pragmatic Bayesianism*.

This is the quest of tractable methods to allow both computers and human brains to perform good approximations of Bayes' rule. As we shall see, the key to do so is to combine adequately computer science with probability theory. This book will discuss numerous promising approaches to do so. It will also provide an extremely wide variety of examples from very diverse fields of knowledge, to train and test our Bayesian thinking.

But perhaps most importantly, in this book, I shall stress the elegance of the properties of *pure Bayesianism*, as well as the excitement of the quest for *pragmatic Bayesianism*. The search for the most reliable paths to knowledge is arguably one of the greatest joys of being human. It is a privilege for me to share this thrilling journey with you, dear reader.

I

Pure Bayesianism

CHAPTER 1

On A Transformative Journey

The theory of probabilities is basically just common sense reduced to calculus; it makes one appreciate with exactness that which accurate minds feel with a sort of instinct, often without being able to account for it. If we consider the analytical methods that the theory gave rise to, the truth of the principles it relies on, the subtle logic that demands its application to solving problems, the public utility goods that is built upon it, and the extensions it has received and can still receive, given its application to the most important questions of natural philosophy and political economics; if we then observe that even in things that cannot be reduced to computation, probability theory allows the most reliable insights to guide us in our judgment, and that it teaches us to steer away from the illusions that often mislead us; we shall see that there is no science more worthy of our meditations, and whose results are more useful.

Pierre-Simon Laplace (1749-1827)

1.1 STUMPED BY A STUDENT

At the end of a lecture in probability and statistics I was giving at the École Polytechnique of Montreal, a trolling student came to test me with a simple-looking puzzle. A man has two kids. At least one is a boy. What is the probability that the other is a boy too?

After a few seconds of thoughts, I successfully gave the right answer, which, as we shall see, is *not* 1/2. The student acquiesced, and moved on to the next puzzle. Suppose you now learn that at least one of the kids is a boy born on a Tuesday. What is the probability that the other kid is a boy too?

This time, though, my answer was wrong. The student had stumped me.

The usual reflex is certainly to regard these two puzzles as mere mathematical games. Sure, there is a right answer. But that answer is only valid in a rigid and restricted mathematical setting. Solving these puzzles is useful in exercises or exams at school. But it's *only* mathematics.

Yet, the puzzle of the troll student is just an ultra-simplified version of many questions that we face in our daily lives. Should I believe a medical diagnosis? Is the presumption of innocence justified? Do judges racially discriminate? Is terrorism worrying? Can one generalize from one example? From a thousand? A million? Is the argument of authority worth anything? Are financial markets trustworthy? Are GMOs harmful? Why would science be *more* right than pseudosciences? Are robots about to conquer the world? Is capitalism wrong? Does God exist? What's good and what's bad?

For most, such questions have absolutely nothing to do with mathematics. And indeed, math alone is insufficient to address such questions. World hunger will not be solved by only proving theorems. Nevertheless, math likely has a lot to offer. It can help better structure our thinking, identify key challenges, and provide unexpected solutions. This is why many endeavours are more and more mathematized - including humanitary aid[1].

Despite the flourishing of mathematical models, it seems that most of us still want to distinguish the "real world" from academic courses that schools force us to take. In particular, the real world, it is often said, far transcends the framework of mathematics. As a result, mathematical theorems never seem to *really* apply to reality. How stupid must one be to think that mathematics has anything to say about the equality of rights[2]?

Sadly, rejecting the usefulness of mathematics is not merely a bad-student reflex. Even years after failing the troll student puzzle, I had not yet realized that my mathematical mishap revealed my inability to correctly reason about the real world. I had not understood that a better understanding of the puzzle would be key to better analyze my traveling friends' advice to plan my next trip - we'll get there.

---

[1] *A Set-Partitioning Formulation for Community Healthcare Network Design in Underserved Areas.* M Cherkesly, ME Rancourt & K Smilowitz (2017).

[2] *Measuring unfairness feeling in allocation problems.* Omega. LN Hoang, F Soumis & G Zaccour (2016).

1.2 MY PATH TOWARDS BAYESIANISM

Granted, I did solve the troll student puzzle later that day, after some obscure and mysterious computations. But it was only three years later, in early 2016, when I investigated the frequentist-Bayesian debate, that I really took the time to meditate about the puzzle. Most importantly, at last, I finally took it out of its confined mathematical setting.

In particular, for the three years that followed, nearly once a day, I kept thinking about the magical equation that solves this puzzle. To my greatest pleasure, this mysterious equation started to reveal its secrets to me. Slowly but surely, this brilliant equation was seducing me. I began to see it everywhere. Months after months, my mind got flooded with the sublime elegance of this untameable equation. It was too much. I *had* to write about it. And I *had* to do this well. This is how, towards the end of 2016, I began the writing of the book you have just started.

The untameable equation I am talking about is what I like to pompously call the *equation of knowledge*. But mathematicians, statisticians, and computer scientists better know it as Bayes' rule.

Bayes' rule is a mathematical theorem of remarkable simplicity. It's a compact equation, which is often taught in high school. It has a one-life proof, and only relies on multiplication, division, and the notion of probability. In particular, it seems vastly easier to learn than many other concepts in mathematics that high school and university students are asked to master.

And yet I'd claim that even the best mathematicians do not understand Bayes' rule - and there is even some mathematics that explains our inability to grasp this equation! More modestly, there is absolutely no doubt that *I* still do not understand Bayes' rule. Indeed, if I did, I would have immediately seen how the fact that at least one kid is a boy born on a Tuesday affects the likely gender of his sibling. I would have instantly given some relevant answer to the *troll* student. He would not have stumped me.

Over the last two years, I have been torturing my mind so as to never fail like this again. I want to know, understand, and feel Bayes' rule. I have already learned a lot, and I am still learning so much! I meditate on Bayes' rule almost every day, as if it were some sort of God I had to devote parts of my days to. And what a pleasure this is! Far from being a repetitive strain, these meditations have continuously fed my curiosity, as they have been discreetly whispering many of the unexpected implications of Bayes' rule. One after the other.

After long months of thinking, I ended up concluding that few ideas were as deep as Bayes' rule. I fell in love with Bayes' rule to the point where I now gladly claim that "rationality" essentially boils down to applying Bayes' rule - in which case no one is rational! This is the foundation of what might be call Bayesian philosophy, or *Bayesianism*.

1.3 A UNIFIED PHILOSOPHY OF KNOWLEDGE

Since I have not yet had the time to present Bayes' rule, for now, I will be intentionally vague about what Bayesianism is. But basically, if I had to sum it up in three clumsy phrases, I would give the following definition. Bayesianism supposes that any model, theory, or conception of "reality" is mere belief, fiction, or poetry; in particular, "all models are wrong". Empirical data must then force us to adjust the importance, or *credence*, that we assign to the different models. Crucially, the way credences are adjusted must obey Bayes' rule as rigorously as possible.

I have long rejected the relevancy of this philosophy of knowledge. It seems to discredit any concept of reality or truth, that many scientists cherish. Yet, it seems to perfectly fit what physics Nobel laureate Richard Feynman once said[3]: "I can live with doubt and uncertainty, and without knowing. I think it's much more interesting to live not knowing than to have answers that may be wrong. I have approximate answers, I have possible beliefs and different degrees of certainty about different things. But I am not absolutely sure of anything. And there are many things I don't know anything about. But I don't *have* to know an answer. I don't feel frightened by not knowing things."

You might fancy this viewpoint. Or you might want to reject altogether this approach to knowledge. Yet, before rejecting or adhering to Bayesianism, I can only encourage you to first take the time to meditate Bayes' rule and its consequences.

In this book, sadly, the main guide that I'll be has a very incomplete understanding of Bayes' rule. To help us in our thoughts, I will invoke a (female) fictitious character, the *pure Bayesian*, and we will try to imagine how this *pure Bayesian* behaves in different contexts. More than myself, it's this *pure Bayesian* that we shall put to the test. This is what we shall do again and again in this book. We shall repeat thought experiments which will be challenges that the *pure Bayesian* will have to face. And we shall carefully scrutinize, judge, and criticize the behaviour of the *pure Bayesian* - although these criticisms will often quickly turn into that of our intuition and of our relentless overconfidence.

---

[3] *The Feynman Series - Beauty*. Reid Gower (2011).

Now, the first Bayesian in history worthy of this name, the great Pierre-Simon Laplace, only had a partial description of the *pure Bayesian*. But over half a century ago, all computations, thoughts, and predictions of the *pure Bayesian* were rigorously described by the brilliant Ray Solomonoff. Unfortunately, as we shall discuss it in length, the *pure Bayesian* that Solomonoff described seems to necessarily violate the laws of physics (in particular the Church-Turing thesis[4]).

This forces us to restrict ourselves to some approximate Bayesianism, which I shall call *pragmatic*. *Pragmatic Bayesianism*, which differs from *pure Bayesianism* by its need of (fast) computability, will be incarnated by another fictitious (male) character, which I shall call the *pragmatic Bayesian*. Unfortunately (or not!), my description of the *pragmatic Bayesian* will be very incomplete, as pragmatic Bayesianism is still a huge and very open field of research - and it's not clear whether it can one day be fully closed.

As you are probably starting to guess, understanding the *pure Bayesian* and the *pragmatic Bayesian* is no easy task. To do so, we will have to discuss numerous fundamental concepts of mathematics, logic, statistics, computer science, artificial intelligence, and even notions of physics, biology, neuroscience, psychology, and economy. We will have to explain logarithms, contraposition, p-values, Solomonoff complexity, and neural networks, as well as entropy, Darwinian evolution, false memory, cognitive biases, and financial bubbles. What's more, we shall also invoke several cases from the history of science to test our two fictitious heroes.

I know. This is a lot to take in to understand Bayes' rule.

The good news is that I love explaining modern science - I have my own (French) YouTube channel called Science4All! Thus, rather than reading this book as a treatise in philosophy, I invite you to (also) read it as a science and mathematics popularisation book. In fact, on our way to Bayesianism, I will not hesitate to take some detours through the world of science, with the secret goal to tease you and make you want to find out more about scientific theories!

But let's get back to philosophy for now. As you can guess, I have surrendered to the appeals of Bayesianism. After long months of meditation, Bayesianism seduced me to the point where I felt the need to write about it. I kept being marvelled by the intelligence of the *pure Bayesian*. And I now aspire to resemble her more and more. Even long after the beginning of the writing of this book, I have kept discovering, again and

---

[4] *The Universal Turing Machine*. ZettaBytes, EPFL. R Guerraoui (2016).

again, the uncountable breathtaking wonders of what has since become my favorite mathematical equations of all.

When I started this book, I was an enthusiastic Bayesian. By now, I have become a convinced Bayesian. I would even call myself an extremist Bayesian, especially compared to others that call themselves Bayesians as well. But more importantly, I would like to become a *competent* Bayesian some day. I dream about the day I'll be able to apply Bayes' rule, as I have become convinced that this is the only way to finally be a rational being!

Ironically, the emotional momentum that Bayes' rule has given me sounds like irrational delirium. I cannot deny it. You may justifiably frown at me. You should be frowning at me. Indeed, I'm even pretty sure I am suffering from a huge cognitive bias caused by a sacralization of Bayes' rule. After all, it's impossible for me to be indifferent to the many secrets of Bayes' rule that I have managed to uncover myself - even though many others uncovered these secrets half a century before me.

Having said that, conscious of this bias, I promise I have fought - and I still do - against the *pure Bayesian*. I have kept trying to prove her wrong; I have kept trying to win a debate against her. In vain.

1.4   AN ALTERNATIVE TO THE SCIENTIFIC METHOD

In mathematics, when a conjecture seems to hold, we usually try to prove it to make it a theorem. Well, this is almost the case of Bayesianism!

As we shall see, for instance, the Jaynes-Cox theorem proves that Bayesianism is the only generalization of Aristotelian logic able to deal with plausibility in a coherent manner. Solomonoff's completeness theorem proves that the *pure Bayesian* will eventually identify all patterns in a data set. What's more, the theorem of expected gains given additional information shows that the *pure Bayesian* never loses by acquiring more data. Finally, statistical decision theory shows that Bayesian inferences are essentially the only admissible learning rules, in the sense that a learning rule is not dominated by any other, if and only if it boils down to applying Bayes' rule[5].

Many additional theorems supporting Bayesianism are unfortunately not discussed in this book. For instance, The Teller[6]-Skyrms[7] theorem

---

[5]These theorems are explained in chapters 3, 7, 9 and 12.

[6]*Conditionalisation and observation.* Synthese. P Teller (1973).

[7]*Dynamic Coherence and Probability Kinematics.* Philosophy of Science. B Skyrms (1987).

asserts that only a Bayesian is never extorted by a "Dutch book" scheme. Joyce's theorem[8] proves that we gain by making our beliefs follow the laws of probability, as prescribed by Bayesianism. Many of these theorems are nicely illustrated by the famous two envelope paradox[9].

Unfortunately though, I have had to state these theorems in a rough manner, as they rely on definitions and theorems that are hard to explain briefly. This is a major problem. In fact, any purist who wants to reject Bayesianism will be able to question and reject the hypotheses of the theorems. I do not claim that the theorems *prove* Bayesianism.

More generally, it seems impossible to *rationally* convince oneself that Bayesianism is the *right* philosophy of knowledge, the right theory of theories, or the right definition of rationality. After all, to convince ourselves of the relevancy of a concept, we need to first have in hand a philosophy of knowledge that measures the relevancy of concepts. To theorise theories, we need a theory that judges and discriminates theories of theories. To discuss rationality rationally, we need a rational definition of rationality.... We have a snake-biting-its-tail problem.

This difficulty is absolutely not specific to Bayes' rule. Any philosophy of knowledge seems doomed to suffer from self-reference. Besides, mathematicians have struggled for centuries to avoid self-referencing theories. Without much success (thanks a lot, Gödel!).

For instance, a supporter of Popper's philosophy, which is sometimes regarded as a description of the *scientific method*, will want to found knowledge on *falsifiability*. Yet, the very requirement of falsifiability does not seem falsifiable. Popper's philosophy seems inconsistent. Or, at least, it does not seem possible to accept Popper's philosophy according to Popper's criteria. This has led many to draw a line between philosophy and science, or between science or theology. This is called the *demarcation problem*. Yet, if you really think about it, this imaginary line is a pure (undesirable?) artefact of Popper's philosophy[10].

When it comes to self-consistency, the *pure Bayesian* performs better. Indeed, while she cannot prove the validity of her philosophy outside her framework, the *pure Bayesian* - for whom, as we shall see, all is belief - seems able to discuss Bayesianism without contradicting itself. Even better, I have applied Bayes' rule to my credences on Bayesianism.

---

[8] *A Nonpragmatic Vindication of Probabilism.* Philosophy of Science. J Joyce (1998).

[9] *Solve the Two Envelopes Fallacy.* Looking Glass Universe. M Yoganathan (2017).

[10] *Beyond Falsifiability: Normal Science in a Multiverse.* S Carroll (2018).

My heuristic computations have only increased my belief in Bayesian philosophy[11].

But there are two other more convincing arguments that have led me to favour Bayesianism over any other philosophy of knowledge. The first is the universality of Bayesianism. As opposed to Popper's philosophy which restricted the range of (scientific) knowledge, for instance by insisting on the reproducibility of scientific experiments[12], Bayesianism has no restriction on its range of applicability. Any phenomenon, whether it belongs to sociology, history, or theology, can be analyzed through the prism of Bayesianism. Bayesianism is a *universal* philosophy of knowledge.

The second argument consists of the rigor, the concision, and the clarity of Bayesianism. Indeed, Bayesianism defines inference rules[13] so precise that applying (even approximately) these rules seems to be sufficient to learn "well enough" about the world. This is a computer scientist's dream. The computer scientist would then only have to push the start button to enable the machine to reach its goal by simply following instructions. Of course, this is above all a description of artificial intelligence! And it's definitely not an accident if, over the last three decades, Bayes' rule has been at the heart of many research breakthroughs in this domain.

This rigidity of Bayesianism heavily contrasts with the *malleability* of most common versions of the scientific method. Indeed, many approaches often consider a sort of statistical toolbox, from which statistical tools may be (cherry-)picked for data analysis. Unfortunately, it has been argued that this allowed scientists to bias their conclusions[14], especially under publishing, financial, or ideological incentives. This has been argued to have led to a blow-up of misleading scientific publications.

More recently, through the work of researchers like Josh Tenenbaum, Karl Friston, and Stanislas Dehaene, Bayesianism has also become an essential theoretical framework to understand how our own intelligence works. In particular, in 2012, Dehaene gave a series of lectures

---

[11]This is related to so-called *hierarchical* Bayesian models, which we shall discuss in chapter 19. Note, however, that the *pure Bayesian* cannot technically discuss pure Bayesianism, at least in Solomonoff's setting, since she should only consider *computable* theories. But as she shall discuss in length in chapter 7, *pure Bayesianism* is *incomputable*.

[12]Reproducibility can be seen as a condition imposed by frequentism.

[13]We shall soon see what this means.

[14]*Medical Nihilism*. Oxford University Press. J Stegenga (2018).

at the prestigious Collège de France entitled *The statistician brain: The Bayesian revolution in cognitive sciences.* "Many biologists are skeptical with the idea that, in neuroscience, there may be general theories", he said. "[But] it really seems that [Bayesianism] yields a theoretical framework which can be applied in an extremely general manner [...] The very existence of general patterns in the architecture of the brain seems to be explained by the hypothesis [according to which] the brain is organized to compute statistical Bayesian inferences."

(Pragmatic) Bayesianism seems to be Nature's solution to natural intelligence[15].

1.5   THE OBJECTIVITY MYTH

Mysteriously enough, though, Bayesianism has long been rejected by several generations of first rank scientists. Why is that? Were the great scientists irrational? What caused the rejection of Bayesianism? And if this rejection is unjustified, what was the fallacy of these great scientists?

It turns out that the two centuries of epistemological war that this book hopes to put an end to boil down to the concept of *objectivity.* Better, the opposition between *subjective Bayesians* and *objective frequentists* can be summed up by the following questions: *What is a probability?*

I have a personal connection to this fascinating question. It was given to me in an oral exam for the entrance to the École Normale Supérieure (ENS) in Paris. This exam was supposed to be the presentation of a year-long project. I was quite proud of mine. I had modelled soccer games, estimated the levels of teams, and simulated different competitions[16]. In particular, based on two years of sports results, my simulations concluded that Portugal, France, and Italy were the three main favorites of the 2006 World Cup. Their probabilities of winning were 20%, 15%, and 10%. Not bad, given that the three teams would end up, respectively, 4th, 2nd, and 1st in the competition!

Examiners of the École Centrale and the École des Mines really enjoyed my work. They gave me 19 points out of 20. However, my simulations did not get the ENS examiners excited. They quickly stopped me. What they wanted to know is whether I could define what probabilities are.

---

[15] *Les algorithmes du vivant.* TEDxSaclay. LN Hoang (2018).

[16] *A model of football games.* Science4All. LN Hoang (2013).

My answer was frequentist. I claimed that the probability of an event was its limit frequency, when an experiment is repeated an infinite number of times. In particular, any empirical frequency would thus be an approximation of some fundamental and objective probability. Frequentists or not, the purists at ENS did not appreciate my efforts. They expected me to rediscover a mathematical definition of probabilities, for instance as unitary measures of sigma-algebras. I got 6 out of 20.

But let's not mourn my fate. Let's focus on what the *pure Bayesian* would call a naive mistake.

I was born a frequentist. I grew up searching for truths, whether mathematical or scientific. I accepted the existence and superiority of *objective* results. Even in 2013, when the troll student challenged me, the major part of the course I was teaching was essentially frequentist - and I thought that these were the *right* statistics to teach! Besides, my own model of soccer games was a classical example of a frequentist approach which, like Stein paradox[17], would have gained by acquiring some Bayesian flavour.

But crucially, the very nature of the probabilities I was manipulating could not be frequencies! The frequency with which France wins the 2006 World Cup is not 15%. This frequency is 0. There has been and there will have been only one 2006 World Cup. And France lost it.

But if the 15% predicted by the model was clearly not a frequency, what is it? Can we still say that it is a probability?

Yes, says the *pure Bayesian*. It's the probability that France wins the World Cup *according to the mathematical model*. In particular, this probability is *subjective*; it's the opinion of the model. But crucially, all probabilities are like this. According to the *pure Bayesian*, no probability is *objective*; and whoever disagrees confuses his subjective desires with a reality to force upon others. Probabilities are *model-dependent*.

Think about it. Any method to search for and organize knowledge is doomed to be biased by the mere choice of this method rather than another - especially if one starts invoking the imprecise Ockham razor, already "established" scientific knowledge, or the very problematic *p-values*. Worse, the way we look at, manage, and select our data inevitably biases the conclusions derived from the data analysis. As we shall discuss in length, facts are often incredibly misleading[18].

---

[17]We shall discuss Stein paradox in chapter 13.
[18]*How statistics can be misleading*. TED-Ed. M Liddell (2016).

What's more, the explicit mention of the method that was followed is insufficient. As data scientists using *machine learning* to extract useful information from *Big Data* quickly learn, the absence of human intervention is absolutely no guarantee of objectivity. Humans or machines, it seems that we always *have* to reason within our models of the world. This shows, the *pure Bayesian* claims that knowledge is necessarily *subjective*. It depends on the algorithm used to compute that knowledge.

This should make you feel uneasy. Bayesianism seems to lead to relativism. If all knowledge is subjective, does this mean that all opinions equally matter? Of course not. We may each see our own red; this does not mean that all opinions about the presence of red in the US flag are equally reliable.

In particular, those who apply Bayes' rule to the same data will end up giving their credences to the same models, especially if there are lots of data. What's more, even with relatively few data, the models that will win the credences of Bayesians will be way more relevant and *useful* than the favorite models of those who, exposed to the same data, have not applied Bayes' rule.

Note that Bayesianism (especially its pragmatic version) is not a substitute to modelling; it is rather a meta-model whose purpose is to discern useful models. The foundation of Bayesianism is in fact very well summed up by Bayesian George Box's holy quote: "All models are wrong, some are useful". I will often be using it! Whether this quote is "true" or not, I've found it incredibly useful to shortcut endless debates which seemed doomed to go nowhere - and thus to greatly bore me.

The *pure Bayesian* much prefers to judge the *usefulness* of models. Especially their *predictive* usefulness. Not their truth. Yet, according to her, judging adequately the usefulness of models can only be done through Bayes' rule.

1.6   THE GOALS OF THE BOOK

While I do intend to share and explain my enthusiasm for Bayesianism, and while I have the secret hope that this will make mathematicians, philosophers, and scientists question what they thought they knew about their disciplines, the goal of this book is actually not to convert you to Bayesianism. What I would like above all is to share with you some of the marvels that have made me fall in love with Bayesianism. And I am willing to bet - a typical Bayesian reflex - that you will be surprised and, I hope, seduced, by the many astounding consequences of Bayes' rule, as well as by its ubiquity in applied mathematics, in our own intuitive thinking, and in the organization of our societies.

Bayesianism explains why the scientific community is far more reliable than any of its members and why our brains are constantly victims of the anchor effect. It explains why it's more desirable to combine incompatible models and why Ockham's razor is an essential tool. It could even be the key to understanding the working of our memory and the usefulness of our dreams. Just as biologist Theodosius Dobzhansky once asserted that "nothing in biology makes sense except in the light of evolution", I would claim that a spectacularly wide range of mechanisms can only be understood through Bayesian lenses.

My discovery of Bayesianism has also been for me the chance to finally measure the extent of my ignorance. This is thanks to the language of probability theory that allows to quantify uncertainty. But most importantly, my inability to apply Bayes' rule, even in the simplest cases like the troll student puzzle, has forced me to acknowledge how bad I am at thinking. I have often had an irrational and unjustified confidence in my intuition, sometimes accompanied with a mysterious distrust of Bayes' rule. But after losing so many debates against the *pure Bayesian*, my Bayesian journey has forced me to acknowledge my unwavering overconfidence. In fact, this will be a major objective of this book. We will fight our overconfidence and measure the extent of our ignorance.

The rest of the book is roughly divided in four parts. In the first part, from chapters 2 to 7, we will straightforwardly tackle Bayes' rule and *pure Bayesianism*. Next, chapters 8 to 13 will reveal the hidden and unexpected presence of Bayesian principles in many phenomena. Then, chapters 14 to 19 will study *pragmatic Bayesianism* and its essential tools. Finally, the last three chapters will be a bit different. Chapter 20 will analyze the antirealist consequences of Bayesianism. Chapter 21 will track down the origins of my personal beliefs to better question our widespread overconfidence. Last but not least, chapter 22 will study consequences of Bayesianism on moral philosophy.

Unfortunately, this book, like any finite book, is absolutely not exhaustive. I apologize for its uncountable deficiencies. In particular, I will not take the time to compare in details Bayesianism to alternative philosophies of knowledge. My goal is more modest: I would like to help you understand the key aspects of Bayesianism. Or, at least, of what I understood. Indeed, like this book, my brain is finite too. Please forgive the extent of my ignorance. I will try to mention all the Bayesian reasonings I find worth mentioning, but I will necessarily omit what I do not know and what I have mistakenly considered unimportant.

In addition to my cognitive limitations, the depth of the book will also be limited by my desire to make it accessible to a wide audience. No prerequisite is assumed. As a result, I will not be nearly as rigorous as the *pure Bayesian* would want me to be - although I will do my best so as not to lead you to any misinterpretation. This is a popularization book.

Nevertheless, there is a good chance that you will not understand everything. Since I really wanted to present some of the most convincing arguments in favor of Bayesianism, I have chosen to provide sections of high mathematical level. These sections are "starred". Be warned. Even doctors in mathematics will struggle to understand all the notions of this book.

Do not rush the reading of the book. Take the time to ponder it. But do not give up either. The book is *not* increasingly difficult. You should be able to find pleasure in any chapter without having read the previous ones - even though it's probably better to read the chapters in the right order. This is not the textbook of a course. There will be no exam. You do not have to understand it all. I would even advise you to skip difficult paragraphs and carry on the reading. My goal, after all, is absolutely not to transform you into experts of Bayesianism.

What I would like above all is for you to search for beauty and pleasure in Bayesian thinking, as well as in the sciences useful to understand and illustrate Bayesianism. I would like you to behave like an explorer who has just arrived in unchartered territory and has plenty of intriguing wildlife, landscapes, and cultures to discover; and who will not necessarily spend much time learning all the subtleties of the language of local people. I would like you to enjoy *your* journey.

If you roughly follow my footsteps, I hope to fill you with enthusiasm, fascination, and questioning. This is the main goal of the book.

FURTHER READING

Probability Theory: The Logic of Science. Washington University. ET Jaynes (1996).

Rationality: From AI to Zombies. Machine Intelligence Research Institute. E Yudkowsky (2015).

Thinking in Bets: Making Smarter Decisions When You Don't Have All the Facts. Portfolio. A Duke (2018).

Medical Nihilism. Oxford University Press. J Stegenga (2018).

Conditionalisation and Observation. Synthese. P Teller (1973).

Dynamic Coherence and Probability Kinematics. Philosophy of Science. B Skyrms (1987).

A Nonpragmatic Vindication of Probabilism. Philosophy of Science. J Joyce (1998).

Beyond Falsifiability: Normal Science in a Multiverse. S Carroll (2018).

A Set-Partitioning Formulation for Community Healthcare Network Design in Underserved Areas. M Cherkesly, ME. Rancourt & K Smilowitz (2017).

Measuring Unfairness Feeling in Allocation Problems. Omega. LN Hoang, F Soumis & G Zaccour (2016).

The Feynman Series - Beauty. Reid Gower (2011).

Bayes: How one Equation Changed the Way I Think. J Galef (2013).

Think Rationally via Bayes' Rule. Big Think. J Galef (2013).

Solve The Two Envelopes Fallacy. Looking Glass Universe. M Yoganathan (2017).

How Statistics Can Be Misleading. TED-Ed. M Liddell (2016).

The Universal Turing Machine. ZettaBytes, EPFL. R Guerraoui (2016).

Les Algorithmes du Vivant. TEDxSaclay. LN Hoang (2018).

A Model of Football Games. Science4All. LN Hoang (2013).

CHAPTER 2

Bayes' Theorem

One of the biggest paradigm shift in my thinking, and also for a lot of people I know, has been learning about Bayes' rule.

Julia Galef (1983-)

We are spontaneously irrational creatures unable to correctly revise our beliefs, and understanding Bayes' rule can really help us improve.

Thibaut Giraud (1986-)

2.1 THE TROLL STUDENT PUZZLE

Let's get back to the troll student puzzle. A man has two kids. At least one is a boy. What is the probability that the other is a boy too? I invite you to try to solve the problem by yourself. Even if you do not succeed, the intellectual effort you'll put in will likely be useful for the sequel.

I'll now discuss the solution. The simplest approach to solve the puzzle consists of listing all possible cases. Let's call Alex and Billie the two kids. There are four possibilities:

- Alex and Billie are boys.

- Alex is a boy. Billie is a girl.

- Alex is a girl. Billie is a boy.

- Alex and Billie are girls.

These four possibilities seem equally likely *a priori*, that is, before we learn that at least one is a boy. Actually, they aren't. Biologists would add that 51% of newborn babies are actually boys - a discovery made by Laplace based on Bayesian computations! But let's simplify and assume that the probability that a kid is a boy is *a priori* 50%. Also we will assume that this probability is independent from the sibling's gender.

But by now, we have learned that at least Alex or Billie is a boy. The three first cases are consistent with this added information. But the fourth is not. We can thus safely strike the fourth case.

But now, given that Alex or Billie is a boy, asserting that the other is a boy too corresponds exactly to saying that both Alex and Billie are boys. This is the only case where at least one kid is a boy, and the other is a boy too. In other words, what we want to compute is the same as the probability that both kids are boys, given that at least one is.

This is only one out of the three remaining cases. Thus, the searched probability is $1/3$. Not $1/2$. Surprised?

The first time I saw this proof - it was way before the troll student challenged me with it - I remember I was not convinced. Our reasoning is not that *clearly* valid. Can we really strike the fourth case, and still consider that the first three cases remain equally likely?

I could get you out of this mess and give you the right way to address this problem right now - which, of course, relies on Bayes' rule. But I think it's worth struggling for now.

2.2   THE MONTY HALL PROBLEM

Let's move on to the Monty Hall problem. This grand classic of probability theory is inspired from a TV game called *Let's Make a Deal* which was hosted by presenter Monty Hall in the 1960s. At the end of the game, a player had to choose one of three curtains. There was a car behind one of the curtains, and goats behind the two other curtains. Once the player chose a curtain, Monty Hall built up suspense. Behind one of the two curtains that the player did not choose, there is a goat. Monty Hall reveals one such curtain.

There are now two curtains left. One hides a car, the other a goat. Monty Hall then proposes to the player to switch curtains. Should the player stick to his initial gut feeling, or should he be willing to change his mind?

Like in the troll student problem, it seems that, here again, we only struck one of the possibilities. It's thus tempting to think that the two remaining curtains remain equally likely. It's tempting to believe that changing curtains or not makes no difference.

If this is what you think, then you are making a mistake that numerous first-rank mathematicians did before you. No need to feel shame! The Monty Hall problem has confused a lot of very smart people. In 1990, when Marilyn vos Savant proposed a correct solution to this problem in the journal *Parade*, 10,000 readers, including 1,000 PhD graduates, wrote to the journal and asserted that vos Savant got it wrong.

Even the world-class mathematician Paul Erdös, the man who has published the most in the history of mathematics, refused to believe vos Savant's rigorous proof. It's only when faced with simulation results that, to his dismay, Erdös reckoned he was wrong. The great Erdös did not understand Bayes' rule. He was not alone.

I was 13 when I first discovered the Monty Hall problem. I did not know Bayes' rule. Nevertheless, there was a convincing reasoning which was accessible to me. Indeed, if you know ahead of time that you will not switch curtains, then all happens as if Monty Hall did not build suspense by revealing a goat curtain. The probability of finding the car would thus be the probability that the curtain you initially chose hid the car. This probability is 1/3. Your chance of winning when not switching curtains is one third. Weirdly enough, though, while I was quite convinced by this reasoning, I was still unable to determine the probability of winning by switching curtains.

Yet, if you lose by keeping your curtain, it means that the other curtain hides the car. After all, this other curtain is the one that Monty Hall suspiciously left unrevealed. In fact, what happens is that 2 out of 3 times, there is a goat behind your initial curtain. Whenever this is the case, once there are two curtains left, the car will necessarily be behind the other curtain. You then win by switching curtains. Two times out of 3.

The math is indisputable. You double your chances of winning by switching curtains! The *pure Bayesian* would win twice as often as he who has not thought through the problem carefully and sticks to his initial choice.

If you are still not convinced by this reasoning, I invite you to do the experiment yourself, as Erdös did. In an excellent BBC documentary, the mathematician Marcus du Sautoy posed the Monty Hall problem to comedian Alan Davies. Alan Davies thought he had his chances in a repeated Monty Hall game with no curtain switching, as opposed to du Sautoy who always switched curtains. After 20 attempts, Davies only won twice. Du Sautoy won 16 times. Granted, these figures do not match the 1/3 and 2/3 of the Bayesian theory - this is an instance of the law of small numbers! - but this definitely helped Alan Davies realize he was wrong. And that he should have paid more attention to du Sautoy's explanations.

In this case, Alan Davies only lost a bit of pride. He might have lost some more pride had he known that pigeons were able to faster understand the Monty Hall problem[1]. But, sometimes, the misunderstanding of Bayes' rule can have dramatic consequences. Sally Clark learned it in the most tragic way there is.

2.3 THE TRIAL OF SALLY CLARK

In 1996, Sally Clark's firstborn baby died after two weeks. A year later, this happened again with her second baby. Sally Clark got sued for double homicide. The pediatrician Sir Roy Meadow was asked to testify. He asserted that the probability of two infant deaths by natural cause was one out of 73 million. This testimony led to the conviction of Sally Clark.

However, three years later, the mishap of Dr. Alan Williams, who was in charge of the autopsy, was exposed. He failed to report the conclusions of his analyses. The second baby did in fact die of a natural cause. Sally Clark was released. But not without consequences. She suffered from major psychiatric issues and died four years later by alcohol poisoning.

Apart from the misstep of Dr. Williams, the cause of Sally Clark's poor fate was a misunderstanding of Bayes' rule, known as the *prosecutor's fallacy*. The judge - and maybe you too - confused the probability of a double death by natural cause with the probability of Sally Clark's innocence. Yet, the rarity of an incriminating evidence is not necessarily incriminating.

We'll come back to the prosecutor's fallacy in great length later on, as it is a fallacy that can be found in most of the classical interpretations of the scientific method. However, it's good to start discussing it right now. The rarity of an incriminating evidence may simply be a consequence of the rarity of the suspect's case. As a matter of fact, Sally Clark's case is exceptionally rare. Therefore, the probability of a double death by natural cause, which is necessarily smaller than the probability of Sally Clark's case, *has* to be small too[2].

---

[1] *Are Birds Smarter Than Mathematicians? Pigeons (Columba livia) Perform Optimally on a Version of the Monty Hall Dilemma.* Journal of Comparative Psychology. W Herbranson & J Schroeder (2010).

[2] Indeed, double deaths by natural cause form a subset of all double deaths. Yet, the probability of a subset is always smaller than that of the whole set.

In fact, approximate computations by mathematics Professor Ray Hill of Salford University showed that the probability of a double death by natural cause, however small, still remains 5 to 10 times larger than the probability of a double homicide. In other words, Bayes' rule forces us to nevertheless prefer the hypothesis of natural deaths over that of double homicide - Hill also pointed out a gross mistake in pediatrician Meadow's computations which did not account for the correlation between the death of the two newborn babies, which made the Bayesian conclusion even more favorable to the natural death hypothesis.

2.4   THE LEGAL CONVICTION OF BAYESIANISM

Sally Clark's trial has been regarded as one of the worst mistakes in the history of British justice. Yet, rather than starting a mass education of Bayes' rule, this tragic story led to an increased mistrust of statistics. In 2010, a UK judge even forbade its use in courts of law[3]. Bayesianism has become illegal! The equation of knowledge is prohibited by justice!

But it's hard to blame the judge. Given Erdös' struggle with Bayes' rule, should we really expect judges and jurors to found their reasonings on this equation?

While the *pure Bayesian* can navigate through the complexity of the judiciary system, it's certainly important to avoid flooding courthouses with misleading statistics that no one can interpret correctly. In a paper published in *The Annual Review of Statistics and Its Applications*, Fenton, Neil, and Berger also add that "there is another rarely articulated but now dominant reason for [Bayes' rule's] continued limited use: it is because most examples of the Bayesian approach have oversimplified the underlying legal arguments being modelled in order to ensure the computations can be carried out manually". Aware of this difficulty, the authors propose a more sophisticated Bayesian approach, based on Bayesian networks, which we'll get to later on. For now, it is worthwhile to observe the irrationality of the judiciary system and the extreme difficulty to correct it.

The oversimplified Bayesian computation is absolutely not an always-satisfying answer. And the correct Bayesian computation is far too complex to be carried out, even by the best of us. The goal of this book is thus absolutely not to transform you into Bayesian computing machines. It would be a lost battle.

---

[3] *A formula for justice.* The Guardian. A Saini (2011).

However, I believe that the simple examples can definitely be understood by all, providing you really take the time to ponder them. Such examples can then serve as reference points or as training for approximate Bayesian thinking in more pragmatic cases. I won't mold you into a *pure Bayesian*. But I hope to make you a better thinker. I aim to help you make your intuitions more Bayesian.

I also hope to convince you of our limits and our inability to think in a purely Bayesian manner - yet, as we shall see, the *pure Bayesian* regards Bayesian thinking as the *only* rational thinking. The realization of our struggles to apply Bayes' rule, as well as of its key role in epistemology, should undoubtedly make us more willing to doubt any of our convictions. This, I think, is the right reaction when discovering how bad we are at thinking. We should accept the fact that we are constantly overconfident. We must decrease our credences in our intuitions and in our non-Bayesian thoughts.

2.5 BAYES' THEOREM

Enough chatting. It's time to tell you about my favourite mathematical equation. Let's present Bayes' rule. To do so, we'll consider a fourth example.

Imagine that a test asserts that you have Ebola. Since you have just come back from Nigeria, this surely sounds bad. But tests are never fully reliable. So you ask about the reliability of this particular test. You are told that when the patient is healthy, the rate of correct tests is 90%. Should you start writing your will?

The *pure Bayesian*'s answer is pretty univocal: no. Even in sub-Saharan Africa where this terrible disease was most deadly, no more than one person out of 10,000 died of Ebola. Thus, a priori, you who only had a brief stay in Nigeria do not have more than one chance out of 10,000 to be sick with Ebola. This probability is succinctly denoted $\mathbb{P}[●]$. It is called the *prior*.

Suppose now that you learn that the test concludes that you are sick[4]. What you should care about now is the probability that you have Ebola, given that the test is bad news. We will write it as $\mathbb{P}[●|🏴]$, where 🏴 denotes the fact that the test yields bad news. By opposition, we shall

---

[4]This is usually called a positive test, but the terminology may be confusing, so I'll avoid it.

use the symbol 👍 to denote the case where the medical test indicates that you do not have Ebola.

What does the so-called conditional probability $\mathbb{P}[😊|😟]$ mean? The theory of probability postulates that this conditional probability is related to the probabilities of events 😊 and 😟 as follows:

$$\mathbb{P}[😊|😟] = \frac{\mathbb{P}[😊 \text{ and } 😟]}{\mathbb{P}[😟]}.$$

In other words, the probability of being sick given that you have a bad-news test result is the ratio of cases where you are sick and have bad news, among all the cases where you end up with bad news.

Note that, these days, even the most anti-Bayesian statistician accepts this postulate. In fact, this is just a definition of conditional probability. Like any definition, it cannot be *false*. However, it's questionable whether it's a *relevant* (and useful) definition of conditional probabilities. In particular, does it match natural language? And is it the right way to think?

The Bayesian's leap of faith is that not only is this definition close to natural language, but also and most importantly, it's *the* right way to think. *To be Bayesian is to rest all knowledge upon the language of conditional probabilities.*

Following the *pure Bayesian*'s footsteps, let's accept that the conditional probability $\mathbb{P}[😊|😟]$ is indeed the probability of being sick given that the test indicates so. However, this is not the quantity that was communicated to you. The 90% figure you were given is the probability of a correct test if you do not have Ebola. In other words, 90% is the probability $\mathbb{P}[👍|😇]$ to have a good-news test result, given that you are healthy (the symbol 😇 indicates your healthiness). The 10% left correspond to the probability $\mathbb{P}[😟|😇]$ of a bad-news test given that you are healthy.

To determine the probability of being sick given a bad-news test, we need to prove and use Bayes theorem. To do so, let's write the definition of $\mathbb{P}[😟|😊] = \mathbb{P}[😟 \text{ and } 😊]/\mathbb{P}[😊]$. Do you notice it? The numerator is the same as in the definition of the inverse probability $\mathbb{P}[😊|😟]$. In particular, the probability of the two simultaneous events is $\mathbb{P}[😟 \text{ and } 😊] = \mathbb{P}[😊]\mathbb{P}[😟|😊]$. In other words, the probability of being sick and having a bad-news test is equal to the probability of first getting sick, and then obtaining a bad-news test when sick.

We are almost done with the proof of Bayes theorem. It now suffices to use the equation above in the definition of the conditional probability

$\mathbb{P}[😷|🌡]$. We obtain the most important equation of the philosophy of knowledge that this book presents, also known as *Bayes' rule*. Please take the time to notice its calligraphic elegance and the pattern that the symbols follow.

$$\mathbb{P}\left[😷 \mid 🌡\right] = \frac{\mathbb{P}[🌡|😷]\ \mathbb{P}[😷]}{\mathbb{P}[🌡]}.$$

In words, the probability of having Ebola given a bad-news test is derived by multiplying the probability of a bad-news test if sick (which requires some thinking!) by the prior probability of being sick, divided by the probability of a bad-news test.

As announced in the first chapter, all you need are multiplications and divisions! How simple is that?

Of course, what makes this equation hard is not the computations it requires, but rather its interpretation - at least in the simple examples of this chapter. It's extremely easy (and tempting!) to misinterpret one of the terms of the equation. I strongly encourage you to take the time to think them through.

2.6   THE COMPONENTS OF BAYES' RULE

In the right expression, the probability $\mathbb{P}[😷]$ is the *prior*, sometimes known as the *base rate*. It's what we could (or rather, should) believe before learning the result of the test. In our case, we estimated it by comparing the number of known Ebola victims to the population of sub-Saharan countries. But this is merely a rough estimate. Besides, we did not even account for the length of your stay, which surely is a major thing to take into consideration to estimate the prior. Just as important is the frequency of interactions with local people, as well as exposure to sick individuals. All these effects are incredibly hard to quantify. We'll work with our rough estimate here.

The other quantity in the numerator in the right expression is the probability $\mathbb{P}[🌡|😷]$ of a bad-news test given that we are sick. This term requires a bit of imagination. It requires us to leave the real world to imagine an alternative one where we definitely got infected by Ebola. In this alternative world, would we get a bad-news result to the test? This is the question that $\mathbb{P}[🌡|😷]$ answers.

Contrary to us, the *pure Bayesian* not only can think along the lines of others' worldview, it's what she does day in day out! This is the art of *thought experiments*. These are essential components of Bayesianism.

Without such thought experiments, it would be impossible to estimate terms like $\mathbb{P}[\text{🏴}|\text{☻}]$. It would be impossible to apply Bayes' rule. According to the *pure Bayesian*, it would be irrational to dismiss thought experiments derived from alternative theories.

Unfortunately, many of us too often refuse to temporarily accept the counterintuitive premises of a theory to explore its consequences. What rather too often happens is that debates will be opposing sides who see the world only through the prism of their epistemology, their worldview, their theology, or their moral principles. Without any effort to start a reasoning on some common ground (and then doing the same for some other common grounds), such debates are doomed to degenerate. Their fallacy is to skip the computation of thought experiment terms like $\mathbb{P}[\text{🏴}|\text{☻}]$. But without such terms, Bayes' rule can't be applied.

Such quantities are usually called *likelihoods* by statisticians. However, this terminology can be misleading. It's important not to forget that the likelihood is the probability of the observed data according to some worldview, which is very distinct from the likelihood of the worldview itself. As I will often repeat it, *the likelihood of the data is not the credence of the theory*. To avoid confusions, I'll rather talk about *thought experiment terms* - although I'll sometimes conform to the commonly accepted terminology.

Last is the denominator of the right expression, that is, the probability $\mathbb{P}[\text{🏴}]$ of a bad-news test. This is a monster term. It's the greatest difficulty of Bayes' rule. It's this term that keeps probability researchers awake at night (or their artificial intelligences). It is called the *marginal*, or the *partition function*. It's this horror that is hardest for me to visualize and understand - although, in the simple cases of this chapter, this quantity turns out to be relatively simple to compute.

To compute the probability $\mathbb{P}[\text{🏴}]$ of a bad-news test, we can distinguish two cases. The first is the case where the test is bad news because we have Ebola. The second is when the test is bad news because the test got it wrong. In each of the two cases, we need to multiply the prior probability by the probability that this prior (having or not Ebola) leads to bad news. In other words, we will use what is known as the *law of total probabilities*[5]:

---

[5]Like Bayes' rule, the law of total probabilities can be derived from the definition of conditional probabilities, in addition to the fact that the probability of incompatible events is the sum of the probability of the events.

$$\mathbb{P}[\pmb{?}] = \mathbb{P}[\pmb{?}|\pmb{\bullet}]\,\mathbb{P}[\pmb{\bullet}] + \mathbb{P}[\pmb{?}|\pmb{\odot}]\,\mathbb{P}[\pmb{\odot}].$$

In particular, the *partition function* $\mathbb{P}[\pmb{?}]$ requires the computation of two *thought experiment terms* that correspond to two different cases. This is what makes its computation so hard. It requires the ability to perform the subtle cerebral gymnastics of combining different reasonings in mutually incompatible versions of reality. This is probably what makes Bayes' rule so hard to apply and understand.

2.7   BAYES TO THE RESCUE OF DIAGNOSIS

We can now finally combine the law of total probabilities to Bayes' rule, yielding

$$\mathbb{P}[\pmb{\bullet}|\pmb{?}] = \frac{\mathbb{P}[\pmb{?}|\pmb{\bullet}]\,\mathbb{P}[\pmb{\bullet}]}{\mathbb{P}[\pmb{?}|\pmb{\bullet}]\,P[\pmb{\bullet}] + \mathbb{P}[\pmb{?}|\pmb{\odot}]\,\mathbb{P}[\pmb{\odot}]}.$$

We now know nearly all quantities of the right equations. We saw that $\mathbb{P}[\pmb{\bullet}]$ could be estimated by 1 out of 10,000 (this is actually an upper bound), from which we derive the fact that $\mathbb{P}[\pmb{\odot}]$ is (at least) 9,999 out of 10,000. Moreover, we notice that $\mathbb{P}[\pmb{?}|\pmb{\odot}]$ corresponds to the error rate of the test on healthy patients. We saw that this is 10%. Finally, the term $\mathbb{P}[\pmb{?}|\pmb{\bullet}]$ is the reliability of the test on sick patients. Let's just note that this term is a probability, and is thus at most 1. Combining it all, we conclude with the following computation:

$$\mathbb{P}[\pmb{\bullet}|\pmb{?}] \approx \frac{1 \cdot 0.0001}{1 \cdot 0.0001 + 0.1 \cdot 0.9999} \approx 0.001.$$

In other words, now that you know the result of the test, you are estimating the probability of having Ebola at 1 out of 1,000. This probability is still very small. In comparison, in the US, the probability of dying from a traffic accident in a year is estimated to be around 1 in 10,000. Surely, you do not need to write your will just yet.

But what happened? Why on earth is the final result so low? How can we guide our intuition to help it better understand the result without blindly trusting the computations? I highly invite you to ponder this yourself.

It's worth focusing on the partition function $\mathbb{P}[\pmb{?}]$ in the denominator. We saw that this quantity could be decomposed into two cases, depending on whether we have Ebola or not. The two cases are absolutely not equally likely. In fact, the case where we have Ebola is 1,000 times less likely than the case where we had a bad-news test because of an error of the test. The difference between these two explanations is so large that we can basically neglect the former case when computing the partition function.

As a result, Bayes' rule is merely a ratio between an unchanged numerator, which is about having Ebola, and a denominator, which is about having a wrong bad-news test[6]. Bayes' rule compares these two alternatives. The smallness of the final result corresponds to saying that the case of having Ebola is much less likely than the case of a wrong bad-news test.

Another useful interpretation of Bayes' rule is as a transfer of credences. Theories whose *thought experiment terms* are small will lose credences to theories whose *thought experiment terms* are large. In our case, $\mathbb{P}[🤮|😊] = 10\,\%$ is 10 times smaller than $\mathbb{P}[🤮|☻] \approx 1$. The credence of ☻ will thus increase by stealing some of 😊. It will gain a factor 10 with respect to 😊. However, *a priori*, ☻ was around 10,000 times less likely than 😊. Therefore, *a posteriori*, ☻ will be around 1,000 times less likely than 😊[7].

In practice, these days, doctors will try to protect you from an unnecessary scare, by combining several (if possible) independent medical tests, and will give you the results only if the combination of all tests concludes that you really are likely to have Ebola. In particular, this means that doctors will try to decrease as much as possible the probability $\mathbb{P}[🤮|😊]$ of a bad-news test for healthy patients.

2.8   BAYES TO THE RESCUE OF SALLY CLARK

To better understand the intriguing Bayes' rule, let's now apply it to Sally Clark's case. Recall that her two babies died. She was then convicted for double homicide. The relevant quantity is the probability of her innocence, given that her two babies died. Here is how Bayes' rule applies here:

$$\mathbb{P}[😇|⚰] = \frac{\mathbb{P}[⚰|😇]\mathbb{P}[😇]}{\mathbb{P}[⚰|😇]\mathbb{P}[😇] + \mathbb{P}[⚰|😈]\mathbb{P}[😈]}.$$

---

[6]This is usually called a "false positive".

[7]Formally, what I described corresponds to the following computation of so-called *odds*:

$$\frac{\mathbb{P}[☻|🤮]}{\mathbb{P}[😊|🤮]} = \frac{\mathbb{P}[🤮|☻]}{\mathbb{P}[🤮|😊]}\frac{\mathbb{P}[☻]}{\mathbb{P}[😊]} = 10 \cdot \frac{\mathbb{P}[☻]}{\mathbb{P}[😊]}.$$

Like earlier, there are three key quantities we need to think through: the *prior* $\mathbb{P}[\text{😇}]$ on the innocence of Sally Clark, and the two competing *thought experiment terms* $\mathbb{P}[\text{⚰}|\text{😇}]$ and $\mathbb{P}[\text{⚰}|\text{😈}]$ which predict the probability of double deaths depending on whether Sally Clark is innocent.

Let's start with the prior. It must be large. This prior is the probability that a person drawn randomly has not killed its two babies. Yet, almost all human beings have not killed their babies! In fact, Doctor Hill estimated the prior probability of Sally Clark's guiltiness at around one out of 500 million.

This remark explains why our societies value the *presumption of innocence*. This is because it's the *right* prior. In the absence of incriminating evidence, and especially for rare crimes, any person is way more likely to be innocent than guilty. However, the presumption of innocence should not be used beyond its scope of applicability. It only corresponds to the prior in the absence of incriminating evidence for rare crimes. If many pieces of evidence are found, then the (posterior) probability of innocence can decrease. It may or may not become smaller than the probability of guiltiness.

The *pure Bayesian* would force us to apply Bayes' rule to better understand the adequate level of suspicion given the incriminating evidence. Most importantly, unless she needs to explain the computations that led to her beliefs, she would not misuse the presumption of innocence which assumes that none or few incriminating evidence has been found.

Let's move on to *thought experiment terms*. The probability $\mathbb{P}[\text{⚰}|\text{😇}]$ of the double death of babies assuming Sally Clark's innocence corresponds to the deaths by natural cause. This is the (underestimated) quantity of one out of 70 million that we mentioned earlier. Finally, the probability $\mathbb{P}[\text{⚰}|\text{😈}]$ of double death, assuming that Sally Clark is guilty, is 1.

We can now combine it all, yielding the following result[8]:

$$\mathbb{P}[\text{😇}|\text{⚰}] \approx \frac{1/70 \, millions \cdot 1}{1/70 \, millions \cdot 1 + 1 \cdot 1/500 \, millions} \approx 0.88.$$

In other words, even given our underestimation of natural death, the probability of innocence of Sally Clark remains large. In particular, it seems unreasonable to convict Sally Clark, since it's still more likely that she is innocent than guilty - even though, as we shall discuss in the last

---

[8]I used the approximation $\mathbb{P}[\text{😇}] \approx 1 - 1/500 \, million \approx 1$.

chapter of this book, this decision, like any decision making, falls beyond the realm of Bayesianism.

So far, we have applied Bayes' rule to two concrete examples. As often in applied mathematics, these two concrete cases are in fact far too complicated to really allow us to grasp Bayes' rule. After all, the *prior* $\mathbb{P}[\bullet]$ on having Ebola and the *thought experiment probability* of natural deaths $\mathbb{P}[\text{⌂}|\text{※}]$ are in fact extremely hard to correctly estimate. It's important to keep in mind that, consequently, final results are necessarily approximate. "All models are wrong".

The *pure Bayesian* has no certainty about the numerical results we obtained. In fact, she must take into account her uncertainty about quantities like $\mathbb{P}[\text{⌂}]$ and $\mathbb{P}[\text{⌂}|\text{※}]$. Now, if pressed by a judge who demands a single result, she would take the average result she obtained, weighted by degrees of beliefs. Nevertheless, she would feel the need to point out that the probability of having Ebola given a bad-news test can only be very low, while the result in Sally Clark's case is less robust, and thus less conclusive. In particular, more search for evidence might be desirable in the latter case.

2.9 BAYES TO THE RESCUE OF THE TROLL STUDENT PROBLEM

Let's now finally use Bayes' rule to solve the troll student puzzle, which is probably its simplest use case. Recall that out of two kids, Alex and Billie, at least one is a boy. What is the probability that the other is a boy too? We already saw that this boiled down to the probability that Alex and Billie are both boys, given that Alex or Billie is a boy.

To simplify notations, denote $A\male$ the fact that Alex is a boy and $B\male$ that Billie is a boy. Bayes' rule then writes

$$\mathbb{P}[A\male \text{ and } B\male | A\male \text{ or } B\male] = \frac{\mathbb{P}[A\male \text{ or } B\male | A\male \text{ and } B\male]\mathbb{P}[A\male \text{ and } B\male]}{\mathbb{P}[A\male \text{ or } B\male]}.$$

The *prior* $\mathbb{P}[A\male \text{ and } B\male]$ is the probability that Alex and Billie are both boys *a priori*. It is $1/4$.

The *thought experiment probability* $\mathbb{P}[A\male \text{ or } B\male | A\male \text{ and } B\male]$ is the probability that Alex or Billie is a boy, given that both are boys. This case is straightforward. Indeed, assuming that Alex and Billie are boys, of course at least one of them is a boy. This *thought experiment term* thus equals 1.

Finally, there is the partition function $\mathbb{P}[A\male \text{ or } B\male]$ in the denominator. This corresponds to 3 cases out of 4.

Interestingly, we can now justify the validity of the reasoning we presented at the beginning of this chapter. If the three hypotheses boy-boy, boy-girl, and girl-boy remain equally likely after the removal of the girl-girl hypothesis, it's because the *thought experiment terms* of the three hypotheses were all equal to 1. This is why there is no transfer of credence between these three hypotheses[9].

We now have all the necessary inputs. We can just use them in Bayes' rule. The searched probability is then $1 \cdot {}^1/_4/{}^3/_4$, which is $^1/_3$. This matches the unrigorous computation we performed at the beginning of this chapter.

2.10   A FEW WORDS OF ENCOURAGEMENT

The reasonings we had to perform to apply Bayes' rule are not easy. For each case, we needed a whole explanatory page. This may scare even the more math-friendly among you. Yes, Bayes' rule is hard to apply, even to simple cases. It's even harder to understand it. Like maths in general, the level of abstraction and sophistication of the equation of knowledge may repel the least courageous among us.

I can only encourage you not to give up. Like maths, Bayes' rule is hard to understand for all of us. Even the greatest mathematicians have had to undergo huge pain to learn to apply Bayes' rule in cases as simple as the Monty Hall problem. But this does not mean that you cannot take advantage of this remarkable equation. However, to get there, you will have to struggle. You will need to fight. And you will have to find the courage not to give up. The price to pay is a huge intellectual effort; but the reward is absolutely fantastic. The sequel will indeed try to convince you that the ability to finally reason correctly and to understand the world is at the end of the journey.

However, reading this book alone will not suffice. Mathematics is not learned by reading religiously. Mathematics requires practice. It requires letting the spirit manipulate abstract objects and seek a clarification of mathematical concepts by oneself. You will need to think mathematically to become good in mathematics. I thus highly encourage you to do and

---

[9]A variant of the problem consists of drawing one of the kids at random, and to observe that it's a boy. In this case, the thought experiment terms of cases boy-girl and girl-boy is no longer 1 (it's now $^1/_2$), which induces a transfer of credences between the 3 hypotheses and leads to a different conclusion.

redo the Bayesian reasonings of this chapter every now and then, in the shower, in transports, or when hiking. And when you feel ready, I invite you to tackle the Monty Hall problem, and then the second part of the troll student puzzle[10] - whose right answer[11] is 13/27.

So, fight. And do not give up. But the most important advice I can give you is to seek enjoyment above all else. Search for the pleasure in the elegance of Bayesian reasoning. It's this pleasure that I'll insist on in the next chapters. In particular, there is something absolutely fascinating in the fact that Bayes' rule is both so compact, fundamental, and tricky to apply and understand. For instance, we have already discovered that it explained the inconclusiveness of excellent medical tests and the relevancy of the presumption of innocence in law! And we are merely at the beginning of our journey!

The elegance of Bayes' rules is the reason why this amazing equation has made me so happy and has led to so many exciting thoughts. This is why I have come to regard Bayes' rule as the most beautiful equation in all of mathematics.

FURTHER READING

Are Birds Smarter Than Mathematicians? Pigeons (Columba livia) Perform Optimally on a Version of the Monty Hall Dilemma. Journal of Comparative Psychology. W Herbranson & J Schroeder (2010).

Bayes and the Law. Annual Review of Statistics and Its Application. N Fenton, M Neil & D Berger (2016).

Conditional Probabilities: Know What You Learn. Science4All. LN Hoang (2013).

A Formula for Justice. The Guardian. A Saini (2011).

Boy or Girl Paradox. Wikipedia (2019).

Alan and Marcus Go Forth and Multiply. BBC (2009).

The Monty Hall Problem. Singingbanana. J Grime (2009).

---

[10] *This May Be The Most Counter-Intuitive Paradox I've Ever Seen: Can you spot the error?* MajorPrep (2019).

[11] Actually, the troll student puzzle is even (much) trickier than what I have presented here, as discussed on Wikipedia or on my (French) YouTube channel....

The Monty Hall Problem - Explained. AsapSCIENCE (2012).

Bayes: How one Equation Changed the Way I Think. J Galef (2013).

A Visual Guide to Bayesian Thinking. J Galef (2015).

Your Brain Is not a Bayes Net (and Why that Matters). J Galef (2016).

Fundamentals: Bayes' Theorem. Critical Thinking. Wireless Philosophy. I Olasov (2016).

Monty Hall Problem. Numberphile. L Goldberg (2014).

Are You REALLY Sick? (false positives). Numberphile. L Goldberg (2016).

The Bayesian Trap. Veritasium. D. Muller (2017).

The Monty Hall Problem. D!NG. M. Stevens (2019).

This May Be the Most Counter-Intuitive Paradox I've Ever Seen: Can You Spot the Error? MajorPrep (2019).

The Boy or Girl Probability Paradox Resolved: It Was Never Really a Paradox. MajorPrep (2019).

CHAPTER 3

Logically Speaking...

Logic takes us nearer to heaven than other studies.

Bertrand Russell (1872-1970)

Every allowed extension of Aristotelian logic to plausibility theory is isomorphic to Bayesian probability theory.

Petri Myllymäki

3.1 TWO THINKING PROCESSES

Imagine you were told: "if a card is a queen on one side, then it is blue on the other side". In other words, we consider the hypothesis "♛ → B". Four cards are on the table, as depicted in Figure 3.1. The first is a queen, the second a 10, the third is blue and the fourth is red. Which card(s) should you flip to test the hypothesis?

Figure 3.1 From left to right, cards are a queen, a ten, blue, and red.

This question was asked to many individuals. Peter Wason found out that the error rate was huge[1]. Only 4% out of millions of respondents gave the right answer. I invite you to ponder the question. Do not fall into the trap(s).

The philosophy of science and knowledge is traditionally divided into two very distinct thinking processes: *deduction* and *induction*. It's often taught that a research scientist must combine both processes in a so-called *hypothetico-deductive* approach. This is the approach that you should have partly adopted to solve the puzzle above. Starting from a hypothesis, you should have derived consequences, and you should have tested them.

Like any other mathematician, I have fallen in love with deductive reasoning. But, perhaps because I am a mathematician, I have often found the inductive part of the scientific reasoning unsatisfactory. I have always been a bit skeptical of the hypothetico-deductive approach. I have never been fully seduced by usual descriptions of the *scientific method*. I have often found such descriptions *ad hoc*, unsound, and very distant from researchers' everyday thinking. Worse, I have often had the impression that the defenders of the "scientific method" had a political goal in mind, namely distinguishing "sciences" from "pseudo-sciences". Like Karl Popper. I have often felt that this effort of explaining the "scientific method" was due to some sort of scientific hooliganism, that is, a will to defend the scientific community by all means, including irrational arguments. The defense of science often felt like motivated reasoning to me.

Of course, this does not mean that I support pseudo-sciences. I actually believe that the distinction between the reliability of science and that of pseudo-science is an extremely important thing to clarify and explain. In fact, I'll stress it later on. But I prefer to say it right now. My discomfort with the "scientific method" does not imply my rejection of scientific conclusions - and definitely not my acception of pseudo-scientific alternatives. In particular, we shall see later on that a Bayesian principle forces us to put great credences on scientific consensuses.

But before getting there, let's get back to the distinction between deduction and induction. This distinction will be obvious to any well-trained scientist today. Yet, curiously enough, the *pure Bayesian* does not make it. For her, there is only one kind of reasoning, which consists

---

[1] *Reasoning about a Rule.* The Quarterly Journal of Experimental Psychology. P Wason (1968).

of applying Bayes' rule. In particular, all our deductive system is only a particular case of Bayes' rule, while induction, as it's often presented, is only a fallacious approximation of Bayes' rule.

The day I discovered this fact, I was stunned. It's this discovery, more than any other, which convinced me to write this book!

In this chapter, we'll only focus on deduction. We'll address induction in the next chapter. Here, we shall see that deduction, also known as *logic*, is in fact way more sophisticated, counterintuitive, and obscure than what one might naively believe. In fact, contrary to what even well-educated scientists might think, there are *several* deductive logics. And we shall see that Bayesian logic has nothing to envy what is taught today.

3.2 THE RULES OF LOGIC

The classical example of logical deduction is Aristotle's syllogism. This syllogism considers the two following premises:

- All men are mortal.

- Socrates is a man.

Aristotle posits that, from these two premises, the following conclusion follows[2]:

- Therefore Socrates is mortal.

Aristotle's logic seems indisputable. It seems straightforward, natural, and unquestionably valid. Years ago, when a friend dared me to doubt it, I forfeited.

Nevertheless, Aristotle's syllogism has inspired many philosophers, logicians, and mathematicians. They analyzed it and broke it into so-called *inference rules*. In particular, they decomposed Aristotle's syllogism into two steps of logical deduction: *universal instantiation* and *modus ponens*. To understand these two inference rules, it's best to start with a simpler example than Aristotle's syllogism. Consider these two events:

---

[2]In fact, Aristotle does not seem to have considered such syllogisms, as his theory did not aim to treat particular cases (and thus would reject reasonings based on the second premise). This contrasts with stoicians who developed the propositional logic we shall discuss later on. See *Computation, Proof, Machine: Mathematics Enters a New Age*. Cambridge University Press. G Dowek (2015).

☂: It rains.

☂: I took my umbrella.

Each event can be true or false. With these two events, called *literals*, we can make up new events, called *formulas*. For instance, we can construct formulas like "not ☂", "☂ or ☂" or "☂ and ☂", or still more complicated formulas like "(not ☂) or ☂". To understand the formulas, it's useful to construct truth tables that express the logical values of the formulas, depending on the values of literals.

Table 3.1 Truth table of "it rains or I took my umbrella"

	☂	not ☂
☂	✓	✓
not ☂	✓	✗

For instance, in Table 3.1, the middle row, which corresponds to ☂, is the case where it rains. Meanwhile, the middle column is the case ☂ where I took my umbrella. The middle entry of the table thus corresponds to the logical value of "☂ or ☂" when ☂ and ☂ are both true. This entry asserts that if ☂ is true and ☂ is true, then "☂ or ☂" is true. I invite you to take the time to analyze the truth table yourself, and to determine truth tables of other logical formulas.

So far we only studied formulas with two literals. But we can go further and consider formulas with three, eight, or a large number of literals. The truth tables must then be way larger to list all possible combinations of values of literals. I invite you to compute by yourself the number of entries of these giant truth tables, as well as the number of different truth tables. But I definitely do *not* recommend you to list all the different truth tables for three literals or more. There are 256 truth tables for 3 literals. And there are about as many truth tables for 8 literals as the number of particles in the universe!

One formula is particular useful in practice. It's the formula "(not ☂) or ☂". This formula is more commonly denoted "☂ → ☂". It intuitively says "☂ implies ☂" or "whenever ☂, there is ☂", or "if ☂ then ☂". I invite you to think about the truth table of this formula provided in Table 3.2, as it may appear surprising if you have not thought about it long enough.

What makes the formula ☂ → ☂ particularly important is that it's at the heart of logical deduction. Logical deduction consists of starting

Table 3.2 Truth table of "if it rains then I take my umbrella"

	☂	not ☂
🌧	✓	✗
not 🌧	✓	✓

from premises, like 🌧, and deriving conclusions, like ☂. If this implication is true and if the premises are true, then so are the conclusions. More formally, we have the logical formula $((🌧 \to ☂)$ and $🌧) \to ☂$. This formula is called *modus ponens*[3].

I invite you to replace the implication arrows by the definition in terms of **or**, **and**, and **not** we gave earlier. By playing around with the logical formula, or by computing the truth table, you should conclude that, no matter what the values of 🌧 and ☂, the *modus ponens* is always true. We say that the *modus ponens* is a tautology, since it is true for all logical values of the literals.

Contrary to its meaning in common language, in logic, a tautology need not be "obvious". It's not necessarily a truism or a triviality. Like *modus ponens*, certain tautologies are hard to notice for most of us, and require much thinking[4].

3.3 ARE ALL QUEENS BLUE?

Let's get back to Wason's experiment. Recall that the hypothesis to test was 👑 → B. Four cards were given. The first was a queen, the second a ten, the third a blue card, and the fourth a red one. Which card(s) should be flipped to test the hypothesis?

The first card is a queen. *Modus ponens* allows a prediction. Indeed, we have the tautology "$((👑 \to B)$ and $👑) \to B$". If the hypothesis to test is true, and given that the first card is a queen, then its back is blue. The hypothesis thus predicts that the back of the first card must be blue. If it's not, then we would have an evidence that falsifies the hypothesis.

---

[3]In fact, it's more complicated than that. To be rigorous, we would need to distinguish the *then* and *and* of the meta-language of logic, from the symbols → and **and** of the language in which logic is stated.

[4]In fact, in a sense, the fundamental P versus NP conjecture, often regarded as the most prestigious problem in computer science, formalizes the intuition that tautologies are not always trivialities.

For the second card, however, no conclusion can be derived from the premise "((♛ → B) and not ♛)". The card may very well be "neither B nor ♛" or "B but not ♛". Similarly, in the case of the third card, nothing about ♛ can be derived from the premise "((♛ → B) and B)". The cases where the card is a queen and where it is not a queen are both compatible with the hypothesis and the fact that it is blue. Take the time to convince yourself of this. One way to do so is to decompose ♛ → B into (not ♛) or B, and to compute the truth tables of these logical formulas.

Finally, the hypothesis does make a univocal prediction regarding the fourth card. Indeed, if a non-blue card is a queen, then we would have an example of a card which is a queen, but whose other side is not blue. It would contradict the hypothesis. We should thus flip this last card to test the hypothesis.

As for the first card, this case corresponds to a tautology. This tautology is "((🍄 → 🌲) and not 🌲) → not 🍄". It is called *modus tollens*. This tautology can also be rewritten as "(🍄 → 🌲) → (not 🌲 → not 🍄)". In other words, the implication " 🍄 → 🌲" implies the implication "not 🌲 → not 🍄", called its *contrapositive*. In fact, any implication is equivalent to its contrapositive.

The equivalence between an implication and its contrapositive is one of the many counterintuitive tautologies of logic. For instance, consider the hypothesis that says "all ravens are black". The contrapositive of the hypothesis asserts that "everything that is not black is not a raven". Yet, since any hypothesis is equivalent to its contrapositive, confirming the contrapositive confirms the hypothesis. In particular, and as a result, each red apple confirms the all-ravens-are-black hypothesis! This highly counterintuitive conclusion is what is known as *Hempel's raven paradox*[5]. In my experience, even doctors in mathematics have trouble anticipating and accepting it!

Another example of logical tautology is *logical disjunction*, which corresponds to the tautology "((🍄 → 🌲) and (not 🍄 → 🌲)) → 🌲". We can also mention *reasoning by contradiction*, also known as *reductio ad absurdum*, whose associated tautology is "((🍄 → 🌲) and (🍄 → not 🌲)) → not 🍄".

---

[5] *The Raven Paradox - A Hiccup in the Scientific Method.* Up and Atom. J Tan-Holmes (2019).

3.4 QUANTIFIERS AND PREDICATES

The logic of propositions that we discussed so far is very rich and already highly counterintuitive. Nevertheless, its language is too restricted. Indeed, each logical formula of the logic of propositions only involves a finite number of literals. Yet, in mathematics like in science, it's often useful to study the set of all possible events, which may be infinite.

For instance, since there are infinitely many numbers, we can create infinitely many literals with these numbers. Typically, there are infinitely many propositions of the form "n is an even number". One for each integer n. Instead of giving a different name for each of these propositions, we may rather consider a proposition $\mathsf{Even}(n)$ that depends on the number n. We then say that Even is a *predicate*. More complex predicates can be constructed, for instance $\mathsf{Addition}(m, n, p)$ to mean "$m + n = p$".

A predicate cannot be *true*. What can be said about a predicate is whether it is *always true*, *always false*, *at least true once*, or *at least false once*. These four assertions correspond to different quantifiers. For instance, we say that P is always true if, for all values of n, $P(n)$ is true. This phrase is compactly denoted "$\forall n P(n)$". The symbol "$\forall$" can be read "for all". It is the *universal quantifier*. Similarly, if P is always false, we write "$\forall n (\mathsf{not}\, P(n))$". Finally, the fact that $P(n)$ is true (respectively, false) for at least one value of n is denoted $\exists n P(n)$ (respectively, $\exists n (\mathsf{not} P(n))$). The symbol "$\exists$" is called an *existential quantifier*. It can be read "there exists".

Crucially, a formula whose predicate variables are all universally or existentially quantified becomes a proposition, which can be true or false. For instance, the phrase "n is even" is neither true nor false. However, quantified phrases "$\forall n\, \mathsf{Even}(n)$" and "$\exists n\, \mathsf{Even}(n)$" are true or false. I'll let you determine which is which.

We can then combine the different symbols of logic in diverse manners to construct all sorts of interesting propositions, like "$\forall n (\mathsf{Even}(n) \rightarrow \mathsf{not}\, \mathsf{Even}(n + 1))$", "$\forall n \exists m (m > n)$" or "$\forall n \forall p \exists q \exists r (n = pq + r)$ and $0 \leq r < p$". In its purest form, the job of a mathematician is to determine (interesting) formulas of predicate logic which are tautologies. When a nontrivial tautology is uncovered, we call it a theorem.

3.5 ARISTOTLE'S SYLLOGISM REINTERPRETED

We can now get back to Aristotle's syllogism. The objects of study in this syllogism are not numbers, but beings. The first premise asserted "all men are mortals". It thus describes a relation between two predicates about beings. In the language of logic, this premise can be rewritten as $\forall x(\mathsf{Man}(x) \rightarrow \mathsf{Mortal}(x))$, where $\mathsf{Man}(x)$ means that x is a man, and $\mathsf{Mortal}(x)$ that x is mortal.

The second premise asserts that "Socrates is a man". It is a premise about the value of the predicate Man for a particular value of x, namely Socrates. It can thus be rewritten as the fact that $\mathsf{Man}(\mathsf{Socrates})$ is assumed to be true.

To conclude, we want to invoke the implication $\mathsf{Man}(x) \rightarrow \mathsf{Mortal}(x)$ for the particular case where x is Socrates. To do so, logicians have introduced an inference rule called *universal instantiation*. Applied to Aristotle's syllogism, this rule says that, if $\mathsf{Socrates}$ is an object of our theory and given the premise $\forall x(\mathsf{Man}(x) \rightarrow \mathsf{Mortal}(x))$, then the logical formula $(\mathsf{Man}(\mathsf{Socrates}) \rightarrow \mathsf{Mortal}(\mathsf{Socrates}))$ is true. Indeed, it was what is obtained when the undetermined variable x is replaced by the object $\mathsf{Socrates}$ of the theory.

We can now conclude with *modus ponens*. Since the implication $(\mathsf{Man}(\mathsf{Socrates}) \rightarrow \mathsf{Mortal}(\mathsf{Socrates}))$ is true, and since the premise $\mathsf{Man}(\mathsf{Socrates})$ is true, then the conclusion of the implication must be true too. We obtain the conclusion $\mathsf{Mortal}(\mathsf{Socrates})$, which is indeed Aristotle's conclusion. We have just derived the validity of Aristotle's logic from the foundations of logic.

You are probably wondering if this has not been an over-explanation of a triviality. Did we not already know that Aristotle's logic was valid? I guess we did. But it's interesting to note that this validity relied on the acceptation of inference rules. Now, it may seem unsound to reject these inference rules. Yet, the rigor of the logician must force him to question his inference rules. And weirdly enough, some logicians, known as *intuitionists* or *constructivists*, now reject some inference rules whose truth tables are full of truth values. By opposition, classical logicians are sometimes known as *Platonists*.

The divergence between Platonists and intuitionists is particularly explicit in their interpretations of Gödel's incompleteness theorem. But to understand this, we need to a make a small detour. We first need to discuss axiomatization.

3.6 AXIOMATIZATION

To determine the truth value of a formula in the logic of propositions, we needed the truth values of its literals. However, in the case of the logic of predicates, we cannot afford to spend an infinite amount of time listing the logical values of predicates for all objects of the theory. The preferred approach is axiomatic. In other words, like in the case of Aristotle's syllogism, we start from fundamental premises called axioms. We then aim at deriving their logical consequences. Formally, mathematics then boils down to determining tautologies of the form Axioms → Theorem.

Let's consider the example of Peano's axioms, which define a theory of natural numbers like 0, 1, 2, 3.... The first axiom asserts the existence of an object, usually denoted 0. Then, the second axiom roughly asserts that any number has a successor. There are many other axioms provided by Peano. But I will not detail them here[6]. What's almost magical is that an impressive list of mathematical theorems can be deduced from Peano's axioms.

However, Peano's axioms are limited to natural numbers. Yet many interesting mathematical objects are not natural numbers. There are also real numbers, geometric curves, or probabilities. These days, many mathematicians prefer to use the Zermelo-Fraenkel (ZF) axioms, which they sometimes complement with the axiom of choice (C). Nearly all known mathematical theorems can then be written as $ZFC →$ Theorem.

Now, could all such true theorems be proved? Unfortunately, no. The devastating *Gödel incompleteness theorem* applies to all axiomatic systems that generalize Peano's axioms. Better, Gödel's theorem applies to all theories that use the logic of predicates, rely on a finite (or computable) set of axioms, and are able to talk about the addition and multiplication of natural numbers[7]. Gödel's theorem asserts that, in all such theories, there exist formulas that axioms cannot prove nor disprove[8]. This includes ZF and ZFC.

3.7 PLATONISTS VERSUS INTUITIONISTS

There are essentially two ways to interpret Gödel's theorem. Interestingly, they point out to two quite distinct ways to think about mathematics.

On one hand, the *Platonist* mathematician typically interprets Gödel's theorem as a deficiency of axioms. For the Platonist, natural numbers, or sets, have a reality in some ideal world. In this world, propo-

---

[6] *What Does It Mean to Be a Number? (The Peano Axioms)*. Infinite Series. G Perez-Giz (2018).

[7] There are additional minor technical details I won't mention here.

[8] *Gödel's incompleteness theorem*. Numberphile. M Du Sautoy (2017).

sitions necessarily either have a true or false value. Unfortunately, the finiteness of words and symbols restricts us to only a partial description of the ideal world. There are truths in this ideal world that our finite axioms cannot capture. For the Platonist, Gödel's theorem shows that there are true theorems without proof.

On the other hand, an *intuitionist* mathematician has a quite different interpretation of the theorem. For the intuitionist, mathematics is a construction game. For instance, Peano's first axiom is a tool to construct the number 0. The second axiom is then a sort of machine that inputs a natural number and uses it to construct another natural number.

Better, especially in type theory which is an alternative to the logic of predicates[9], the intuitionist considers that mathematical proofs are objects of his theories. Proofs therefore must be constructed as well. In particular, the intuitionist's focus is the constructibility of his mathematical objects. Not the truth of theorems. For the intuitionists, Gödel's theorem asserts that in any theory, there are sentences for which no proof, neither in favour nor against, can be constructed. This poses no metaphysical problem to him, as the matter of the truth value of the theorem is of secondary importance.

The crux of the debate between Platonists and intuitionists can be summed up by the law of excluded middle. This law asserts that the proposition "P or not P" is a tautology. It seems that it suffices to compute a truth table to prove it. Indeed, if P is true, then P or not P is true too. Meanwhile, if P is false, then not P is true, and thus P or not P is true too.

However, for the intuitionist, a third possibility remains. P may be neither provable nor disprovable. In this case, we say that P is *undecidable*. But then, if P is undecidable, then neither P nor not P are true. The proposition P or not P is then undecidable too. Indeed, in the absence of proofs of P and of not P, it's impossible to construct a proof of P or not P. Therefore this proposition is *not* a tautology for the intuitionist.

The opposition between Platonists and intuitionists is thus not limited to an interpretation of Gödel's theorem. The intuitionist rejects all nonconstructive proofs that the Platonist has proved. Among the most infamous of such theorems, we can mention the Banach-Tarski

---

[9] *Type Theory: A Modern Computable Paradigm for Math.* Science4All. LN Hoang (2014).

paradox[10], the existence of bases of vector spaces, or the uniqueness of algebraic closures.

3.8 BAYESIAN LOGIC*

What about our *pure Bayesian*? Which logic does she believe in? One of the most exciting discoveries of my Bayesian meditations is that the *pure Bayesian* has her own deductive logic. This logic is neither classical nor intuitionist. We may talk about *Bayesian logic*. Bayesian logic corresponds to particular cases of Bayes' rule.

In this logic, the truth of an event like ❀ is the extreme case where the event occurs with probability 1, that is, when $\mathbb{P}[❀] = 1$. What's more, the fact that ❀ implies 🍄 corresponds to $\mathbb{P}[🍄|❀] = 1$. In Bayesian words, ❀ implies 🍄 if and only if the probability of 🍄 given ❀ is 1.

Modus ponens and *modus tollens*, like many other inference rules, are also particular cases of the law of total probabilities and Bayes' rule. Recall that *modus ponens* is the tautology "$((❀ \rightarrow 🍄) \text{ and } ❀) \rightarrow 🍄$". Its Bayesian version consists of assuming $\mathbb{P}[🍄|❀] = 1$ and $\mathbb{P}[❀] = 1$. The *pure Bayesian* then uses these two assumptions to perform computations, which I invite you to do yourself. These lead her to conclude $\mathbb{P}[🍄] = 1$. Similarly, using Bayes' rule, I invite you to prove *modus tollens*, the contraposition and the law of excluded middle.

Bayesian logic thus does not seem inferior to commonly accepted logics. However, it's not quite equivalent to such logics. In particular, there is a slight difference between the classical implication $❀ \rightarrow 🍄$ and the Bayesian equality $\mathbb{P}[🍄|❀] = 1$, namely in the case where ❀ is false. Indeed, when ❀ is false, the logical formula $❀ \rightarrow 🍄$ is true, even when 🍄 is false. However, if $\mathbb{P}[❀] - 0$ then Bayesian logic concludes that $\mathbb{P}[🍄|❀]$ is in fact ill-defined.

What's amusing is that the Bayesian interpretation seems more natural than its classical logic counterpart. Indeed, the phrase "if France won the 2006 World Cup, then pigs would fly" is true in classical logic. However, the truth of this phrase does not seem to match common sense. It seems more tempting to say that this phrase has no truth value; that it's nonsense. This is what Bayesian logic asserts. It says that $\mathbb{P}[🍄|❀]$ makes no sense when $\mathbb{P}[❀] = 0$.

Bayesian logic also naturally extends to the logic of predicate. However, once again, there are slight differences with classical logic. To

---

[10] *The Banach-Tarsky Paradox.* VSauce. M Stevens (2015).

understand this distinction, it's useful to first assume that the objects of the theory are randomly drawn. Consider a probability law to draw objects of the theory. The universal quantifier $\forall x A(x)$ then corresponds to the identity[11] $\mathbb{P}[A(x)] = 1$, when x is thereby randomly drawn. In Bayesian logic, this identity can be more succinctly written $\mathbb{P}[A] = 1$. The existential quantifier $\exists x A(x)$ then translates into $\mathbb{P}[A] > 0$.

Universal instantiation then has a Bayesian equivalent that can be derived from Bayes' rule. Indeed, if $\mathbb{P}[A] = 1$, then for any object y of the theory[12], we have $\mathbb{P}[A(y)] = \mathbb{P}[A|y] = 1$. However, Bayesian existential instantiation is slightly different from the classical one. In Bayesian logic, if $\mathbb{P}[A] > 0$, all that can be said is that there exists an object y such that $A(y)$ has a strictly positive probability to occur, that is, $\mathbb{P}[A|y] > 0$.

3.9 BEYOND TRUE OR FALSE

The magic of Bayesian logic, however, is to allow for the generalization of classical logic to manipulations of different degrees of certainty. In fact, it can be shown that it's the only logic capable of doing so[13]. This is what the Jaynes-Cox theorem and its variants assert[14], which derive Bayesian logic from certain natural prerequisites on any logic of plausibilities.

I would even claim that Bayesian logic allows to understand why so many classical inference rules seem so counterintuitive. To do so, it's useful to imitate some algorithms in *machine learning*, like the Boltzmann machines we shall discuss later on, that forbid themselves from considering probabilities equal to 0 or 1. After all, in practice, when we discuss the real world, it's reasonable to never exclude anything.

For instance, if the phrase "if France won the 2006 World Cup, then pigs would fly" seems false, it may be because we may not be fully certain of the fact that France did not win the 2006 World Cup. Maybe we slept

---

[11]In general, in measure theory, the equivalence I am presenting here is actually wrong, as some (if not all!) objects of the theory may have a zero probability to be drawn (I would need to define complicated sigma-algebras to explain this, and other funky stuff...). To simplify, throughout most of this book, we shall assume that Bayesian (prior) probabilities are defined over countable sets, with full support, so that any object x has a strictly positive probability to be drawn.

[12]With a nonzero probability to be drawn.

[13]Fuzzy logic is sometimes proposed to be such a candidate, but the fuzziness of fuzzy logic does not describe probability (and thus does not correspond to *epistemic* uncertainty).

[14]*From Propositional Logic to Plausible Reasoning: A Uniqueness Theorem.* International Journal of Approximate Reasoning. K Van Horn (2017).

the day before the final and dreamt the years that followed. Or maybe our memory is confusing the 2006 World Cup with the 2000 European championship - we'll get back to the unreliability of our memories! And who knows if Italy's victory will not one day be invalidated because of doping or some other reason?

For the *pure Bayesian*, the defeat of France at the 2006 World Cup is almost a sure fact, but we cannot fully exclude the fact that it may be wrong. As a result, the phrase "if France won the 2006 World Cup, then pigs would fly" would be incorrect, as the probability that pigs fly assuming France won the 2006 World Cup would not equal 1. In other words, this apparent paradox vanishes as soon as we refuse to fall into the binary logic that traps us into classifying everything as either true or false!

And indeed, much is clarified by the use of degrees of credences. For instance, one of the many Bayesian thoughts that greatly excited my neurons was the explanation of our uneasiness with contraposition. Indeed, a hypothesis is true if and only if its contrapositive is. However, a hypothesis may be very likely true, even though its contrapositive is not.

This is precisely what we uncovered in the case of Sally Clark! When a mother is innocent, her babies will very likely *not* die at their birth. In other words, $\mathbb{P}[\text{not} \, 👶 | 😇]$ is nearly 1. However, if the babies die at birth, then the mother nevertheless remains likely innocent. In other words, the contrapositive probability $\mathbb{P}[😈 | 👶]$ is close to 0. We thus uncovered an explanation of why contraposition is not intuitive: outside of the Platonist world to which binary logic applies, it's much more reasonable to reason with degrees of credence. Given this, contraposition no longer holds.

Going beyond the true-or-false dichotomy also allows to justify some intuitive reasonings that classical logic condemns. Suppose $🌧 \rightarrow ☂$. In other words, suppose that, whenever it rains, I take my umbrella. Then, it's tempting to assert that if it does not rain, then I am less likely to take my umbrella. This intuitive remark, which does not have any grounds in classical logic, is actually a Bayesian theorem. This theorem asserts that if $\mathbb{P}[☂ | 🌧] = 1$, then $\mathbb{P}[☂ | \text{not} \, 🌧] \leq \mathbb{P}[☂]$. More generally, surely enough, the absence of evidence is no proof of absence. But *the absence of evidence can only increase the suspicion of absence*. Though often not by much.

Embracing Bayesian logic and its uncertainties also clarifies the mystery of the black raven. Indeed, a Bayesian analysis shows that each red apple does indeed confirm that "all ravens are black". However, this confirmation is extremely slight, much, much slighter than the observation of a black raven, mostly because the number of objects which are not ravens is vastly larger than the number of ravens[15]. Confirming the contrapositive does confirm the original implication, but this effect may be so small that it's vastly negligible. It's the case of the black raven paradox!

This conclusion has an immediate corollary regarding the all-queens-are-blue problem. The two cards that we did not consider useful to flip to reject the hypothesis ♛ → B are actually useful. However, the confirmation or partial rejection that they would imply is so tiny that it's essentially negligible.

More generally, when it comes to the confirmation or rejection of scientific theory, the binary language of classical logic seems inappropriate. It ignores the extent of confirmation or the magnitude of rejection. Besides, as Eliezer Yudkowsky often argues, the Bayesian computation of the credences of theories is not much of a walk. And it is definitely not a small set of leaps of faith. *Learning is a dance.* Like stock markets or average temperature on Earth, the credences of the *pure Bayesian* always fluctuate as data pour in. During learning, even the credences of the best theories will not increase monotonically. Even the best theories will very likely undergo numerous small losses of credences, since, by chance, some observations will perfectly fit concurrent theories. However, in the long run, if a theory really is more relevant than any other, then its credence will follow an upward trend.

Unfortunately, the rigidity of the *scientific method* as described by many is unable to capture the dance of the credences. The more appropriate language rather seems to be that of the (conditional) probabilities of the *pure Bayesian*.

3.10   THE COHABITATION OF INCOMPATIBLE THEORIES

In particular, the wonder of the *pure Bayesian*'s probability language is to allow to think along several theories at once and to combine their predictions. This technique has had great success in *machine learning*. It is known as *ensembling* or *bagging*. Its practical efficiency is astonishing. Combining different incompatible theories often yields better predictions than any individual theory does!

---

[15]This actually also depends on the selection bias of the red apple. Did you pick it because it was red? Or was it because it was a non-raven? Here, I'm assuming that it was because it was red.

This approach can be described as follows. The *pure Bayesian* thinks along a theory "Theory" by conditioning probabilities by Theory. For instance, in the framework Theory, we may know that ☂ implies ☂ and that ☂ occurs with probability $1/2$. We then deduce that $\mathbb{P}[\text{☂ and ☂}|\text{Theory}] = \mathbb{P}[\text{☂}|\text{☂ and Theory}] \cdot \mathbb{P}[\text{☂}|\text{Theory}] = 1 \cdot 1/2 = 1/2$.

The *pure Bayesian* will compute certain probabilities in many different theories. Reconsider Sally Clark's example. The *pure Bayesian* assigns different values to probabilities $\mathbb{P}[\text{⚰}|\text{☂ and Theory}]$ of double deaths by natural causes and $\mathbb{P}[\text{⚖}|\text{Theory}]$ of Sally Clark's *a priori* guilt, depending on her different theories Theory. This leads her to different conclusions $\mathbb{P}[\text{☂}|\text{⚰ and Theory}]$ about the posterior probability of Sally Clark's innocence, given the double death of her babies.

If the judge now demands a single definite result, the *pure Bayesian* will compute the average result of her computations. Not all results will be equally weighted. In fact, each result will be weighted by the *pure Bayesian*'s credence in the theory that led to the result. Formally, this averaging is the law of total probabilities, which is the computation

$$\mathbb{P}[\text{☂}|\text{⚰}] = \sum_{\text{Theory}} \mathbb{P}[\text{Theory}|\text{⚰}]\,\mathbb{P}[\text{☂}|\text{⚰ and Theory}],$$

where the symbol $\sum$ means that the right term is a sum of (many) terms of the form $\mathbb{P}[\text{Theory}|\text{⚰}]\,\mathbb{P}[\text{☂}|\text{⚰ and Theory}]$, for different theories Theory.

Then comes the problem of computing the probabilities $\mathbb{P}[\text{Theory}|\text{⚰}]$. These probabilities are the credences of the *pure Bayesian* in different theories. For the averaging to be indeed an averaging, these probabilities evidently need to add up to 1.

But most importantly, these probabilities are not arbitrary. In fact, their computation is the main focus of this book. As often, the computation relies on Bayes' rule. And it's what we'll get to in the next chapter.

FURTHER READING

Probability Theory: The Logic of Science. Washington University. ET Jaynes (1996).

Homotopy Type Theory: Univalent Foundations of Mathematics. Institute for Advanced Studies. The Univalent Foundations Program (2013).

Reasoning about a Rule. The Quarterly Journal of Experimental Psychology. P Wason (1968).

From Propositional Logic to Plausible Reasoning: A Uniqueness Theorem. International Journal of Approximate Reasoning. K Van Horn (2017).

What Does It Mean to Be a Number? (The Peano Axioms). PBS Infinite Series. G Perez-Giz (2018).

Gödel's Incompleteness Theorem. Numberphile. M Du Sautoy (2017).

The Raven Paradox - A Hiccup in the Scientific Method. Up and Atom. J Tan-Holmes (2019).

5 Stages of Accepting Constructive Mathematics. Institute for Advanced Studies. A Bauer (2014).

The Banach-Tarsky Paradox. VSauce. M Stevens (2015).

Computer Science ∩ Mathematics (Type Theory). Computerphile. T Altenkirch (2017).

The Netflix Prize. ZettaBytes, EPFL. AM Kermarrec (2017).

Type Theory: A Modern Computable Paradigm for Math. Science4All. LN Hoang (2014).

Homotopy Type Theory and Inductive Types. Science4All. LN Hoang (2014).

Univalent Foundations of Mathematics. Science4All. LN Hoang (2014).

CHAPTER 4

Let's Generalize!

All knowledge degenerates into probability; and this probability is greater or less, according to our experience of the veracity or deceitfulness of our understanding, and according to the simplicity or intricacy of the question.

David Hume (1711-1776)

Our brain has this annoying habit to think that [...], if under some hypothesis results are unlikely, then the hypothesis is unlikely. This is false.

Christophe Michel (1974-)

4.1 THE SCOTTISH BLACK SHEEP

A biologist, a physicist, and a mathematician go on holidays in Scotland for the first time. While they are still in the train to Edinburgh, they see a black sheep. The biologist says: "Incredible! Sheeps are black in Scotland!" Annoyed, the physicist replies: "All that can be said is that there is at least a black sheep in Scotland". The mathematician, placid, adds: "Actually, all that can be said is that at least half a sheep is black in Scotland".

I like this joke. If the biologist may have overgeneralized his observation, the physicist seems too prudent, while the mathematician's rigidity seems laughable. If the half sheep you see is black, it seems unreasonable not to generalize its blackness to its other half.

We might even extend the joke by adding a philosopher who, upon hearing these thoughts, would add: "How do you know we really are in Scotland? You might be dreaming at home. Worse, a demon might have implanted your memories. While you think you live on Earth, the demon might be playing tricks to your mind. Or perhaps you are living in a simulation, and none of what surrounds you is real."

The problem posed by this joke turns out to be one of the greatest problems of epistemology in particular, and of philosophy in general. This problem is the *problem of induction.*

4.2   A BRIEF HISTORY OF EPISTEMOLOGY

The most influential thinker on this matter may be David Hume, an 18th century Scottish philosopher. Other philosophers of his time thought that they had proved necessary truths. For instance, Descartes seemed to have proved the existence of God, by pondering the essence of perfection. It is more perfect to exist than not to exist, he said. Yet God is the absolutely perfect being. Thus God exists. Unfortunately, such a reasoning is full of logical fallacies, deficient axiomatization, and wishful thinking.

A contrario, in his masterpieces *A Treatise on Human Nature*, and then *An Enquiry Concerning Human Understanding*, Hume argued that it was impossible to *deduce* absolute and general facts from empirical observations only. Empiricism cannot lead to absolute truths. For instance, no matter how many observations we make, we can never conclude that the sun will always rise up. Univocal past observations cannot allow univocal predictions about the future.

Nevertheless, Hume also argued that such generalizations are often quite correct, or at least, *useful* to make. And according to Hume, the reason why such generalizations are sensible is because of a *uniformity principle* that Nature seems to satisfy. The laws of Nature do not seem to change. And even if they do, like the nuclear activity of the Sun, they seem to evolve sufficiently slowly to allow for generalizations to a not-too-distant future.

In particular, the uniformity principle (which we shall partially justify by the Church-Turing thesis) allows predictions on what is likely to occur. In a stroke of genius, Hume foresaw the central role that the theory of probabilities had to play in the resolution of the induction problem. If Laplace is the father of Bayesianism, Hume is arguably its grandfather.

Unfortunately, Hume's Bayesian seeds did not blossom. For the two centuries that followed, few tried to formalize and mathematize his ideas. Worse, in 1934, Karl Popper took a very different direction by publishing *The Logic of Scientific Discovery*. In this book, Popper describes what he believes to be *the* philosophy of science. According to Popper, above all else, any scientific theory must be *falsifiable* by experiments. This is known as the *principle of falsifiability*. A theory is scientific if and only if it imposes constraints on conceivable experimental observations that make it subject to a possible falsification. Scientists must then repeat experiments to reject the scientific theory, or to corroborate it. But according to Popper, a corroboration is *not* a validation of the theory.

4.3 A BRIEF HISTORY OF PLANETOLOGY

However, today's science philosophers often underline the fact that Popper's beautiful principles do not quite match the reality of scientific research. Let's illustrate it with planetology.

In 1821, astronomer Alexis Bouvard noticed anomalies in the trajectory of Uranus. This distant planet, the seventh farthest planet of the Solar system, did not seem to obey Newton's laws of gravity. If Popper's philosophy was the philosophy of scientists, Bouvard would have had to reject Newton's laws. This is not what happened.

Bouvard, followed by John Couch Adams and then Urbain le Verrier, preferred to postulate the existence of an eighth planet in the Solar system. Bouvard, Adams, and le Verrier preferred to believe in the existence of unobserved entities, rather than listening to Popper's philosophy (although Popper was not yet born). The *pure Bayesian* would say that Bouvard's credence in Newtonian mechanics was simply greater than his credence in the absence of an eighth planet.

Weirdly enough, our three theorists were right! After complex computations, Adams and le Verrier even managed to determine the precise location of this eighth planet. Adams had the poor idea to ask astronomers of the Cambridge observatory to make the observation. Rather than complying, Sir Airy raised doubts about Adam's computations. On the other hand, le Verrier, baffled by the lack of enthusiasm in France, rather called the Berlin observatory. On the same evening, Johann Gottfried Galle confirmed le Verrier's astounding prediction and discovered Neptune[1]!

You might think that the moral of this story is that one should never question Newton. Here again, the history of science seems to have been playing tricks with us. The same le Verrier also studied the trajectory of

---

[1] *Pluto is NOT (not?) a Planet.* Science4All. LN Hoang (2014).

Mercury. The anomalies of Mercury led him to predict the existence of a zeroth planet in the Solar system. He called it Vulcan.

However, for the decades that followed, no one observed Vulcan. Maybe the planet was too close to the Sun to be detected, as the extreme luminosity of the Sun made everything nearby appear dark. Or perhaps should we now question Newton. The young physicist that dared this audacious thought was a certain Albert Einstein.

After eight long years of meditations, full of doubtful sophisms, rough computations, and strokes of genius, Einstein eventually published a revolutionary new theory of space, time, and gravity. The starting point of his thought was somehow both obscure and brilliant. In 1907, while still only a patent clerk, Einstein dared a thought that got mocked by many of my YouTube viewers: what if gravity was not a force? What if gravity was only an illusion[2]? What if gravity was just an artefact, due to the upwards acceleration of the ground and our egocentrism that makes us reason in the non-inertial reference system of the ground?

This thought, called the *equivalence principle*, is what Einstein would call the "happiest thought of [his] life". But it may not have been the one that gave him the most dangerous heart palpitations. Eight years later, in 1915, Einstein established new equations of gravity that relied on a new (non-Euclidean!) conception of space and time. But more importantly, in November 1915, Einstein proved that his mysterious (but wonderfully elegant) equations perfectly explained the anomalies of Mercury! He was now convinced: his theory was *right*. In winter 1916, a few months later, his theory was now taught by the great mathematician David Hilbert at the University of Göttinghen.

But how could Einstein and Hilbert, perhaps the two greatest geniuses of their time, believe in a theory that no observation had yet confirmed? Indeed, it would be only be four years later that experimental observations of optical aberrations of a solar eclipse would confirm Einstein's theory. A student asked the German scholar what he would have done otherwise. Einstein answered: "Then I would feel sorry for the good Lord. The theory is correct."

Albert Einstein, the superstar of science, he who had great interest in the works of philosophers like Kant and Mach, would not have accepted a falsification of his theories by experiments. Einstein's epistemology did not seem to quite fit Popper's philosophy.

---

[2] *Is Gravity An Illusion?* PBS Space Time. G Perez-Giz (2015).

4.4 SCIENCE AGAINST POPPER?

The examples of the inadequacy of Popper's philosophy I have provided so far are far from being exhaustive. It suffices to pay attention to the details of the history of science to realize that Popper's methodology is not the norm. From Pasteur's biogenesis to Darwin's evolution, from Newton's Principia Mathematica to Mendeleïev's periodic table, from quantum mechanics to string theory, the inventors of these theories seem to have been convinced by their theories much before they had been tested as Popper would require.

One of the latest examples to date is the case of faster-than-light neutrinos. In 2011, the OPERA experiment announced the detection of such neutrinos. The problem is that such neutrinos contradict the foundations of special relativity. For almost all physicists whose credences in special relativity is huge, it was much easier to imagine the presence of a measurement error.

This is actually one of the many cases where an experiment has been rejected by theory. This may sound paradoxical, or contrary to the scientific method. But as we shall see in the next chapter, for the *pure Bayesian*, the physicists' reasoning is actually perfectly sound.

Popper himself knew that his philosophy could not be applied *stricto sensu*. After all, experimental results are subject to measurement errors and random perturbations. Popper's philosophy thus needed to be adapted to take into account the statistical fluctuations that unavoidably accompany scientific experiments.

4.5 FREQUENTISM*

The heroes of the formalization of the theory of statistical errors are named Karl Pearson, Egon Pearson (his son), Jerzy Neyman, and, most importantly, Ronald Fisher. In the 1920s, these great minds developed a statistical theory that has invaded the world of science since. Their theory is dubbed *frequentism*, as it assumes that probabilities are measures of frequencies. For a *frequentist*, understanding probabilities was mostly understanding how errors cancel when sample sizes were sufficiently large.

Frequentists' appealing argument, like Popper's, was the *objectivity* of their methods. In particular, Pearson, Neyman, and Fisher were proud to provide methods that were both rigorous, predetermined, and widely applicable. As opposed to the Bayesian methods that *frequentists* denigrated, the frequentist method did not allow proponents of theories to bias the conclusions of their experiments with ill-defined priors.

One central object of the philosophy of *frequentists* was the *p-value* method, also known as the statistical null-hypothesis test. This consists of testing the plausibility of any theory T. For the frequentists, there is a sort of plausibility presumption[3]. According to Popper's philosophy, the statistical test will then attempt to reject the plausibility of the theory T through experimentation. Let's call d the experimental data. If the data d are highly unlikely under theory T, then *frequentists* propose to reject T.

Now, translating this into more mathematical terms raises a major issue, though: if we consider highly precise values of the data, then any datum is highly unlikely. Indeed, if I obtain $d = 0,1583197412 \pm 10^{-10}$, and if my theory asserts that I should get a number between 0 and 1, then the probability of obtaining exactly d down to the 10th digit is roughly one out of a few billion. So it'd be tempting to reject the theory. Except that, weirdly enough, any such precise measurement would have led to such a rejection of the theory!

As a result, *frequentists* propose to map any datum d to the set D of all data that are "even more unlikely than d", according to theory T. Typically, if the theory T says that we should obtain $d \approx 0$, and if we obtain a datum $d > 0$, then "D worse than d" will typically be the set D of data whose values are larger than d, or the set of data whose distances to 0 are at least d.

The infamous *p-value* of datum d, associated to theory T and to the statistical test that we considered, is then defined as the probability p of obtaining some datum D like d or worse, assuming T true. In other words, we define p as

$$p = \mathbb{P}[D \text{ worse than } d | T].$$

This *p-value*, or variants of this quantity, is sometimes known as *likelihood*. If we compare it to Bayes' rule, we see that this quantity resembles the *thought experiment term* that we discussed in chapter 2. This term measures to what extent theory T explains the data "well".

Intuitively, the smaller p is, the more the datum d seems inconsistent with theory T, and the more it's tempting to reject theory T. Fisher proposed to reject theories whose *p-values* took values smaller than 5%.

---

[3] A trolling Bayesian would call it a (bad) subjective prior!

These days, when technologies allow for the collection of billions, or millions of billions of data points, like in some physics experiments, the threshold is often brought down to 0.00003%. Results with *p-values* smaller than a predetermined threshold are called *statistically significant*[4].

Whatever the details, Fisher's principle has been undeniably amazingly fruitful in 20th-century science. For instance, in 2012, the *Large Hadron Collider* at CERN announced the detection of the Higgs boson. In fact, what should technically have been said is that the CERN showed that it was highly unlikely to observe what it observed (or worse), assuming the Higgs boson did not exist in the standard model of particle physics. In other words, the probability of CERN data (or worse), given that the Higgs boson does not exist, is less than 0.00003%. This led CERN scientists to reject the inexistence of the Higgs boson. Or as the media said, they accepted its existence.

Frequentists' methods have ruled almost unchallenged in 20th-century science. Fisher played a central role in this wide adoption. His brisk criticisms of Bayes' rule, and his unwavering willingness to repress any opinion opposed to his genius, made any alternative to his *frequentist* philosophy taboo. He claimed: "The theory of inverse probability [also known as Bayes' rule] is founded upon an error, and must be wholly rejected". If the history of statistics were to be romanticized in a Manichaean manner, and if the *pure Bayesian* were the hero of this telling, then Fisher, more than any other *frequentist*, would be the great villain.

Having said that, setting aside his arrogance, nervousness, and condescendance, as well as his eugenic and racist convictions, Fisher is nevertheless a first-rank mathematician and one of the most influential thinkers of the 20th century. Thanks to his rigor and genius, 20th-century sciences, especially so-called soft sciences, have tremendously improved and gained a lot of credibility.

Nevertheless, our *pure Bayesian* has numerous objections. In fact, the *frequentist* approach makes little sense to her. Why should we accept the plausibility presumption? Why assume *a priori* that all theories are equally valid? Are there not some simpler or more structured theories? And are such theories not more promising? Should we not take into account past successes of theories? Why should we consider data worse than *d*? Is it not some *ad hoc* hack? Is there always a *natural* way to

---

[4] *How p-values help us test hypothesis.* Crash Course Statistics. A Hill (2018).

determine the set of data worse than d? Why would a reject be final? What are we to do if all theories have been rejected? Should we not rather compare the successes of concurrent theories? Why use the threshold of 5%? Why 0.00003%? Is it not completely arbitrary, if not *subjective*? Don't frequentist methods eventually require the *subjective* expertise of a statistician? And wouldn't this statistician be exposed to cognitive biases? What if only a very small amount of data is available? Should we leave the unanswered question completely open? How can we treat life, whose only known exemplary is terrestrial? What about the universe of which there is only one version? How can we deal with the case of the Scottish black sheep?

4.6 STATISTICIANS AGAINST THE *p-VALUE*

The *pure Bayesian* is not the only one to criticize Popper's and the *frequentists'* epistemology. Especially since the beginning of the 21st century, the *p-value* has received a lot of bad press[5]. A major cause of its bad reputation is the selection bias due to the publication of conclusive works only. Worse, we witness more and more *p-hacking* strategies, which we'll get back to later on. No matter the causes, we observe an overabundance of false published results. Statistician John Ioannidis estimates[6] that most published research findings may be wrong[7].

In fact, the presence of errors in scientific publications is likely far larger, especially if we consider non-statistical errors. Computer scientist Leslie Lamport even suggests that one mathematical publication out of three contains a false theorem[8], despite peer review!

Even more surprisingly, if the *p-value* was taken very seriously, then *all* scientific theories should be rejected. Even true ones. Indeed, according to many descriptions of the scientific method, any theory must be tested, again and again. Yet, with the 0.00003% threshold, each experiment has a 0.00003% probability to reject a true theory that it is testing. As a result, even a true theory will be rejected after a few million experiments. And it will be so thousands of times after billions of experiements! Clearly, *if our scientific theories still hold, it's only because they have not been tested enough.* But as long as these theories keep being tested again

---

[5] *P-Value Problems.* Crash Course Statistics. A Hill (2018).
[6] *Why Most Published Research Findings Are False.* PLoS Med. J Ioannidis (2005).
[7] *Is Most Published Research Wrong?* Veritasium. D Müller (2016).
[8] *Comment écrire une démonstration au 21ème siècle.* Math Park. Institut Henri Poincaré. L Lamport (2016).

and again, the day of rejection will come. Unavoidably. Weird, right? The scientific method is doomed to reject all true theories[9].

Statisticians have been more and more vehement over the years. In 2010, Tom Siegfried explained his skepticism[10]: "It's science's dirtiest secret: the 'scientific method' of testing hypothesis by statistical analysis stands on a flimsy foundation". In 2014, Regina Nuzzo added[11]: "[The problem] lay in the surprisingly slippery nature of the *p-value*, which is neither as reliable nor as objective as most scientists assume".

Brisk and statistically grounded criticisms by numerous statisticians built up in 2016 with a statement[12] of the American Statistical Association (ASA): "The statistical community has been deeply concerned about issues of reproducibility and replicability of scientific conclusions[13]. Without getting into definitions and distinctions of these terms, we observe that much confusion and even doubt about the validity of science is arising. [...] Misunderstanding or misuse of statistical inference is only one cause of the 'reproducibility crisis', but to our community, it is an important one".

And all of this culminated in March 2019, with a publication in *Nature* entitled *Scientists Rise Up Against Statistical Significance*, co-signed by 800 scientists and statisticians. At the same moment, the *American Stastician* released a special edition called *Moving to a World Beyond "$p < 0.05$"*, which compiles 43 articles that highlight the flaws, misuses, and troubles with p-values. These articles contain criticisms like "a p-value [...] should not be enshrined as the 'mother of all statistics' for scientific decision-making", "in my proposal, the terms 'significant" and 'nonsignificant' would not be used at all", or "it is remarkable that the most widely used approach - null hypothesis significance testing [also known as the p-value method] - has been subjected to devastating criticism for so long to so little effect". Ioannidis writes: "the use of P values has become an epidemic affecting the majority of scientific disciplines. Decisive action is needed both from the statistical and wider scientific community."

---

[9]For a 5% threshold, around 20 experiments (only!) suffice to reject a true theory. And about 100 experiments will suffice for a 1% threshold.

[10]*Odds are, it's wrong*. Science News. T Siegfried (2010).

[11]*Statistical Errors*. Nature. R Nuzzo (2014).

[12]*The ASA's Statement on p-values: Context, Process, and Purpose*. The American Statistician. R Wasserstein and N. Lazar (2016).

[13]*The Replication Crisis*. Crash Course Statistics. A Hill (2018).

The editors of the journal conclude: "So, let's do it. Let's move beyond 'statistically significant,' even if upheaval and disruption are inevitable for the time being. It's worth it. In a world beyond '$p < 0.05$,' by breaking free from the bonds of statistical significance, statistics in science and policy will become more significant than ever."

4.7   p-HACKING

Of course, there are those that do not sufficiently understand the *p-value*. But there are also those who understand it too well and see the opportunity to use it to boost their careers. This is aggravated by the *"publish or perish"* atmosphere that scientists have to fight through. Yet, to publish, too often, it's necessary, if not sufficient, to obtain (surprising) *p-values* below the 5% threshold. Obtaining such values to reject theories is quite improbable. But not that improbable. If we seek to reject valid theories, the probability of such *p-values* is in fact, by construction, precisely 5%. This means that if I launch 20 distinct unrelated experiments, then, in average, one out of the 20 experiments will yield a publishable *p-value*! In other words, it suffices to multiply experiments to obtain scientifically publishable results. This is *p-hacking*[14].

The danger of *p-hacking* is particularly well illustrated by Randall Munroe's xkcd cartoon entitled *Significant*. Munroe imagined a suspicion that some jelly bean causes acne. A scientist runs an experiment and concludes that the theory T according to which the jelly bean causes acne has a *p-value* larger than 5%. In other words, the experiment does not allow to reject T. So far, so good.

But a second rumor says that only a certain color of jelly bean causes acne. However, there are 20 colors of jelly beans, and thus 20 experiments to run. Not surprisingly, one of the 20 experiments, the one for green jelly beans, for instance, yields a *p-value* less than 5%. This allows to reject the hypothesis according to which green jelly beans do not cause acne. The next day, newspapers' headlines read: "science proved that green jelly beans cause acne with 95% confidence[15]!"

At the scale of all sciences produced on Earth, the number of independent scientific experiments far, far, far exceed 20. Tabloids will easily find thousands of surprising scientific publications to discuss. It's thus not surprising that a large proportion of the publications end up rejected by other works, if not retracted by their authors.

---

[14]*P-Hacking*. Crash Course Statistics. A Hill (2018).

[15]What's more, annoyingly, such too-common phrases confuse the posterior credence of the hypothesis with the likelihood of the data assuming the hypothesis.

Aside from the multiplicity of experiments, there is another method that is just as efficient to obtain scientifically conclusive results: it suffices to cumulate experimental data until the data are conclusive. As counterintuitive as it may sound, it has been proved that all theories end up rejected by the *p-value* method if experimental data are cumulated until a conclusion can be reached[16]! In other words, if your data do not yet allow to reject some theory, be patient. Just collect more data and you will eventually be scientifically allowed to reject the theory.

Do you see the problem? By choosing when to stop the data collection, a huge selection bias is introduced. The worst thing is that, a posteriori, if you publish all of the collected data without detailing the way the number of data points has been determined, then it will be impossible to find holes in your research. Your publication will appear to satisfy the norms of the *p-value* "scientific method". Yet, cumulating data until a statistical conclusion is reached is, unfortunately, a widely adopted approach[17].

One might think that this phenomenon is restricted to social sciences, and perhaps biology and chemistry. This is not the case. Even in physics, and even when the extreme threshold of 0.00003% is used, artefacts of statistical analysis and the multiplicity of scientific experiments have led to statistical errors, as was the case of the 2003 discovery of pentaquarks. This discovery was even confirmed by other independent experiments that tortured their data to reach their conclusions. The discovery was eventually rejected by the scientific community that failed to replicate the original experiment[18].

The absolutely disastrous consequence of the deficiencies of Fisher's "scientific method" by statistical analysis is the loss of confidence in the validity of science. As statistician David Colquhoun put it[19]: "Every false positive that gets published in the peer-reviewed literature provides ammunition for those who oppose science. That has never been more

---

[16]The number of necessary data to be almost sure to reject a hypothesis is however exponential. Nevertheless, Johari, Pelekis, and Walsh have shown that the probability of rejection already increases in a surprising manner even for reasonable amounts of collected data.

[17]Johari and his collaborators propose a variant of the *p-value* to correct the issue. *Always Valid Inference: Bringing Sequential Analysis to a/b Testing*. R Johari, L Pekelis & D Walsh (2015).

[18]The pentaquark seems to have finally been really discovered in 2015.

[19]*The False Positive Risk: A Proposal Concerning What to Do About p-Values*. The American Statistician. D Colquhoun (2019).

important. There is now strong pressure from within science as well as from those outside it, to improve reliability of the published results."

4.8   WHAT A STATISTICS TEXTBOOK SAYS

It's interesting to see what a renowned statistics textbook says on the topic. In his textbook, statistician Larry Wasserman writes: "Results from observational studies start to become believable when:

(i) the results are replicated in many studies,

(ii) each of the studies controlled for plausible confounding variables,

(iii) there is a plausible scientific explanation for the existence of a causal relationship."

We'll discuss *confounding variables* in chapter 13. For now, just note that it's an additional extremely dreadful difficulty that is orthogonal to the numerous deficiencies of the *p-value*.

What I want to insist on is the vagueness that accompanies Wasserman's description of the art of statistics. The quote is full of voluntary imprecise terms, like "start to become", "many studies", "plausible", "scientific explanation", "causal relationship". The word "believable" even sounds like an invitation to Bayesianism!

Note that this is not a criticism of Larry Wasserman or of his book. His lesson is excellent. In fact, it seems that any sufficiently well-written *frequentist* textbook must stress the unavoidable vagueness of applied statistics and to insist on the necessary prudence of any interpretation of statistical data.

Unfortunately, this vagueness also has undesirable consequences. It seems to imply that science can never be conclusive. It leads to the impression that nothing can be proved beyond reasonable doubt in science, which has led many to doubt the effects of vaccination, the existence of climate change, and the danger of tobacco. Even I cringe when a scientist claims to have *proved* the existence of the Higgs boson, or that there is *zero doubt* that the universe is quantum. Our *pure Bayesian* too.

More worryingly, the vagueness opens the door to *malleability*. Malleability is the ability for the scientist to choose how to perform his data analysis. In an ideal world where scientists only tried to extract knowledge from data, this would not be a problem. However, in practice, unfortunately, even scientists are prone to alternate incentives. They might feel the pressure to publish, which incentivizes surprising findings.

Also, like all of us, they might suffer from *motivated reasoning*, typically by trying to justify past successes they had, or by favoring findings consistent with their ideologies. Finally, there are also cases of political or financial corruption.

As Ioannidis notes[20]: "Strong selection biases can make almost everything (seem) statistically significant and it is very likely that these biases do operate in many, probably most scientific fields that use P values, especially with lenient $P < 0.05$ thresholds for claiming success. Implausibly, 96% of the biomedical literature that uses P values in the abstract or in the full text claims statistically significant results."

Philosopher Jacob Stegenga adds[21]: "when evidence can be bent in one direction or another because of the malleability of methods, such bending is very often toward favoring medical interventions, and away from truth". This leads him to a very drastic claim, which he calls *medical nihilism*. This is the view that "even when presented with evidence for a hypothesis regarding the effectiveness of a medical intervention, we ought to have low confidence in that hypothesis[22]".

Interestingly, just like Wasserman, Ioannidis and Stegenga rely on a language that is heavily filled with degrees of credence. This also turns out to be the *American Statistician* editorial's main advice: "Accept uncertainty". The heuristic, or simplified, version of the language of uncertainty is that of words like "unlikely", "plausible", "believable", "highly likely", and "beyond any reasonable doubt". The rigorous translation of these vague degrees of credence is then inescapably a language similar (or *isomorphic*) to Bayesian probabilities.

4.9 THE EQUATION OF KNOWLEDGE

For the *pure Bayesian*, any philosophy of knowledge must boil down to the computation of Bayesian credences. To know is to assign adequate credences to different theories. And to do so, there is a magical formula. As you guessed, this magical formula is Bayes' rule. In particular, the most fundamental version of Bayes' rule, the one that really ought to be called the *equation of knowledge*, is, I believe, the following astounding equation:

---

[20] *What Have We (Not) Learnt from Millions of Scientific Papers with P Values?* The American Statistician. J Ioannidis (2019).

[21] *Medical Nihilism.* Oxford University Press. J Stegenga (2018).

[22] Note that Stegenga invokes Bayesian arguments to reach this harsh conclusion.

$$\mathbb{P}[\mathsf{Theory}|\mathsf{Data}] = \frac{\mathbb{P}[\mathsf{Data}|\mathsf{Theory}]\ \mathbb{P}[\mathsf{Theory}]}{\mathbb{P}[\mathsf{Data}|\mathsf{Theory}]\mathbb{P}[\mathsf{Theory}] + \sum\limits_{\mathsf{Alter}} \mathbb{P}[\mathsf{Data}|\mathsf{Alter}]\mathbb{P}[\mathsf{Alter}]},$$

where **Alter** designates theories that are alternatives to **Theory**. I refer you to chapter 2 to better familiarize yourself with this equation.

I highly encourage you to longly meditate, again and again, this absolutely remarkable equation. In particular, I invite you to ponder how the *pure Bayesian* uses this equation to address the objections she raised against Popper's philosophy.

Let's now analyze the equation of knowledge! To begin, like the frequentist method, Bayes' rule involves the *likelihood* $\mathbb{P}[\mathsf{Data}|\mathsf{Theory}]$ of a data, given the theory. Such a term is, of course, essential for the *pure Bayesian* too, as it measures the ability of a theory to explain observed data. However, for the *pure Bayesian*, this term, which is similar in spirit to the *p-value*, is only part of the equation.

The other important term is the *prior* $\mathbb{P}[\mathsf{Theory}]$. This term is absolutely fundamental for the *pure Bayesian*. It allows her to distinguish theories *a priori*, typically based on their simplicity. As we shall see in future chapters, such criteria are essential for everyday learning, especially when training data are scarce. Better, we shall show in chapters 7 and 12 that the combination of Bayesianism and theoretical computer science implies Ockham's razor, a celebrated philosophical principle that asserts that simpler theories are more believable!

Finally, there is the denominator of the above equation, called the *partition function*. It equals $\mathbb{P}[\mathsf{Data}]$, which we decomposed using the law of total probabilities. One component of the partition function is the numerator itself. But the partition function also consists of similar terms associated to alternative theories. It basically simulates a competition between the different theories - and thereby guarantees that the sum of all credences always equals 1.

As a result, the *pure Bayesian* assigns a large credence to a theory if and only if this theory is vastly more believable than its competition. This means that she will not give a large credence to a theory that explains easy-to-explain observations very well, if other simpler theories explain the observations just as well or better. Conversely, in the case of hard-to-predict phenomena, a theory that gains the *pure Bayesian*'s credences will not have to perfectly explain the phenomena, especially if no other theory does so.

Another thing to discuss is what I mean here by **Data**. This variable does not designate the set of scientific experimental results only. It designates all sensory data that the *pure Bayesian* has had access to in all her life. In particular, this means that no experiment should be analyzed in an isolated manner.

Besides, despite the frequentist protocols that they use, scientists actually rely much more on an accumulation of data than what many descriptions of the "scientific method" often suggest. This is why scientific articles always begin with a long literature review that shows how the presented work contributes to a larger body of work. Scientists arguably reason more according to the principles of Bayesianism than to those of Popper.

4.10 CUMULATIVE LEARNING

In practice, we do not determine the credences of a theory by recalling all the data collected in all our life - notably since, as we shall discuss, our limited memory is quite deficient. Our learning is rather cumulative. Well, Bayes' rule also allows to integrate new collected data to refine our credences. This process is known as *Bayesian inference*. Each inference, or Bayesian update, relies on the partial decomposition of Bayes' rule that follows:

$$\mathbb{P}[T|\text{News and } D] = \frac{\mathbb{P}[\text{News}|T \text{ and } D]\mathbb{P}[T|D]}{\mathbb{P}[\text{News}|D]},$$

where, for more concision, we denoted T for **Theory** and D for **Data**. The particularly interesting case is then the one where the new data News has been gained independently from past data D, in which case[23] we can rewrite Bayesian inference as follows:

$$\mathbb{P}[T|\text{News and } D] = \frac{\mathbb{P}[\text{News}|T]\mathbb{P}[T|D]}{\mathbb{P}[\text{News}|T]\mathbb{P}[T|D] + \sum_{A \neq T} \mathbb{P}[\text{News}|A]\mathbb{P}[A|D]}.$$

Now, what's remarkable is that the above formula is the same as the earlier Bayes' rule, except that the prior here is a probability given already collected data D. In other words, in practice, before acquiring new data News, the *pragmatic Bayesian* will replace the fundamental

---

[23]It is actually necessary that, in all theories T or Alter, News be independent from D.

prior $\mathbb{P}[T]$ by his current credence $\mathbb{P}[T|D]$ in theory T, and his fundamental priors $\mathbb{P}[A]$ on the alternatives by his current credences $\mathbb{P}[A|D]$. It's these current credences that the *pragmatic Bayesian* will use to apply Bayes' rule. "*Today's prior is yesterday's posterior*", as explained by Bayesian Dennis Lindley. What's more, in future chapters, we shall see that this principle turns out to be at the heart of both Darwinian evolution, the reliability of the scientific consensus, and online machine learning algorithms.

These computations also allow to show that, according the *pure Bayesian*, it's very much possible to study history in a rational manner. This is by opposition to certain rigid definitions of what *science* ought to be. Indeed, according to some, history, but also phylogeny and cosmology, are non-scientific disciplines because they do not lend themselves to the reproducibility of experiments. It's noteworthy that such a thought is a pure artefact of Popper's philosophy and frequentist statistics.

For the *pure Bayesian*, there is no fundamental line of demarcation to be drawn between the fields that aim to uncover the past and those that aim to study time-invariant laws. In both cases, what's needed is the collection of relevant data combined to Bayesian inferences to determine the adequate credences on different theories. *If you've got data, then you can apply Bayes' rule!*

In particular, for the *pure Bayesian*, what distinguishes "sciences" from "pseudosciences" is not the falsifiability of the hypotheses of the fields, but the extent to which (an approximation of) Bayes' rule is performed by proponents. Scientists apply Bayes' rule a lot better - and as we'll see, the scientific community applies it even better than any of its members!

4.11   BACK TO EINSTEIN

To better understand the *pure Bayesian*'s philosophy, let's get back to the Einstein conundrum. The first thing to note is that, while too abstract to be understood by almost any physicist of his time, Einstein's general relativity is actually remarkably simple. Indeed, it boils down to saying that 4-dimension spacetime has a curvature determined by equation $G_{\mu\nu} \propto T_{\mu\nu}$. Details do not matter here. What matters is that this equation has only a unique parameter, which is the coefficient of proportionality of the equation. This coefficient describes the intensity with which matter $T_{\mu\nu}$ affects the curvature $G_{\mu\nu}$ of spacetime.

Now, Newton's law of gravity $M_1 \vec{a} \propto m_1 m_2 \vec{r}/r^3$ also contains a proportionality coefficient which describes the intensity with which two massive bodies attract one another through gravitational pull. However, in addition, to fit observations, Newton's gravity assumes that the inertial mass M_1 that fights the acceleration of objects is equal (or proportional) to the gravitational mass m_1 which causes the force of gravity. Thus, even if Newton's law seems easier to conceptualize and even if it lends itself better to computations, Newton's law is actually more arbitrary than Einstein's theory.

These days, we even know the extent to which Einstein's theory is anything but arbitrary. We have proved that the linear combinations of the metric tensor and Einstein's tensor were the only tensors of valency 2 that only consist of derivatives of the metric tensor of at most the second order[24]. Unfortunately, this mathematical theorem is beyond the scope of this book. But importantly, Einstein's equation is absolutely not arbitrary. It is the unavoidable consequence of the simple postulate that says that the spacetime curvature is the cause of what is commonly known as "gravity". This explains why Hilbert independently discovered Einstein's theory. This unavoidable aspect of Einstein's theory makes it greatly more likely *a priori* than Newton's. This is why, even without any observational data, it's not unreasonable to prefer Einstein's theory over Newton's.

When Einstein determined his equation, he did not yet know that it was an inevitable fallout of his initial postulate. But the elegance of his equation, combined with years of frustration of work on the wrong equations, probably led him to suspect the inevitability of his equation. It's thus not surprising that his credence in his new theory at this point was already superior to that of Newton's, or at least sufficiently larger to hope to overthrow Newton. But there is, of course, an empirical datum in particular that would give him huge confidence in his theory.

First, note that Einstein's general relativity behaves like Newton's law whenever gravity is "weak". In fact, in our entire Solar system, to which data at that time was limited, the only location where gravity is sufficiently strong to make Einstein's equations diverge from Newton's was near the Sun. Yet, Newton's theory seemed to fail there, as it got Mercury's trajectory wrong. Granted, a planet like Vulcan might explain Mercury's trajectory. But observers had been searching for Vulcan for

---

[24] *The Einstein tensor and its generalizations.* Journal of Mathematical Physics. D Lovelock (1971).

decades in vain. The absence of evidence, even if it's not the proof of the absence (another Popperian concept!), can only affect the credences on Vulcan negatively.

However, Einstein's theory successfully explains Mercury's trajectory with great precision, in the highly more likely case where no unobserved planet affects its trajectory. Even if the prior credence on Einstein's theory was roughly comparable to that on Newtonian mechanics, the fact that Einstein's theory perfectly explains Mercury's trajectory without any additional unlikely hypothesis leaves no doubt to the *pure Bayesian*: Bayes' rule implies a vastly greater credence in Einstein's theory than in Newton's.

Einstein and Hilbert surely intuited this Bayesian computation. As soon as November 1915, contrary to the entirety of the scientific community, both were already convinced that general relativity was by far the most believable theory of gravity. The *pure Bayesian* agrees.

FURTHER READING

A Treatise of Human Nature. Courier Corporation. D Hume (1738).

An Enquiry Concerning Human Understanding. London: A. Millar. D Hume (1738).

The Logic of Scientific Discovery. Routledge. K Popper (2005).

All of Statistics: A Concise Course in Statistical Inference. Springer Science & Business Media. L Wasserman (2013).

The Big Picture: On the Origin of Life, Meaning and the Universe Itself. Dutton. S Carroll (2016).

Statistical Methods for Research Workers. Genesis Publishing Pvt Ltd. R Fisher (1925).

On the Problem of the Most Efficient Tests of Statistical Hypotheses. Breakthroughs in Statistics. J Neyman & E Pearson (1933).

The Einstein Tensor and its Generalizations. Journal of Mathematical Physics. D Lovelock (1971).

Can I take a Peek? Continuous Monitoring of Online a/b Tests. ICWWW. R Johari (2015).

Always Valid Inference: Bringing Sequential Analysis to a/b Testing. R Johari, L Pekelis & D Walsh (2015).

Why Most Published Research Findings Are False. PLoS Med. J Ioannidis (2005).

Odds Are, It's Wrong. Science News. T Siegfried (2010).

Revised Standards for Statistical Evidence. PNAS. V Johnson (2013).

Statistical Errors. Nature. R Nuzzo (2014).

Editorial. Basic Applied Social Psychology. D Trafinow & M Marks (2015).

The Reproducibility Crisis in Science: A Statistical Counterattack. Significance. R Peng (2015).

The ASA's Statement on p-Values: Context, Process, and Purpose. The American Statistician. R Wasserstein & N Lazar (2016).

Scientists Rise up against Statistical Significance. Nature. V Amrhein, S Greenland & B McShane (2019).

Moving to a World Beyond "p < 0.05". The American Statistician. R Wasserstein, A Schirm & N Lazar (2019).

The False Positive Risk: A Proposal Concerning What to Do About p-Values. The American Statistician. D Colquhoun (2019).

What Have We (Not) Learnt from Millions of Scientific Papers with P Values? The American Statistician. J Ioannidis (2019).

Significant. xkcd. R Munroe.

Pluto is NOT (not?) a Planet. Science4All. LN Hoang (2014).

Hypothesis Test with Statistics: Get it Right! Science4All. LN Hoang (2013).

Is Most Published Research Wrong? Veritasium. D Muller (2016).

Scientific Studies. Last Week Tonight. J Oliver (2017).

How p-Values Help Us Test Hypothesis. Crash Course Statistics. A Hill (2018).

P-Value Problems. Crash Course Statistics. A Hill (2018).

The Replication Crisis. Crash Course Statistics. A Hill (2018).

P-Hacking. Crash Course Statistics. A Hill (2018).

CHAPTER 5

All Hail Prejudices

All human knowledge begins with intuitions, proceeds from thence to concepts, and ends with ideas.

Immanuel Kant (1724-1804)

Under Bayes' theorem, no theory is perfect. Rather, it is a work in progress, always subject to further refinement and testing.

Nate Silver (1978-)

5.1 THE LINDA PROBLEM

Linda is 31 years old, single, outspoken, and very bright. She majored in philosophy. As a student, she was deeply concerned with issues of discrimination and social justice, and also participated in anti-nuclear demonstrations. Which is more probable?

1. Linda is a bank teller.

2. Linda is a bank teller and is active in the feminist movement.

Please take the time to think this puzzle through. I invite you to give your answer out loud or to write it down before continuing to read.

This problem, famously known as the *Linda problem*, was proposed by Amos Tversky and Daniel Kahneman, two psychologists that aimed at better understanding the way people think. Thanks to this work and others, Kahneman would win the 2002 Nobel prize in economics. In his excellent book *Thinking Fast and Slow*, he wrote that, had Tversky not died in 1996, the two researchers would have likely shared the Nobel prize.

If the Linda problem has become well known, it's due to its huge error rate. Between 85% and 91% of people that Tversky and Kahneman interrogated gave the wrong answer, in different replications of the experiment. Our brilliant human brains are doing here much worse than a chimpanzee whose answers are random.

Some critics noted the ambiguity of answer 1, which might suggest that Linda is not active in the feminist movement. However, once answer 1 has been clarified and replaced by "Linda is a bank teller and she is or is not active in the feminist movement", the error rate was still 57% - still worse than the chimpanzee's.

If you have not solved the problem, you might be surprised that there is a *right* and a *wrong* answer. Well, it is the case. Indeed, answer 2 turns out to be a special case of answer 1. In other words, if 2 is true, then so is 1. In Venn diagram terms, the set of cases where 2 is true is a subset of the set of cases where 1 is true. In probability terms, this boils down to the inequality $\mathbb{P}[\mathsf{Bank\ and\ Feminist}] \leq \mathbb{P}[\mathsf{Bank}]$. The probability that two events both occur is always less than the probability that one does. The mathematics is indisputable. Answer 1 is the *right* answer.

If you did not get the right answer, you might take comfort in the fact that you belong to the majority. Few successfully do the mathematical reasoning. Tversky and Kahneman argue that, instead, people reason by association. The question they ask themselves is not that of the probabilities of answers 1 and 2, but that of their representativity with respect to the preamble. Answer 2 seems more "representative" of 31-year-old single women, who have studied philosophy and participated in movements against discrimination.

5.2 PREJUDICES TO THE RESCUE OF LINDA*

There may be a more enlightening way to understand the incredibly high error rate of the Linda problem, which leads us back to the debate opposing frequentists to Bayesians. According to a *pure frequentist*, a hypothesis test consists of studying the probability of data given the hypothesis. Here, the data we have about Linda is the preamble. The hypotheses to be considered are answers 1 and 2. The *pure frequentist* will thus focus on probabilities $\mathbb{P}[\mathsf{Preamble}|1]$ and $\mathbb{P}[\mathsf{Preamble}|2]$, where 1 and 2 designate answers 1 and 2, respectively.

It's then reasonable to think that the preamble is more likely assuming 2 than assuming 1. Formally, this corresponds to inequality $\mathbb{P}[\mathsf{Preamble}|1] \leq \mathbb{P}[\mathsf{Preamble}|2]$. In statistical terms, the preamble is more *likely* for answer 2. The *pure frequentist* would say that answer 2 is the *maximum of likelihood.*

Unfortunately, the "scientific method" that we discussed in the previous chapter often focuses solely on these *likelihoods*. The terminology does not help. In fact, it is deeply misleading. It leads to the confusion between the *likelihood* of the data given the hypotheses, and the *credence* of the hypotheses given the data. The *pure Bayesian* regards this confusion as the crux of the frequentist fallacy.

As we saw, these *likelihoods*, which I called *thought experiment terms*, are only part of Bayes' rule. Instead, according to the *pure Bayesian*, the important quantity is the inverse probability. In the case of the Linda problem, this inverse probability is that of the answers given the preamble. For answer 1, given our definition of conditional probabilities, this inverse probability is computed as follows:

$$\mathbb{P}[1|\text{Preamble}] = \frac{\mathbb{P}[1 \text{ and Preamble}]}{\mathbb{P}[\text{Preamble}]} = \frac{\mathbb{P}[\text{Bank and Preamble}]}{\mathbb{P}[\text{Preamble}]}.$$

Similarly, the probability of answer 2 given what we know is

$$\mathbb{P}[2|\text{Preamble}] - \frac{\mathbb{P}[\text{Bank and Feminist and Preamble}]}{\mathbb{P}[\text{Preamble}]}.$$

The *pure Bayesian* now concludes using the fact that the probability that two events both occur is less than the probability that one does. As a result, we necessarily have $\mathbb{P}[\text{Bank and Preamble}] \geq \mathbb{P}[\text{Bank and Feminist and Preamble}]$. This is what allows the *pure Bayesian* to give the right answer. As opposed to the *pure frequentist*.

Better, we can compute the extent to which the *pure Bayesian* finds answer 1 more likely than answer 2, given the preamble. Indeed, by playing with the laws of probability - which I invite you to do again and again at home - we see that her computation is equivalent to what follows:

$$\mathbb{P}[2|\text{Preamble}] = \mathbb{P}[\text{Feminist}|\text{Preamble and Bank}] \cdot \mathbb{P}[1|\text{Preamble}].$$

In other words, no matter what the *pure Bayesian*'s assumptions are, they will have to obey the laws of probability. As a result, the *pure Bayesian* will necessarily conclude that, given the preamble, the probability of hypothesis 2 is $\mathbb{P}[\text{Feminist}|\text{Preamble and Bank}]$-times that of hypothesis 1. Since all probabilities are at most 1, we conclude that, given the preamble and no matter which (Bayesian) model is considered, hypothesis 2 will *always* be less likely than hypothesis 1. And the comparison of the probabilities of the two hypotheses boils down to the computation of a *thought experiment term* - namely the probability that Linda is active in the feminist movement, given the preamble and the fact she is a bank teller.

5.3   LONG LIVE PREJUDICES

You may have felt like the *pure Bayesian*'s reasoning was too cumbersome. After all, it sufficed to note that answer 2 implied answer 1 to determine the right answer to the Linda problem. You would be entirely right. The reason why I presented the *pure Bayesian*'s reasoning is not so that you better understand the Linda problem; it's to better illustrate how the *pure Bayesian*'s reasoning diverges from the *pure frequentist*'s. Namely, the *pure Bayesian* is unhappy with the computation of the *likelihood* alone. Her analysis also includes her prior belief in answers 1 and 2.

Formally, the quantity that the *pure Bayesian* cares about is not the likelihood $\mathbb{P}[\text{Preamble}|1]$, but the inverse probability $\mathbb{P}[1|\text{Preamble}]$, which may be called the credence of theory 1. This can be derived from the likelihood using Bayes' rule:

$$\mathbb{P}[1|\text{Preamble}] = \frac{\mathbb{P}[\text{Preamble}|1]\mathbb{P}[1]}{\mathbb{P}[\text{Preamble}]}.$$

In particular, the *likelihood* $\mathbb{P}[\text{Preamble}|1]$ must *always* be accompanied with the *prior* $\mathbb{P}[1]$.

Now, in this book, I will take the risk of replacing the technical terminology of *priors* by a synonym with a negative connotation: *prejudices*. I am doing so out of intellectual honesty. Given my pro-Bayesian bias, I want to try to defend a negative version of Bayesianism, and to convince you that even this version is appealing - by opposition to a *bullshit* approach that would consist of playing with the connotation of words for better persuasion. Please keep in mind what *prejudice* literally means: it's a judgment prior to observation.

Of course, this can also be described by positively connotated phrases, like the *current state of knowledge*. But I'll stick with *prejudices* here.

Prejudices are at the heart of the opposition between Bayesians and frequentists. *Prejudices* are the main reason why in the last two centuries Bayes' rule has been rejected. Scientists wanted their work to be *objective*. Yet, *prejudices* seemed necessarily *subjective*. As a result, frequentists and most scientists regarded the subjectivity of *prejudices* as the fatal deficiency of Bayesianism.

However, while they are subjective, *prejudices* are definitely *not* arbitrary, especially if they have been computed through Bayes' rule and if they obey the laws of probability! In fact, like in the case of the Linda problem, the *pure Bayesian* considers that *prejudices* are the key assets of Bayesian reasoning - at least if these *prejudices* have been determined by Bayes' rule. According to the *pure Bayesian, prejudices* are essential to reason correctly. *Prejudices form a cornerstone of rationality.* This is the most controversial claim of Bayesianism.

5.4   xkcd's SUN

Such a counterintuitive and criticizable claim definitely requires more than the example of Linda. So let's now discuss the formidable thought experiment proposed by Randall Munroe on xkcd[1], which I'll slightly romanticize.

It's almost midnight in Paris, where you currently live. Your intern is in Hawaii. You ask him to call you at midnight. But before that, he will have thrown two dice. If he lands double sixes, then he calls you to tell you that the sun disappeared. Otherwise, he will tell you *if* the sun disappeared. Midnight comes and the intern says that the sun disappeared. Can you conclude that the sun indeed disappeared?

A *p value frequentist* would compute the *p-value* of the theory that we may want to reject. This theory ☀ says that the sun is still here and did not disappear. The corresponding *p-value* thus equals $p = \mathbb{P}[\text{☎}|\text{☀}]$. It's the probability of what you observe[2] - namely the fact ☎ that your intern told you that the sun disappeared - assuming that the sun did not disappear.

However, given ☀, you only could have received the phone call you received if your intern rolled two sixes. The probability p that your intern tells you that the sun disappeared assuming ☀ is thus equal[3] to the

---

[1] *Frequentists vs. Bayesians.* xkcd. R Munroe.

[2] Here, there is no "worse data than observed" to consider.

[3] The rigorous computation here involves the law of total probability, which I encourage you to write down.

probability of two sixes, which is $p = 1/36 \approx 0.028$. In particular, we have $p < 0.05$. The fact that the intern tells you ☏ assuming ☀ is thus highly unlikely. We can conclude: we reject theory ☀. Yet, rejecting ☀ is rejecting the fact that the sun did not disappear: we conclude that the sun disappeared.

Incredible! By following the footsteps of *p-value frequentists*, we end up with an absurd conclusion: the mere fact that the intern tells you that the sun disappeared suffices to conclude that the sun indeed disappeared!

Monroe's cartoon ends with a reply of an amused *pure Bayesian*: "bet you $50 it hasn't". Weirdly enough, the conclusion of the *pure Bayesian* sounds much more sensible than that of he who followed the "scientific method" as prescribed by *p-value frequentists*.

5.5   PREJUDICES TO THE RESCUE OF xkcd

Fortunately, whether they know it or not, scientists have Bayesian reasoning tendencies. None of those to whom I presented this thought experiment asserted that the *pure Bayesian*'s reaction was ridiculously irrational. None said it was *pseudo-scientific*. To understand why, it's worth focusing on how the *pure Bayesian* determines her credences. Why did the *pure Bayesian* keep believing in the existence of the sun, despite the intern's phone call?

As you may expect, the answer lies in the *pure Bayesian*'s prejudices. Her prejudices are not arbitrary. They rely on all past observed data. Let's call LifeExperience the set of all past data of the *pure Bayesian*, and ☏ the fact that the intern announced the disappearance of the sun. Bayes' rule asserts that

$$\mathbb{P}[☀|☏\text{ and LifeExperience}] = \frac{\mathbb{P}[☏|☀\text{ and LifeExperience}]\mathbb{P}[☀|\text{LifeExperience}]}{\mathbb{P}[☏|\text{LifeExperience}]}.$$

The details may be overwhelming. Don't worry. Our goal here is not a full understanding of the equation - though I invite you to ponder it at length. What I want to insist on is the quantity $\mathbb{P}[☀|\text{LifeExperience}]$. It is the *pure Bayesian*'s *prejudice* before the intern's call. It is a *prior* conditioned by the *pure Bayesian*'s life experience. Importantly, this life experience includes the observation of the sun rising all mornings thus far, but also the physics textbooks she read. These textbooks claim that the sun is a plasma ball, energetically fueled by the nuclear fusion of hydrogen nuclei, and whose hydrogen quantity is largely sufficient for the sun to keep shining for billions of years.

In particular, while the *pure Bayesian*'s prejudice definitely *is* subjective, it is not arbitrary! In this case, it relies, among other things, on centuries of scientific research. In other words, all the *pure Bayesians* whose life experiences, albeit distinct, include the observation of thousands of sun rises will agree on one point: the *p-value frequentist*'s conclusion is *wrong*. Indeed, for all these *pure Bayesians*, the probability $\mathbb{P}[\text{no} \; \text{☼} \; |\text{LifeExperience}]$ that the sun disappeared, prior to the intern's call, is ridiculously small. This prejudice is what allows them to reject the sun's disappearance despite the intern's call.

More generally, each piece of data cannot be analyzed independently. *A datum is like a rock. It has value only as part of a larger construction.*

5.6   PREJUDICES TO THE RESCUE OF SALLY CLARK

Let's move on to a third example illustrating the importance of prejudices. Our third example will be the lawsuit of Sally Clark that we discussed in chapter 2. Recall that within two years, Sally Clark sadly lost her two babies. This led to suspicions of a double murder. However, we saw that the important quantity is the probability $\mathbb{P}[\text{☺}|\text{⚰}]$ of her innocence ☺ given the double death ⚰ of her two babies. We applied Bayes' rule:

$$\mathbb{P}[\text{☺}|\text{⚰}] = \frac{\mathbb{P}[\text{⚰}|\text{☺}]\mathbb{P}[\text{☺}]}{\mathbb{P}[\text{⚰}]}.$$

The *p-value* or *likelihood* approach invites us to focus on the *thought experiment term* $\mathbb{P}[\text{⚰}|\text{☺}]$, which measures the difficulty of explaining the double death assuming Sally Clark's innocence. The *prosecutor fallacy*, the one that Professor Sir Roy Meadow was victim of, is a profoundly *frequentist* argument, since it insists on the incredible smallness of this *thought experiment term*. There is no doubt. The *likelihood* of the data, which is the probability $\mathbb{P}[\text{⚰}|\text{☺}]$ that an innocent person suffers the death of her two babies, is ridiculously small. Meadow estimated it at one out of 70 million, which is about 0.000001%. This is below the 0.00003% threshold of physics! Like the prosecutor, a *p-value frequentist* would have to reject the hypothesis ☺ of Sally Clark's innocence. He would have to convict her.

The way mathematics professor Ray Hill contested this decision is based on a profoundly Bayesian argument. For Hill, one *must* apply

Bayes' rule, which means taking into account the presumption of innocence $\mathbb{P}[\text{ }]$. It's only once this *prejudice* has been added to the reasoning that we reach an actual understanding of Sally Clark's case. Only then can we make an informed decision about her deserved sentence.

As opposed to what some defenders of the "scientific method" often claim, prejudices are not a flaw in our reasoning that we need to get rid of. Prejudices are essential to reason well.

5.7   PREJUDICES AGAINST PSEUDO-SCIENCES

This principle that was critical to address the Linda problem, the xkcd sun thought experiment, and Sally Clark's lawsuit finds numerous applications when dealing with highly likely or highly unlikely theories.

Indeed the *pure Bayesian* does not waste her time listening to those who claim to know how to create energy out of nothing. Since the principle of (local) energy conservation is a fundamental principle of all theories of physics, she has a strong prejudice against the possibility of creating free energy. What's more, even assuming energy conservation, it remains very likely that *some* people on Earth are convinced, for the wrong reasons, that they successfully achieved perpetual motion[4] - we'll get back to the Bayesian interpretation of the argument of authority in chapter 10. No need to watch a large number of videos. Prejudices suffice to reject such experiments.

Similarly, the *pure Bayesian* has a strong prejudice against pseudo-scientific theories on paranormal phenomena. Telekinesis, which supposedly allows to bend spoons, and precognitions able to foresee the future both violate fundamental principles of physics that have gained her credences. The numerous known cases of fraud, and the numerous cognitive biases that explain why attendants got convinced, are additional reasons why the *pure Bayesian* will not change her credences, even when defenders of pseudo-scientific theories try to convince her. To be exposed to such defenders is indeed highly likely, even under the hypothesis that pseudo-sciences are wrong.

This does not mean that the *pure Bayesian* cannot change her mind. But for her to do so, you will need to present an extraordinary datum D. For there to be a transfer of credence, $\mathbb{P}[D|A]$ will have to be very large for some alternative theory A, by opposition to the *thought experiment* terms $\mathbb{P}[D|T]$ for all believable theories T - and we shall see in chapter 10 that selection bias often makes $\mathbb{P}[D|T]$ quite large, even when data

---

[4] *Are perpetual motion machines possible?* Physics Girl. D Cowern (2015).

D seem mysterious. More precisely, for alternative theory A to suddenly become as believable as T, the ratio $\mathbb{P}[D|A]/\mathbb{P}[D|T]$ will have to be at least as large as the inverse ratio $\mathbb{P}[T]/\mathbb{P}[A]$ of prior credences[5]. Yet, the monumental pile of scientific knowledge gathered over the last centuries often makes the latter ratio gigantic. Carl Sagan once said: "Any extraordinary claim requires extraordinary proofs." We have just derived this principle from Bayes' rule!

Another way a *Bayesian* may radically change his mind is if he is presented a theory that he has never thought of - in chapter 7, we'll see that this cannot apply to the *pure Bayesian* as she knows all (computable) theories. This alternative theory will have to be theoretically more solid than established theories, and it will have to explain all known physical observations at least as well as established theories do. Finally, it will have to better explain some physical phenomenon, as Einstein's general relativity better explained Mercury's trajectory than Newton's.

On the contrary, the *pure Bayesian* may have large credences on some claims, despite a blatant lack of empirical data. For instance, if I tell her I went for a 6000-meter-high summit in the Himalayas and suffered a lot, she will need no additional data to believe that it wasn't a piece of cake. If the *pure Bayesian* believed my story based on a single testimony, it's because she had a strong (justified) prejudice about my inability to easily climb high summits.

5.8 PREJUDICES TO THE RESCUE OF SCIENCE

Prejudices also have very useful consequences on the way scientists deal with empirical anomalies. For instance, when the OPERA experiment thought that they had detected faster-than-light neutrinos, their announcement was accompanied with general skepticism. Not only by theoretical physicists, but also by the authors of the experiments! Eventually, no one got really surprised when experimental flaws were discovered. The physicists' credences on Einstein's special relativity (which asserts that no particle can move faster than light) is so huge that the hypothesis of an experimental error was far more believable than that of a violation of Einstein's theory[6].

What may seem weirder is that even mathematicians can have strong credences on unproved theorems. For instance, these days, the vast

---

[5]I invite you to derive this from Bayes' rule!

[6]*Why you can't go faster than light (with equations)*. Sixty Symbols. M Merrifield (2017).

majority of number theorists believe in the famous Riemann hypothesis[7], which is often regarded as the most prestigious open problem of mathematics. Their belief is so strong that, these days, a huge number of theorems assume the Riemann hypothesis to be true to explore its consequences.

Similarly, many computer scientists have strong credences on the conjecture $P \neq NP$, which is often regarded as the most prestigious open problem in theoretical computer science. While these credences are not founded on indisputable rigorous mathematical proofs, they are nevertheless somewhat justifiable by (pragmatic) Bayesian arguments. Typically, I highly advise you to read Scott Aaronson's excellent blog entry[8] to better understand the origin of theoreticians' credences on unproved theorems. In fact, Aaronson's credences are so great that the publication of mathematical proofs of $P = NP$ hardly affects his credences on the opposite conclusion. Aaronson seems to find much more believable that these mathematical proofs are erroneous. History has been proving him right thus far.

Identical reasonings apply to more controversial topics. In 2016, the American government decided to authorize the commercialization of GMO mushrooms without any testing procedure[9]. This decision raised serious concerns in Europe, where GMOs have a bad reputation. Authorizing GMOs is already controversial. Doing so without tests seems absolutely outrageous to many!

Yet, this announcement has been accompanied with joy and hope by scientists. Biologist Caixa Gao asserted: "The research community will be very happy with the news". Is the research community behaving like a collective of mad scientists? Are they even aware of potential risks on public health? Are they not creating some Frankenstein monster?

To understand the scientists' viewpoint, it's essential to reason in terms of (founded) prejudices. To begin, it's noteworthy that all organisms are constantly genetically modified. Indeed, at each reproduction, genes of two different sexes are combined to form a new DNA strand which almost surely never existed before. In addition to this mixing, DNA mutations also change the genetic legacy of organisms.

---

[7] *Visualizing the Riemann zeta function and analytic continuation.* 3Blue1Brown. G Sanderson (2016).

[8] *The Scientific Case for $P \neq NP$.* Shtetl-Optimized. S Aaronson (2014).

[9] *Gene-Edited CRISPR Mushroom Escapes US Regulation.* Nature. E Waltz (2016).

In Nature, natural selection then favours some genetically modified organisms over others. In agriculture, a similar, but artificial, selection occurs. For thousands of years, domesticated species, both vegetal and animal, have been undergoing this artificial selection of genetically modified organisms. This has completely altered domesticated lives, thereby transforming agressive wolves into docile chihuahuas and disgusting small wild bananas full of large seeds into the delicious bananas that we eat every day. All the organisms that we are familiar with, by opposition to those that existed a few centuries ago, are genetically modified organisms.

But this is not all. There are also changes to local biodiversity due to, among other things, large-scale monoculture exploitations, importations of species from distant lands, increased use of pesticides, and even more recent ultraviolet irradiations that aim at accelerating the rate of genetic mutations.

All these genomic modifications are fast and hardly controllable. There are great uncertainties with respect to mutated species. This makes the *pure Bayesian* suspicious about their potential harm to public health - even though the meta-analysis over several decades of scientific research suggests that GMOs are no more dangerous for public health than traditional agriculture[10].

On the other hand, since 2012, biology researchers have discovered a way to edit the genomes of organisms. This technology, called CRISPR Cas9, allows editing of individual letters of DNAs. In other words, they allow to know exactly which modifications to genomes have been made. As a result, the *pure Bayesian* considers that GMOs obtained by CRISPR Cas9 through a controlled and purposeful protocol are, *a priori*, much more reliable than the GMOs obtained through methods whose control of DNA modifications is much weaker[11]. This explains why the authorization of the commercialization of CRISPR Cas9 GMOs has not worried biologists much. The *pure Bayesian* has a justified prejudice due to the fact that such GMOs are much less dangerous than the new plant varieties obtained through traditional means.

Whether it comes to GMOs or medical diagnostics, to the Linda problem or to justice, to experimental or theoretical sciences, the *pure*

---

[10] *An overview of the last 10 years of genetically engineered crop safety research.* Critical Reviews in Biotechnology. A Nicolia, A Manzo, F Veronesi & D Rosellini (2013).

[11] Though the control of CRISPR Cas9 modifications does not seem perfect for now.

Bayesian cannot think clearly without prejudices. Prejudices are her secret weapon. They are the origin of the success of her predictions.

5.9   THE BAYESIAN HAS AN OPINION ON *EVERYTHING*

Let's consider a hypothesis H that has not been studied much. Assume that nearly no datum (directly) relevant to testing H has been collected[12]. Should we believe H? Rather than taking the risk of an error, some scientists will assert that it's better to say "I don't know" or "I have no idea". Some Bayesians would add that it's reasonable to adopt a so-called *non-informative* prior. This prior often takes the form of a *uniform* distribution, that is, it favors no hypothesis in particular. In particular, if H is either true or false, then it seems reasonable to assume that, *a priori*, H has a $1/2$-probability to be true - or rather, it should be assigned a prior credence of $1/2$. But this actually seems very unsatisfactory, for several reasons.

The first reason is the most fundamental one. This position is generally incompatible with Bayesianism. Indeed, even when we have only barely studied H, quite often H is related to questions that we did study, and for which numerous data have been gathered. This is typically the case of all the examples of this chapter. Whether it's about Linda, the sun, Sally Clark, conspiracy, theorems, or GMOs, we have all done a lot of thinking and have gathered numerous data on related topics. Worse, Stein's paradox, which we shall discuss in chapter 13, proves the (statistical) *inadmissibility* of the chunking of knowledge. In particular, since the *pure Bayesian*'s reasoning is necessarily *admissible*, it's *impossible* that her life experiences have absolutely no effect on her credences on H. As a result, it's extremely unlikely that her computations lead to precisely $\mathbb{P}[H|\mathsf{LifeExperience}] = 1/2$.

The second reason is about motivation. Saying "I don't know" encourages laziness. As Poincaré once said: "To doubt everything or to believe everything are two solutions equally convenient, as both exempt us from thinking." In particular, determining the prior $\mathbb{P}[H|\mathsf{LifeExperience}]$ is a subtle and difficult computation. But technically, it's just a computation. Granted, unlike the *pure Bayesian*, we cannot perform this computation instantly. Still, it seems irrational not to take the time to find some fairly good approximation of it. Now, it will

---

[12]To fix ideas, you may consider the hypothesis H that says that the Big Bang is a rebound of an older universe.

be important to keep in mind that our heuristic computations are only a rough approximation. In particular, this implies that the validity of the approximation might be questioned by purely theoretical arguments - typically those that show that some overlooked computations actually have a strong impact on the quality of the approximation. Nevertheless, crucially, to find the motivation to perform such tough and tricky computations, we must first convince ourselves that the answer "I don't know" is *not* satisfactory.

The third reason is pedagogical. If we never express our prejudices, then we can never know that they were misconceptions. And we may fail to realize our ignorance and the biases of our unconscious beliefs. To fight overconfidence, it seems much preferable not to discard the chance to highlight the mistakes of our intuitive prejudices. Au contraire, it seems preferable to finish phrases like "I may be wrong, but I would bet that", "it's probably naive, but it seems to me that" or "before discovering X, I thought that". By making our prejudices explicit, we better underline their inadequation with empirical data. It's then much easier to change our mind. And we may then catch the habit of making our beliefs evolve. *Learning is a dance.* So let's dance. This is how we can more easily identify the cases where we can reasonably trust our intuitions, by opposition to those where our intuitions are definitely *not* reliable. In such cases, it will then be easier to defer our judgment to some mathematical model or some renowned and trustworthy authority.

The fourth reason is playful. Yes, because finding out that some prediction is remarkably right is quite pleasing. Just look at physicists who take pleasure repeating experiments whose outcomes are well known! Being right is nice. But being wrong can be even better! In fact, discovering that a very convincing intuition turns out to be flawed can be an intense enjoyment. This is what happened when physicists thought they had detected faster-than-light neutrinos[13], or when mathematicians discovered that consecutive prime numbers did not quite behave randomly[14]! As Isaac Asmiov argues: "The most exciting phrase to hear in science, the one that heralds new discoveries, is not 'Eureka!' [I found it] but 'That's funny'." He who has never lived the ecstasy of the discovery of a counterintuitive fact will never understand the raison d'être of scientists. Yet the key to the ecstasy is not the fact. It's the rejection of some very believable prejudice.

---

[13] *Neutrinos slower than light.* Sixty Symbols. E Copeland & T Padilla (2012).

[14] We shall come back to this in chapter 14.

In practice, though, unfortunately, our social, educative, and professional environment tends to stigmatize errors. We usually get blamed for our errors. This is why we have learned to fear them. And this is probably why saying "I don't know" or $\mathbb{P}[H|\mathsf{LifeExperience}] = 1/2$ is such a popular loophole. But this actually has very harmful consequences, especially when it comes to learning mathematics. Because mathematics is the one field where errors are easiest to spot, to avoid any error, it's tempting to shut up and never reveal our prejudices about mathematical conjectures. Worse, this leads many to lose all their capabilities when they are told that they are given a *math* problem. This has even been dubbed "mathematical anxiety", which has its own Wikipedia page. Far from affecting only "bad" people in maths, this syndrome seems to cause a drop in math skills for all those who fear errors too much.

By contrast, my first-year university math professor, the one who really made me discover the joy of doing math, often used phrases like "I believe it" or "I don't believe it" when judging our mathematical explanations - and they were usually not understatements! It was a deliverance for me to realize that. We *can* treat math lightly. We *can* bet on the truth of a theorem or on the validity of a proof (sketch). And we may be wrong in our bets. But it's actually quite pleasing to fail this way! It's such errors of our intuitions that will allow it to improve - and witnessing the progress of the mathematical intuition is also pretty sweet! In mathematics, perhaps more than in other fields, celebrating the errors of our intuitions seems like an essential first step towards a more sound and efficient learning.

Nevertheless, this glorification of errors may feel cringeworthy to you. A doctor that prescribes the wrong medicine or a surgeon that cuts some vein by mistake should not be celebrated. Similarly, we tend to think that a politician acknowledging his errors should not be promoted. And rather than talking rubbish or polluting debates with stupid remarks, when it comes to public speaking with potentially major consequences, asserting "I don't know" seems like a good rhetoric strategy. For instance, this can be a way to stress the difficulty of the problems at stake.

Sure. But in all these examples, it's important to see that there is a moral (or strategic) dimension that enters in play. Yet, Bayesianism is not a moral philosophy. It thus has nothing to say about what is *morally desirable*, or about one's selfish goals. Bayesianism is a philosophy of science. Its purpose is to organize learning and knowledge. It's for this purpose that it seems desirable to celebrate errors.

To elicit Bayesian opinions, philosopher Thibaut Giraud proposes *epistemic bets*. These are bets that we have to make, because a gun is pointed at us. We may hesitate and "freak out". But according to Giraud, unless estimating exactly $\mathbb{P}[H|\mathsf{LifeExperience}] = 1/2$, it would be irrational to throw a coin rather than to bet H or non H. Computer scientist Alexandre Maurer goes further and imagines a thought experiment where a serial killer proposes a dilemma. Either you choose to bet on H, and if H does not occur, he'll kill you. Or you choose not to bet on H, in which case the serial killer will throw a die. If the die lands on 6, he will then kill you. Will you bet on H? If you really hesitate, this means that you assign a probability of roughly[15] $5/6$ to H.

More generally, Bayesianism *imposes* to have prejudices on *everything*,[16] except moral judgments that are beyond the realm of Bayesianism[17]. The probability $\mathbb{P}[H|\mathsf{LifeExperience}]$ always has a precise value. And it seems unlikely that this value is $1/2$. According to Bayesianism, *sweeping prejudices under the rug would be very irrational*.

5.10   ERRONEOUS PREJUDICES

This glorification of prejudices is likely to surprise you. After all, in common language, the word "prejudice" is very negatively connoted. And for good reasons. Prejudices seem to lead to discrimination, racism, sexism, or, more generally, to the oppression of misunderstood minorities. However, there are two crucial distinctions to be made. The former is between Bayesian prejudices and erroneous prejudices. The latter is between the quest of a better understanding of the world and the use of this understanding.

Let's start by reminding ourselves that nearly all of our prejudices are non-Bayesian, in the sense that they have not been derived from (an approximation of) Bayes' rule. In fact, our prejudices are almost always inconsistent with the laws of probability. We saw it. Even the greatest mathematicians fail to understand and apply Bayes' rule, even in simple cases. Our prejudices are thus poorly justified, especially if we have not taken the time to ponder their origins.

One recurring fallacy is the lack of contextualisation of Bayesian credences. For instance, movies and TV shows often present the stereo-

---

[15]Evidently, the killer may choose a different probability of killing you if you don't bet on H, so as to help you better figure out your epistemic belief!

[16]Or at least, on every future data.

[17]Technically, it's a bit trickier than this. At least in Solomonoff's setting, there should only be probabilities on data and on (computable) probability law on data.

type of Asians witb poor driving skills. Amusingly, this stereotype is justifiable. Indeed, the vast majority of Asians have grown up in developing countries where learning to drive was not a given. In Vietnam where I was born, especially a few years ago, traffic was so chaotic that taking the scooter or the taxi was more practical. It thus seems reasonable to assert that Asians do not drive as well as Americans for many of whom having a car is an essential part of life. However, if you now consider an Asian who grew up in a Western country (like me!), then the difference in driving competence with an American person is much less clear, if not essentially inexistent[18].

Similarly, Andre Kuhn, Professor of Criminology and Criminal Law at universities of Lausanne, Neuchâtel, and Geneva, asserts that foreigners are associated with a higher crime rate. If I ask the *pure Bayesian* to think about a local and a foreigner, she will likely assign a slightly larger credence on the foreigner's criminality than on the local's. However, if I now tell her that the local and the foreigner are both young and male people from a modest socio-economic background, then André Kuhn asserts that the *pure Bayesian* would now assign similar credences on the two individuals' criminality. Indeed, the reason why she thought that the foreigner was slightly more likely to be a criminal is that the foreigner is more likely to be young, male, and from a modest socio-economic background. However, once we have contextualized adequately the setting, the fact that the two individuals were local or foreigner now has a negligible effect on the *pure Bayesian*'s credences on their criminality.

More generally, we tend to like describing, publishing, and reading relations like "A causes B". In probabilistic terms, this may be interpreted as the fact that $\mathbb{P}[B|A]$ is far greater than[19] $\mathbb{P}[B|\text{not } A]$. However, general phrases generally do not apply to individuals. Indeed, each individual A has all sorts of characteristics Z that distinguishes him from the generic individual A. The quantity that applies to this individual is then $\mathbb{P}[B|A, Z]$, not $\mathbb{P}[B|A]$. Yet, the two quantities may be very different. Typically, the body mass index (BMI) is very useful to describe populations, but it's not necessarily conclusive for an individual in particular. The trouble, of course, is that listing all values $\mathbb{P}[B|A, Z]$ for different values of Z is unlikely to make the headlines of the news.

---

[18]In mathematical terms, this corresponds to saying that $\mathbb{P}[A|B \text{ and } C]$ can be very different from $\mathbb{P}[A|B]$.

[19]In chapter 13, we'll see that this is not exactly what Fisher meant by "A causes B", but the criticism below still applies to Fisher's definition.

Worse, such links of causality may be of limited interest in practice. Indeed, the interesting question is often rather to know if an individual Z can hope to obtain B, if he *starts* doing A. Yet, starting to do A, call it $\mathsf{do}(A)$, is not the same as doing A. For instance, bodybuilders can lift heavy weights. But if you start bodybuilding exercises, then you will likely soon be unable to lift anything! The relevant quantity to advise individual Z is then $\mathbb{P}[B|Z, \mathsf{do}(A)]$, not $\mathbb{P}[B|Z, A]$. Unfortunately, estimating $\mathbb{P}[B|Z, \mathsf{do}(A)]$ is often far more difficult than estimating $\mathbb{P}[B|Z, A]$, and even more so than $\mathbb{P}[B|A]$!

Another cause of flawed prejudices is *selection bias*. This is still essentially a matter of contextualisation. Namely, it's important to realize that all of our observations are biased by a context. Thus, they may not be representative of similar observations made in different contexts. When I told an Indonesian guide that I was a mathematician, he looked at me with surprise: "but... you're not old!" Weirdly enough, the most widespread image of a mathematician is that of an old bearded wise man. But this is a biased sample of mathematicians. I replied: "All old mathematicians were once young". I had no need to formalize the proof.

Excessive generalizations, overfitting on irrelevant characteristics, and selection bias are only some of the numerous causes of the incorrectness of our prejudices. In addition, we suffer from numerous other well-known cognitive biases, such as confirmation bias, availability bias, or cognitive ease. In particular, a major cause of the incorrectness of our prejudices is the fact that these prejudices often largely rely on hearsay than on reliable data. Worse, our beliefs depend far more on the affection and admiration we feel for the charisma, rhetoric, and reputation of those that try to persuade us than on the (approximate) computation of Bayes' rule.

A frequent criticism of Bayesians is that they can bias their conclusions by the choice of their prejudices. Indeed, for any conclusion, there seems to exist a prejudice that leads to this conclusion. I gladly concede it. The Bayesian approach can be easily corrupted and turned into an unfair weapon in debates whose purpose is to triumph over the other side. If your goal is persuade and win the support of many, I rather recommend you exploit the art of rhetoric, clickbait, and provocation, and essentially forget about philosophies of science. First-order logic will not win you signatures and votes.

The problem is equally tricky when it comes to companies or organizations with major economic or ideological interests, as in the case of medical tests, crash tests, or quality label tests. In all such cases, we tend to prefer a simple, explicit, and auditable procedure. Yet adequate prejudices are often extremely complex to model, describe, and understand. Unfortunately, the easy interpretability of tests is often incompatible with the foundations of Bayesianism.

Having said this, if you and your interlocutors seek to better understand the world, even if it means using models that you do not fancy, then it seems essential to me that you start by first making your prejudices explicit, and to search for the origins of the prejudices. It's only once the prejudices are clarified, corrected, and made sufficiently Bayesian that you can then serenely move forward, and apply Bayes' rule to new data to converge to better theories.

So, our prejudices are not Bayesian. But the presence of incorrect prejudices is absolutely not an argument against the necessity of Bayesian prejudices. It would be like rejecting deductive logic under the pretext that no one understands contraposition.

5.11 PREJUDICES AND MORAL QUESTIONS

Having said this, the *pure Bayesian* does not exclude the possibility that the origins of an individual nevertheless affect the *pure Bayesian*'s credences on the individual's physiologic or cultural characteristics. Leo Grasset, host of YouTube channel *Dirty Biology*, asserts that[20] "since the conditions of selection vary locally, human populations differ genetically". It would be scientifically unfounded to ignore such differences. It would be irrational, in Bayesian terms, to think that the genetic and social origins of individuals should not affect the credences on their characteristics, competency, and intelligence.

The problem is not the existence of differences. It's rather the fact that moral judgments often accompany different genetic origins. The ability to digest lactose need not create a sense of social or moral superiority. Meanwhile, intellectual quotient and mathematical ability seem to be much more dependent on economical levels and the equality of education than on genomes. More importantly, even if this were not the case, associating greater intelligence with some sort of superiority is a *moral* judgment. A philosophy of knowledge like Bayesianism has no say on such moral judgments.

Since the question of moral judgments and ethics is so crucial, we shall nevertheless discuss it in the last chapter of the book, especially given the fact that Bayesianism has a central role to play in so-called

---

[20] *Des races dans l'humanité ?* Dirty Biology. L Grasset (2016).

consequentialist normative moral philosophies. However, to conclude this chapter, I can tell you right now why showing off your Bayesian credences may be immoral, even when these credences were computed by Bayes' rule. The obvious example is that of the not-so-tasty pie that your friend seems so proud of. You do not have to say what you truly think. More generally, it often seems moral to lie to some other person if this lie helps him feel better - although Kant and his followers would disagree.

In the case of prejudices, there is a more subtle, but equally hurtful example. When I was a kid, I was one of the two Asians in a school of hundreds of pupils - and I did not even know the other Asian kid! I was also the smallest kid in my class. Each question, remark, and joke on my Asian origins or my height was not hurtful per se - though kids did not use flattering words.... The problem, the thing that was upsetting, was the fact that every kid kept repeating the numerous prejudices that all other kids were saying about me. I received a special treatment. I was constantly faced with the same stereotypes. It's the repetition of the exposure to the (yet justified) stereotypes that was upsetting. *No, I do not do kung fu.*

Correcting one prejudice per year is easy. This is what happens to those whose visual traits are ordinary. But correcting twenty times the same prejudice every day without the help of a peer is much more difficult, exhausting, and unpleasant. In fact, I was somewhat amused when I noticed the irritation of my Nepali guide when, for 15 days, he had to explain to other Nepalis, again and again, that no, I am not Nepali.

But there is worse. I am clearly not the most to be pitied. Others are regularly disadvantaged because of (possibly justified) prejudices associated to their physical appearance or their origins. This problem is particularly salient when it's about hiring, justice, or social aid. As we shall stress when discussing game theory, there may be a striking contrast between the incentives of decision-makers and the harmful consequences of their decisions on the victims of prejudices. It's to avoid such harmful consequences that it's essential to ponder moral philosophy, which is what we shall do at the very end of the book.

Like technological innovations, prejudices are a powerful tool that can be used both for good or evil. Having said this, I cannot stress enough the fact that *prejudices are essential to sound thinking*. They are a necessary condition to rationality. According to our *pure Bayesian*, only those who use their prejudices can claim to be rational.

FURTHER READING

Thinking Fast and Slow. Farrar, Straus and Giroux. D Kahneman (2013).

Gene-Edited CRISPR Mushroom Escapes US Regulation. Nature. E Waltz (2016).

Frequentists vs. Bayesians. xkcd. R Munroe.

The Scientific Case for $P \neq NP$. Shtetl-Optimized. S Aaronson (2014).

Neutrinos Slower Than Light. Sixty Symbols. E Copeland & T Padilla (2012).

Are Perpetual Motion Machines Possible? Physics Girl. D Cowern (2015).

Why You Can't Go Faster Than Light (with Equations). Sixty Symbols. M Merrifield (2017).

CHAPTER 6

The Bayesian Prophets

[Bayes' rule] had no practical application in [Bayes'] lifetime, but today, thanks to computers, is routinely used in the modelling of climate change, astrophysics and stock-market analysis.

Bill Bryson (1951-)

Inside every non-Bayesian there is a Bayesian struggling to get out.

Dennis Lindley (1923-2013)

6.1 A THRILLING HISTORY

Frequentists and their guru Ronald Fisher have long persecuted those who, to their eyes, were merely an obscure sect. For two centuries, the few Bayesian believers had to act in secret and did not dare to publicly confess their heretic convictions. Forbidden by frequentists, Bayesianism nearly went extinct on several occasions. However, by relying on Price's and Laplace's ancient texts, a handful of fervid apostles managed to keep the Bayesian flame burning. These prophetic Bayesians adapted the Bayesian dogma to the modern world, whether it's for finance, engineering, or science. More recently, the great universities throughout the world have even started to propose weekly masses to invite believers and learners to read and reread Bayes' precepts to the point where it has become no longer unreasonable to self-proclaim oneself a Bayesian, including within the academic sphere - although people still look at me weirdly when I fully endorse and stand up for Bayesianism.

While the religious metaphor does not help to promote Bayesianism, the history of Bayesianism really is as fascinating as the history of religions. Interestingly, it also yields a great insight into the history of science. As opposed to how it's sometimes told, science is not only a succession of strokes of genius and a triumph of rationality. Power abuse, jealousy, and rivalries have played an equally important role in the evolution of ideas. And what has been rejected by the greatest scholars for centuries might eventually be celebrated by the entire scientific community.

According to science writer Sharon McGrady, Bayesianism precisely underwent such a roller-coaster acceptance. McGrady even dedicated a book to the dramatic history of Bayesianism, and entitled the book *The Theory That Would not Die: How Bayes' Rule Cracked the Enigma Code, Hunted down Russian Submarines, and Emerged Triumphant from Two Centuries of Controversy.*

In this chapter, I propose a brief exploration of the fascinating back alleys of the history of Bayesianism. And to do so, it's worth making a detour by the 17th century, when Blaise Pascal and Pierre de Fermat finally attempted the mathematization of the concept of probability.

6.2 THE ORIGINS OF PROBABILITY

Pascal and Fermat asked how to redistribute the money bet in a game of chance, if the game is interrupted, given the score sheet so far. Typically, imagine two players who each bet 10 dollars in a game of 11 head-or-tail throws with an unbiased coin. Let's assume that the player with the most winning throws, meaning the one with 6 wins or more, wins the total pot of 20 dollars. However, as the score is currently 4 against 0, some event suddenly interrupts the game. What would be the fair way to share the 20 dollars?

Intuitively, he who leads by 4-0 should receive more money, since his chances of eventually winning are larger. To provide a rigorous answer, Pascal and Fermat had to determine how to propagate the uncertainty of each throw. In other words, they knew the cause of the uncertainty of the game (the uncertainty of each individual throw), and had to derive the consequences on the probabilities of eventual wins of the two players, given that the current score is 4-0. They had to find the deductive logic of probabilities. This led them to lay out the bases of the theory of probability, and to introduce notions like the expectation and the binomial law.

But Pascal and Fermat's theory was still very incomplete. The person who really shaped the theory of probability is rather Abraham de Moivre.

After religious persecutions forced him to flee France around the end of the 17th century, de Moivre enjoyed the intellectually stimulating environment of the *Royal Society* in London, where he got to interact with Isaac Newton, John Wallis, and John Locke, among others. There, he published a seminal book entitled *The Doctrine of Chances*. This book sketched the first version of one of the most beautiful theorems of mathematics: the central limit theorem. This theorem derives the law of probability that a random variable obeys if it is an infinite sum of independent random infinitesimal perturbations.

6.3   THE MYSTERIOUS THOMAS BAYES

However, there was a problem with the theory of probability that de Moivre could not solve, and which would echo David Hume's later philosophical thoughts that we discussed in chapter 4. The fundamental problem I am hinting at would be nicknamed the problem of *inverse probability*. But it's actually nothing more than the *problem of induction*. It's the problem of establishing the probability of the causes given the consequences.

This is where Presbyterian minister Thomas Bayes enters the scene. Like any good mathematician faced with a tough question, Bayes focused on a simpler version of it. He imagined a table on which a white ball was (uniformly) randomly put. Bayes turned his back to the table so he had no idea of where the white ball was. He would then have to guess the position of the white ball, or rather its likely positions, given consequences of this position.

More specifically, Bayes' assistant would then put a black ball on the table, still (uniformly) randomly. Again, Bayes could not see the table, so he had no idea of where the black ball might be. But then the assistant gave Bayes some clue. He told Bayes whether the white ball was on the left or on the right of the black ball. This procedure was then repeated with other black balls randomly put on the table by the assistant.

If he knew the position of the white ball, Bayes would have been able to compute the probability of the answers of the assistant. The white ball position would have been (in part) the cause of the assistant's answers. The inverse probability problem consists of determining the likely cause, which here is the white ball position, given the consequences, namely the assistant's answers. As you might have guessed, it's by intuiting the equation that now bears his name that Thomas Bayes solved the white ball position problem[1].

One might think that this closed the inverse probability problem. It did not. Thomas Bayes, like many statisticians that we'll talk about in this chapter, had a puzzling reaction. He did not publish his magical formula. Was he afraid of controversy? This seems unlikely given that he had taken a stand against Georges Berkeley's criticisms of Newton's mathematics. Was he afraid of questioning his religious beliefs? Definitely not, since the theory of inverse probability was motivated by stressing the importance of causes. By backtracking to root causes, the theory was supposed to prove the existence of God.

One more believable hypothesis for Bayes' refusal to publish was simply that Bayes did not see the beauty of his equation. Perhaps he did not even believe it. In any case, experts seemed to agree with high credence on the fact that Bayes was not much of a Bayesian. Amusingly, "Bayesian" is almost surely a misnomer.

In fact, Bayes' rule was only published posthumously. Two years after Bayes' death, in 1763, Richard Price did a monumental work to write down and publish Bayes' ideas. In fact, of the two, Price seemed to be the more Bayesian one. But he too was not that Bayesian. Besides, his efforts were clearly religiously motivated. He wrote: "The purpose I mean is, to shew what reason we have for believing that there are in the constitution of things fixt [sic] laws according to which things happen, and that, therefore, the frame of the world must be the effect of the wisdom and power of an intelligent cause; and thus to confirm the argument taken from final causes for the existence of the Deity".

6.4   LAPLACE, THE FATHER OF BAYESIANISM

In fact, the first person who can be called a Bayesian was not British. It was the Frenchman Pierre-Simon Laplace. Laplace is one of the greatest mathematicians in history, and perhaps the greatest of my heroes. For a long time, he was better known for his work in calculus and applications in astronomy, which he published in a 5-volume masterpiece entitled *Traité de mécanique céleste*. Among other things, this work offered new answers to the question of the stability of the Solar system. Newton had already proved that if the Earth and the Sun were alone in the universe, then they would form a stable system until the end of time.

---

[1] *Bayesian Statistics with Hannah Fry*. standupmath. H Fry & M Parker (2019).

However, once Jupiter was added to the model, the equations became unsolvable. Newton gave up the mathematics and concluded that only a divine intervention could put order in this complex system and stabilize the trajectories of the planets.

Armed with new mathematical tools, including the transform that bears his name, Laplace gave some very good reasons to believe that the Solar system is in fact stable without divine intervention. Upon reading Laplace's *Traité*, general Napoleon Bonaparte supposedly asked: "Newton spoke of God in his book. I have perused yours but failed to find his name even once. Why?" To which Laplace is said to have replied: "I have no need of that hypothesis".

However, Laplace did not rigorously solve the stability problem. He can hardly be blamed, though. Generations of mathematicians have been struggling with this extremely challenging problem, including Carl Friedrich Gauss, Henri Poincaré, Andrey Kolmogorov, Jacques Laskar, and Cédric Villani. Like Poincaré who had to retract one of his memoirs that supposedly proved the stability of the Solar system, the scientific community went back and forth in its beliefs in this stability. These days, Jacques Laskar's simulations seem to have won the credences of the scientific community. These simulations predict the instability of the Solar system in the very, very long term. Rest assured. We have time.

One of the difficulties that Laplace had to face in his astronomy research was the imprecision of the observational data he had to work with. After all, these data had been collected by the Arabs around 1000 AD, by the Romans around 100 AD, by the Greeks around 200 BC, and even by the Chinese around 1100 BC. Yet the measurement instruments of those days were inaccurate. Laplace had to work with unreliable data. Could he still somehow exploit such erroneous data?

Laplace tackled the problem through a typically Bayesian angle. He knew the measures made by the astronomers of the previous centuries, and had to determine their causes - namely the actual positions of the celestial bodies in the sky. Conscious of the structure of the problem, and apparently not aware of Bayes' discovery, Laplace addressed the problem of the inverse probability in its general form. In 1774, Laplace published the *Mémoire sur la probabilité des causes par les événements*. What a fabulous document! He combined de Moivre's previous work, Lagrange's calculus, and his own genius, to state Bayes' rule in all its generality and splendour.

Crucially, Laplace did not restrict himself to astronomy. Later in his life, he would publish his thoughts in two books, where he took mathematics out of its usual domain of application. For instance, Laplace proposed to apply the theory of probability to natural sciences like astronomy, but also social sciences, testimony, medical tests, law courts, population censuses, and to many more other problems. Laplace exploited his new philosophy to the study of the sexes of newborns, which would lead him to conclude with large credence that a newborn is more likely to be a boy than a girl.

For Laplace, probabilistic reasonings were just a mathematization of common sense. Laplace certainly regarded Bayes' rule as the *right* way to think. However, he was also aware of his contemporaries' inability to apply Bayes' rule. His contemporaries' "common sense" were full of fallacies. Part of his books can thereby be interpreted as germs of cognitive science.

Towards the end of his life, Laplace also developed non-Bayesian statistical methods, many of which rely on the central limit that he proved. Laplace understood that the frequentist approach was equivalent to the Bayesian one for sufficiently large data sets. Given that frequentist computations were more manageable, Laplace often rather used them in practical cases. Laplace was a *pragmatic Bayesian*.

6.5 LAPLACE'S SUCCESSION RULE

Let's detail one of the most intriguing computations in Laplace's 1774 *Mémoire*. To illustrate his theory of the inverse probability, Laplace introduced the example of an urn with a large number of black and white tickets. This example is in fact very similar to Bayes' balls - a mathematician would say that the two problems are *isomorphic*. The proportion of white balls is assumed to be unknown. Laplace then draws a random ticket out of the urn. The ticket is white. What can we infer about the proportion of white tickets in the urn? In other words, what caused the white ticket?

The fierce *frequentist* Fisher would have probably mocked the question and asserted that it made no sense. Fisher would say that this is a non-statistical question, if not a non-scientific one.

Not Laplace. Laplace had the brilliant idea to first posit a prejudice about the proportion of white tickets, prior to any observation. He assumed that, *a priori*, the proportion of white tickets in the urn is a (uniformly) random number between 0 and 1. Note that the Laplace's randomness is *not* a real, physical uncertainty. It's merely a representation of Laplace's (subjective) ignorance.

In any case, Laplace went on performing a Bayesian inference to update his prejudices, given the color of the drawn ticket. After applying Bayes' rule (which was actually his own equation!), Laplace inferred that, even a posteriori, the proportion of white tickets was still an unknown random number between 0 and 1. But if he now had to predict the color that a second drawn ticket would be, Laplace would assign a 2/3-probability to it being white.

More generally, if he had drawn p white tickets and q black tickets, Laplace would have assigned a probability of $(p+1)/(p+q+2)$ to the fact that a newly drawn ticket were white. This is now known as Laplace's *succession rule*. It is a consequence of Bayes' rule.

Unfortunately, Laplace's Bayesian computations to derive the succession rule require calculus tools which I do not intend to dwell on in this book - in particular, it requires the computation of an integral[2]. But I highly invite those who can to perform the computations to get a feel for Laplace's thinking. In particular, it's noteworthy that there are two levels of uncertainty: the randomness of a draw from the urn, and the randomness used to represent Laplace's ignorance of the proportion of white tickets. If Laplace's contemporaries spent more time pondering the brilliance of Laplace's solution, the history of science, and that of the philosophy of science, might have taken a very different turn.

For instance, Laplace's succession rule enabled him to finally provide an answer to Hume's question. Given that the Sun has been rising for s consecutive days, should we believe that it will rise again tomorrow? If each day is a ticket, and if the ticket is black when the Sun rises (and white otherwise), then we would have $p = 0$ and $q = s$. Laplace's

---

[2]Let X be the fraction of white tickets. Laplace's computation was the following:

$$\mathbb{E}[X|p, q] = \int_{x=0}^{1} x \frac{\mathbb{P}[p, q|X = x]dx}{\mathbb{P}[p, q]} = \int_{0}^{1} \frac{\binom{p+q}{p}x^{p+1}(1 - x)^q dx}{\int_{z=0}^{1} \binom{p+q}{p}x^p(1 - x)^q dx} = \frac{B(p+1, q)}{B(p, q)},$$

denoting $B(p, q) = \int_{0}^{1} x^p(1 - x)^q dx$. Integration by parts then yields:

$$B(p, q) = \frac{q}{p+1}B(p+1, q-1) = \frac{q(q-1)}{(p+1)(p+2)}B(p+2, q-2) = \ldots = \frac{q!p!}{(p+q+1)!}.$$

The searched expectation then becomes:

$$\mathbb{E}[X|p, q] = \frac{q!(p+1)!}{(p+q+2)!}\frac{(p+q+1)!}{q!p!} = \frac{p+1}{p+q+2}.$$

Bayesian theory then implies that the probability of the sun not rising tomorrow equals $1/(s + 2)$.

Based on the Bible, Laplace then chose a value of s that corresponds to 5000 years. This led him to conclude that the probability of a disappearance of the Sun is about one out of a million. When faced with the absurdity of this result, Laplace added that "this number is far greater for him who, seeing in the totality of phenomena the principle regulating the days and seasons, realizes that nothing at the present moment can arrest the course of it". Any *pure Bayesian* must involve the entire breadth of her knowledge to make predictions.

Despite this remark, Laplace's (incomplete) prediction was unfortunately deeply criticized. It was mocked, again and again, which led many to denigrate Laplace's whole theory of probability. Laplace's unfortunate prediction likely was a major cause of the decline of the Bayesian philosophy for the two centuries that followed. However, weirdly enough, Laplace's succession rule was, in fact, given today's state of knowledge, astoundingly right!

To begin, we need to correct the value of s that Laplace used. We now know that the Sun has been rising every day for 5 billion years. As a result, Laplace's succession rule asserts that the probability that the Sun does not rise tomorrow is around one out of two trillion. In particular, using this quantity, we should now predict that the Sun will stop rising in a few billion years. Yet, curiously enough, astrophysicists now tell us that in 5 billion years, the Sun will become a red giant that will grow so much that it will eat up the Earth. And even if the transformation of the Sun into a giant red did not do the trick, Laskar's simulations suggest that our pale blue dot may leave its orbit in a few billion years. This is astonishing! Modern physics gives us two different reasons to believe that Laplace was spot on!

It's tempting to think that this is merely a cosmic inexplicable coincidence. Well, it's definitely a lucky guess. After all, Laplace's prediction was fundamentally probabilistic - anything could have happened! Besides, this very same reasoning dramatically fails when it comes to predicting the disappearance of the universe[3]. What's more, the

---

[3]Although a 2003 paper by Caldwell, Kamionkowski, and Weinberg entitled *Phantom Energy and Cosmic Doomsday* predicts a *Big Rip* of our universe in 22 billion years, which again fits Laplace's prediction!

interpretation I gave here was not entirely Bayesian[4]. Nevertheless, Laplace's lucky guess was far less a cosmic coincidence than meets the eye.

Imagine we had to guess the lifetime of a human being, based on his age. Laplace's method[5] boils down to predicting he will live as much more as he is old. Of course, this prediction will be wrong if the human being is a newborn or a very old person. However, this is unlikely. Most of the time, we will be drawing a human being who is between 20 and 60 years old, which will lead to predicting that he will have lived between 40 and 120 years.

Even better, assume that all humans live exactly 100 years, and suppose that all age groups are equally present. Then, a probability computation shows that the average prediction of the life expectancy of humans will be exactly 100 years[6]!

This weird phenomenon has been dubbed the *Lindy effect* by writer Albert Goldman, and then by mathematicians Benoît Mandelbrot and Nassim Taleb, in reference to a Deli called Lindy's where comedians gathered to wonder how to survive in show business. Goldman noted that the number of appearances of a comedian to come was roughly proportional to the number of past appearances. Mandelbrot added: "However long a person's past collected works, it will on the average continue for an equal additional amount". Taleb would then justify this empirical observation by the ubiquity of so-called power law distributions, like Zipf's surprising law that predicts that the n-th most frequent letter of a language is roughly n times less likely than the most frequent one[7].

---

[4]We used the *posterior-mean*, that is, the prediction of the (weighted) average of the probabilities of the sun rising. By opposition, the *pure Bayesian* would take the (weighted) average of the predictions of all probabilities, which would lead her to an unbounded expected life of the sun. However, here, expectation would be an unrepresentative summary of the *pure Bayesian*'s posterior belief. In fact, the median of the *pure Bayesian*'s posterior matches the magnitude scale of the *posterior-mean*. Formally, denoting $LIFE$ the lifetime of the sun and p the probability of the sun rising on a particular day, we have computed $\mathbb{E}[LIFE|D, \mathbb{E}_p[p|D]]$. The more Bayesian computation should be that of $\mathbb{E}[LIFE|D] = \mathbb{E}_p[\mathbb{E}_{LIFE}[LIFE|D, p] \mid D]$, while the Bayesian median should be the value x such that $\mathbb{P}[LIFE \leq x|D] = 1/2$.

[5]In fact, Laplace only gave the probability of the sun not rising tomorrow, and did not discuss the life expectancy of the sun.

[6]Indeed, define $X = 100$ and x the age of a randomly drawn human. The prediction of his life expectancy will be $2x$. The average prediction of life expectancy is then $\mathbb{E}[2x] = \int_0^X 2x \frac{dx}{X} = \frac{1}{X}\left[x^2\right]_0^X = X$.

[7]*The Zipf Mystery*. VSauce. M Stevens (2015).

Startlingly enough, Laplace's succession law has numerous practical applications. For instance, it allowed Allies to infer the number of Nazi tanks during the second world war from the serial numbers of captured tanks[8].

6.6 THE GREAT BAYESIAN WINTER

Unfortunately, in the 1800s, science was not there yet. Rather than noting another astounding illustration of the brilliance of the Bayesian reasoning, the great scholars of that time rejected almost entirely Laplace's theory of probability. Mathematician George Chrystal declared: "The laws of [...] Inverse Probability being dead, they should be decently buried out of sight, and not embalmed in textbooks and examination papers. [...] The indiscretions of great men should be quietly allowed to be forgotten."

Others had harsher reactions, and violently criticized Laplace's subjective credences. Philosopher John Stuart Mill called Laplace's philosophy "an aberration of the intellect" and "ignorance [...] coined into science". Apart from punctual uses by Joseph Bertrand to address uncertainty in wartime and by Henri Poincaré to reject the relevancy of some evidence in the Dreyfus lawsuit, Laplace's credences and Bayes' rule were essentially ruled out of the field of science.

This became even more the case at the start of the 20th century, as Egon Pearson, Jerzy Neyman, and Ronald Fisher founded the theory of frequentist statistics. Even though the three geniuses did not get along, all agreed on one point: the time had come to put an end to the subjectivity of Bayes' and Laplace's theories. Fisher insulted such theories, calling them "fallacious rubbish", while Neyman removed any Bayesian notion from his theory of confidence intervals as "the whole theory would look nicer if it were built from the start without reference to Bayesianism and priors". At this point, and nearly all throughout the 20th century, the words "subjective", "prior" and "Bayesian" got banished from almost all departments of statistics.

Bayesianism was not quite dead, though. A few inveterate thinkers like Émile Borel, Frank Ramsey, and Bruno de Finetti still found subjective probability useful to model gambling bets. However, they were mostly ignored in their lifetimes.

Fisher's great Bayesian rival was rather geologist Harold Jeffreys. While Fisher was brilliantly applying his frequentist methods to genetics,

---

[8]Laplace's succession law also appears in the *doomsday argument*, which predicts that "only" around one hundred billions humans will be born from now on.

Jeffreys found himself severely limited when trying to apply frequentism to seismology. Indeed, it turns out to be quite difficult to repeat earthquakes to measure frequencies. Seismographic data were scarce and imprecise. Nevertheless, armed with Bayesian tools, Jeffreys successfully interpreted his data to locate the epicentres of seisms, and even to rightly guess that the Earth center is liquid[9]. Unfortunately, though, Fisher's vehemence won over Jeffreys' placidity.

6.7 BAYES TO THE RESCUE OF ALLIES

When the World War II broke out, in the academic world, statistics were anti-Bayesian. But outside the academic world, statistics were mostly denigrated. When it realized the importance of deciphering Nazi codes, the British government rather hired linguists, art historians, and crossword puzzlers. Fortunately, British mathematicians saw it coming. They declared themselves physicists so that the government would notice them. Statisticians, however, were ignored. And this was probably for the better, as Bayes' rule, by then rejected by "true" statisticians, was to play a key role.

World War II cryptography was of a new kind. It was mechanized. The Nazis were using a coding machine called *Enigma*. *Enigma* was a sort of typewriting machine, whose particularity was to print an encrypted version of what was typed. Even better, to decrypt the code, it sufficed to type the encrypted code.

Well, not exactly. The way the machine was encrypting and decrypting the code depended on the configuration of the machine. Every day, the Nazis were using a different configuration of the machine. Yet, *Enigma* was sold with millions of such configurations. Worse, the Nazi militaries were using additional functionalities to make *Enigma* decoding even harder. There were now hundreds of billions of billions of configurations. There was absolutely no way to test them all.

Slowly but surely, thanks to the influence of Winston Churchill, the British authorities realized that mathematics was key to decrypt Nazi codes. A dream team was assembled at Bletchley Park. This team included Peter Twinn, Gordon Welchmann, Derek Taunt, Bill Tute, Max Newman, Jack Good, and a certain Alan Turing.

Armed with his 1936 theory of computation that we'll discuss in the next chapter, Turing quickly understood how to automate many of the

---

[9]We now know that the outer core is liquid. But the inner core is solid.

necessary computation steps to crack *Enigma*, as staged by the movie *Imitation Game*. This allowed him to construct the *Bombe* machine which, every day, was decrypting the codes of the Nazi army and air force. However, the *Kriegsmarine* was using an even more sophisticated version of *Enigma*. *Bombe* was not sufficiently fast for the *Kriegsmarine*. Even worse, the Nazi authorities were using an even more complex code which, instead of relying on *Enigma*, was using the *Lorenz machine*[10].

One of Turing's greatest challenges was to first convince the British authorities that the *Enigma* of the *Kriegsmarine* and the Lorenz machine even *could* be cracked. He had to convince them that the investment in their deciphering would not be in vain. For a while, the authorities were not convinced. These codes seemed too complicated, and their deciphering too costly in terms of time, human resources, and materials. However, Turing argued that it was worth it.

Churchill eventually agreed. He would later assert that "the only thing that ever really frightened me during the war was the U-boat peril [of the *Kriegsmarine*]". The U-boats had already sunk a large number of supply ships that were coming from America. Captain Jerry Roberts adds that if the situation had continued, "it is entirely possible, even probable, that Britain would have been starved and would have lost the war". Meanwhile, cracking the Lorenz machine would allow knowing directly Adolf Hitler's intentions and strategies - and to discover that he was expecting a landing at Calais rather than in Normandy.

Turing eventually received the green light to start his research. All was left was to find ideas. As you may guess, one major idea was Bayes' rule. Turing determined a heuristic way to quantitatively apply Bayes' rule. Turing's unit of measurement was dubbed the *banburismus* or *ban*, from the name of the city that provided the material needed to automate the computation of *bans*. After the war, the mathematician Claude Shannon, whom Turing met in the US during the war, would formalize a variant of Turing's *bans* and give it a by-now familiar name: *bits*. We'll detail this in chapter 15.

For now, let's get back to Turing and World War II. When an Enigma configuration seemed to partially decode a message, then this configuration won a few *bans*, which basically correspond to some Bayesian credence. By combining the *bans* of different configurations, Turing managed to orient his research towards the more promising configurations. This process, which I have shamefully oversimplified, allowed to greatly

---

[10] *Lorenz: Hitler's "unbreakable" Cipher Machine*. singingbanana. J Grime (2014).

accelerate code breaking. Eventually, Turing, his colleagues, and his machines managed to decrypt a significant fraction of Nazi messages.

Historian Harry Hinsley asserts that without the British mathematician's work, "the war would have been something like two years longer, perhaps three years longer, possibly four years longer than it was." Others even suggest that the outcome might have been uncertain. What is less disputable is the fact that Turing's mathematics, including its clever use of Bayes' rule, saved millions of lives.

However, after the war, all of Turing's work was confidential. And Winston Churchill did everything so that it stayed this way. He ordered the destruction of any document that might suggest that the Nazi codes had been broken. Churchill had buried Bayes' rule (and Turing's machines) six feet under ground.

6.8   BAYESIAN ISLANDS IN A FREQUENTIST OCEAN

When the war was over, the term "Bayesian" was still an insult. During the McCarthyism of the 1950s, while communists were chased from American institutions like witches, an American statistician said about one of his colleagues, half-jokingly, that this colleague was "un-American because [he] was Bayesian, and [...] undermining the United States Government". Another statistician added: "Bayesian statisticians do not stick closely enough to the pattern laid down by Bayes himself: if they would only do as he did and publish posthumously we should all be saved a lot of trouble". University departments of statistics, in particular, were profoundly anti-Bayesian. Jack Good, who exploited Bayes' rule alongside Turing during the war, did try to praise Bayesian methods. But his speeches kept falling on deaf ears.

It's actually far from academia that the Bayesian flame got revived. The charismatic American actuary Arthur Bailey played a big role in this. Estimations of the probabilities of life's mishaps were key to determine the price of insurance. The greater the probability of a risk is, the larger the cost of covering the risk is, and the greater its insurance price should be. However, such probabilities were not founded on Fisher's p-value. They were computed with obscure seemingly groundless formulas. Few actuaries knew where they came from, but all acknowledged that they yielded consistent results. The actuaries' computations worked in practice, but no one knew why! Bailey, who was trained as a frequentist, was horrified by that.

Yet, Bailey eventually discovered that actuaries' strange formulas, like many other magical formulas that we'll discuss in future chapters, were mysteriously similar to Bayes' rule. After a year of skepticism, Bailey eventually embraced the quasi-Bayesian inferences that were used to compute the price of insurances. He would even come to reject his frequentist education and to campaign against Fisher's methods. In 1950, he published an article analyzing the connection between the theory of credibility that actuaries were relying on, and the works of Laplace, Price, and Bayes. He praised subjective probabilities and declared the end of the frequentist tyranny. Unfortunately, Bailey died of a heart attack shortly after attacking Fisher's statistics.

There were two theorists and a half whose theoretical pondering also led to Bayesian statistics. Let's start with the half-Bayesian. Before the war, in 1933, Andrey Kolmogorov finally proposed axioms that the theory of probability could be founded upon. For Kolmogorov, what mattered was not the interpretation of probabilities, but the rules of manipulations of probabilities. When pressed to apply his theory to military strategy, Kolmogorov developed the same reasoning as Bertrand a century before. This reasoning was Bayesian, even though Kolmogorov rather called himself a frequentist.

After the war, the efforts to mathematize the theory of probability, which contrasted with Fisher's more handwavy approaches, led Dennis Lindley and Leonard Savage to reject frequentist statistics. By opposition, Bayes' rule was a straightforward consequence of Kolmogorov's axioms, and was thus perfectly mathematically founded. What's more, in 1958, Lindley published an article that proved the inconsistency of Fisher's so-called *fiduciary induction*. Lindley dared to publicly criticized Fisher; and he was right. Victorious, Lindley became an activist for Bayesianism, declared that all statistics is a special case or an approximation of Bayes' rule, and opened up to 10 departments of Bayesian statistics in England.

Meanwhile, in 1954, Savage published *Foundations of Statistics*, where he fully endorsed the subjective interpretation of probabilities. Savage had a messianic acceptance of Bayes' rule. Savage, more than any other, did not regard this equation as merely one reasoning tool among others. For Savage, it was the *only* reasoning tool. Any trade-off was *irrational* (but may be justified by pragmatism). Savage was religiously Bayesian.

When he was asked whether this questioned the objectivity of science, Savage replied that objectivity is the emergence of a consensus within the scientific community, which appears when sufficiently many data are gathered. However, Savage added that this was surely the only way to define objectivity. For Savage, the frequentist methods are definitely not objective, since they constantly require an interpretation of the statistical results, if not a choice of the frequentist method. What's more, he said that objectivization attempts by Fisher, like his fiduciary induction, were a "bold attempt to make the Bayesian omelet without breaking the Bayesian eggs".

Unfortunately, though, like Bailey, Savage died from a heart attack while crusading in the name of Bayesianism.

6.9 BAYES TO THE RESCUE OF PRACTITIONERS

Outside the circle of theorists, Bayesian statistics found applications in domains where frequentist methods fell short. In particular, Robert Schlaifer and Howard Raiffa built upon the theory of games by von Neumann and Morgenstern, and combined their utility theory to subjective probabilities to develop a theory of decision in the presence of uncertainty. Thereby, Schlaifer and Raiffa transformed the Harvard Business School into a Bayesian " hot house". Soon enough, after the publication of their book, all business schools got into Bayesian statistics, and numerous Nobel prizes of economics were given to Bayesian researchers, like John Harsanyi and Roger Myerson, as we shall discuss it in chapter 9.

The genius of Bayesian statistics was its ability to address settings where data are scarce. When, in 1950, an economist asked statistician David Blackwell how to determine the probability of another world war within the next 5 years, Blackwell replied like any good frequentist student: "Oh, that question just doesn't make sense. Probability applies to a long sequence of repeatable events, and this is clearly a unique situation. The probability is either 0 or 1, but we won't know for five years". The economist was disappointed: "I was afraid you were going to say that. I have spoken to several other statisticians, and they have all told me the same thing". Later, when he realised the predictive deficiencies of frequentist statistics, Blackwell converted to Bayesianism.

Another major application of Bayesian statistics was the study of the harmful effects of tobacco on lung cancer. The hero of this epidemiological study was Jerome Cornfield. Cornfield faced harsh criticisms of anti-Bayesians Neyman and Fisher. Fisher, in particular, rejected Cornfield's arguments on the basis that they did not involve the controlled and repeated experiments that frequentist methods require. Fisher (who received fundings from tobacco industries and denied the harmful effects of tobacco) even proposed the hypothesis that lung cancer was actually linked to a predisposition to smoke tobacco! Eventually, though, like Lindley, Cornfield would have the last word. The scientific community is now unanimous. Tobacco is a major risk factor for lung cancer.

Meanwhile, John Tukey was applying Bayesian statistics to predict the outcomes of presidential elections. In 1960, the race between Nixon and Kennedy was remarkably tight and uncertain. So tight that no TV station dared to claim a final result. At 2 am, Tukey finally gave the green light to the NBC TV channel to announce the victory of Kennedy. But it would only be at 8 am that the channel would find the courage to do so. Tukey's methods remained secret for a long time. Tukey himself, being a professor of statistics, long refused to admit their Bayesian flavour.

More recently, Bayesian methods have found themselves prominent in the prediction of presidential elections, especially as, in 2008, Bayesian Nate Silver became the first in history to correctly predict all outcomes of the 50 American states. Silver's 2016 prediction would be less successful. We'll get back to it later in chapter 15.

Similarly, numerous practitioners that had to deal with the uncertainty of rare events found the solutions of their problems in Bayes' rule. For instance, Norman Rasmussen equipped himself with Bayesian credences to estimate the probability of a major incident in nuclear power plants, while NASA hired a contractor whose Bayesian computations predicted a probability of 1 out of 35 of a major incident in the launch of a rocket. This probability was much greater (and much more realistic!) than the probability of 1 out of 100,000 that NASA predicted.

However, until the 1990s, Bayesian successes remained rare and disparate. And there was a good reason for this. Bayesian computations are long, difficult, and quickly out of reach of the mathematical equations that humans can manage. They often rely on the computation of integrals with no closed form. Bayesianism seemed promising. But it was not always practical. The advent of a more general and applicable theory of computation would change it all.

6.10 BAYES' TRIUMPH, AT LAST!

In the 1960s, Ray Solomonoff, the superstar of our next chapter, cleverly combined Turing's theory of computation to Bayes' rule to anticipate a general formulation of artificial intelligence. Like many others before him, Solomonoff was extremely critical of frequentism and their guru:

"Subjectivity in science has usually been regarded as Evil [...] if it does occur, the results are not 'science' at all. The great statistician, R. A. Fisher, was of this opinion. He wanted to make statistics 'a true science' free of the subjectivity that had been so much a part of its history. I feel that Fisher was seriously wrong in this matter, and that his work in this area has profoundly damaged the understanding of statistics in the scientific community - damage from which it is recovering all too slowly". Unfortunately, Solomonoff's ideas remained purely theoretical for a long time, since he himself had none of the necessary machines to experiment with them.

When computing machines became practical, however, Bayesianism finally lived its messianic renaissance. Frederick Mosteller was one of the first to exploit the new machines to solve challenging Bayesian problems. Then, in the 1980s, the emergence of so-called *Monte Carlo methods*, and especially *Markov Chain Monte Carlo* (MCMC), revolutionized the practical use of Bayes' rule. Instead of performing the exact computation of integrals that formal mathematics could not handle, Monte Carlo methods drew samples to perform the approximate computations of integrals. In particular, the program *Bayesian inference Using Gibbs Sampling* (BUGs) led to the final triumph of Bayesianism, while, more recently, *deep learning* and other *machine learning* methods take advantage of Bayesian priors to augur the most spectacular social changes of human history.

Finally, in the last few decades, Bayes' rule and the Bayesian formalism also seem to revolutionize our understanding of intelligence, whether artificial or human. Computer scientists like Judea Pearl, Geoffrey Hinton, and Michael Jordan, as well as neuroscientists like Josh Tenenbaum, Karl Friston, and Stanislas Dehaene, have all come to regard Bayesianism as an essential pillar of any form of cognition. We'll get back to this.

6.11   BAYES IS UBIQUITOUS

To end this chapter, it's worth listing the mesmerizing wideness of the application range of Bayesian statistics throughout history. Among so many others, we can give the following list of applications: medical diagnostic, genetics, epidemiology, astrophysics, biology, politics, war, cryptography, geology, theology, game, insurance, gambling, decision making, economics, spatial engineering, artificial intelligence, neurosciences....

Now, this is merely what we had the time to discuss in this chapter. But applications far transcend this list. We can add: sports, psychology, archeology, paleontology, education, social networks, automated translation, signal processing, genome sequencing, protein analysis, resource allocations, communication, image processing, publicity, finance, planning, logistics, and so many other fields....

Unfortunately, this chapter is too short to really explore the meanders of the fascinating ups and downs of the history of Bayesianism. Fortunately, there are other excellent available resources, on Wikipedia or Less Wrong[11]. But I especially strongly recommend the excellent book by Sharon McGrayne that I mentioned at the beginning of the chapter. This book shows that the march of science is anything but a peaceful long calm river. Yet despite the swirls, this march seems to unavoidably tend towards epistemic progress; and this progress seems to be the acceptance of Bayesian methods.

FURTHER READING

Mémoire sur la Probabilité des Causes par les Événements. Imprimerie Royale. PS Laplace (1774).

Théorie Analytique des Probabilités. V. Courcier. PS Laplace (1812).

Essai Philosophique sur les Probabilités. Bachelier. PS Laplace (1840).

The Doctrine of Chances: or, A Method of Calculating the Probability of Events in Play. W. Pearson. A de Moivre (1718).

The Foundations of Statistics. Wiley Publications in Statistics. L Savage (1950).

Game Theory and Economic Behavior. Princeton University Press. J von Neumann & O Morgenstern (1944).

Applied Statistical Decision Theory. MIT Press. H Raiffa & R Schlaifer (1944).

The Unfinished Game: Pascal, Fermat, and the Seventeenth-Century Letter That Made the World Modern. Basic Books; First Trade Paper Edition. K Devlin (2010).

The Theory That Would not Die: How Bayes' Rule Cracked the Enigma Code, Hunted down Russian Submarines, and Emerged Triumphant from Two Centuries of Controversy. Yale University Press. S McGrayne (2011).

---

[11]*A History of Bayes' Theorem.* Less Wrong. lukeprog (2011).

The Influence of Ultra in the Second World War. Cambridge Security Group Seminar. H Hinsley (1993).

Credibility Procedures: Laplace's Generalization of Bayes' Rule and the Combination of Collateral Knowledge with Observed Data. New York State Insurance Department. A Bailey (1950).

Smoking and Lung Cancer: Recent Evidence and a Discussion of some Questions. International Journal of Epidemiology. J Cornfield, W Haenszel, C Hammond, A Lilienfeld, M Shimkin & EL Wynder (1959).

Algorithmic Probability: Theory and Applications — Information Theory and Statistical Learning. R Solomonoff (2009).

Mathematicians: Blaise Pascal. singingbanana. J Grime (2009).

Lorenz: Hitler's "Unbreakable" Cipher Machine. singingbanana. J. Grime (2014).

Bayesian Statistics with Hannah Fry. standupmath. H Fry & M Parker (2019).

The Zipf Mystery. VSauce. M Stevens (2015).

Confidence Interval for the Mean. Wandida, EPFL. JY Le Boudec (2016).

A History of Bayes' Theorem. Less Wrong. lukeprog (2011).

CHAPTER 7

Solomonoff's Demon

My earliest interest in this area arose from my fascination with science and mathematics. However, in first studying geometry, my interest was more in how proofs were discovered than in the theorems themselves. Again, in science, my interest was more in how things were discovered than in the contents of the discoveries. The Golden Egg was not as exciting as the goose that laid it.

Ray Solomonoff (1926-2009)

7.1 NEITHER HUMAN NOR MACHINE

I have had the immense pleasure of pondering *pure Bayesianism* by myself. Slowly, but surely, I convinced myself that *the* right philosophy of knowledge had to be a clever marriage of Bayes' rule and theoretical computer science.

And then I read Ray Solomonoff.

It was one of the greatest moments of my life. My mind was blown. I felt that all the pieces of a gigantic puzzle were perfectly clicking in front of my astonished eyes. The Holy Grail of the philosophy of knowledge, which I had been in quest of for years, was somehow being constructed in front of me. It was prodigious! To access knowledge, I realized, it suffices to perform Solomonoff's Bayesian computations - and any alternative will almost surely fail.

Unfortunately, though, *pure Bayesianism* as defined by Solomonoff is an ideal that requires enormous computation power. Our *pure Bayesian* is not human. Worse, her computations even transcend what the laws of physics authorize. In particular, they violate the Church-Turing thesis.

In other words, she cannot be a machine either! The *pure Bayesian* is neither human nor machine! She can only be a demon that defies the laws of physics (assuming the Church-Turing thesis holds).

As a reference to the absolutely brilliant work of Solomonoff of the 1960s, in this chapter, I shall call her *Solomonoff's demon*. But before discussing Solomonoff's demon, we'll need to go back a few decades earlier, and introduce one of the most fundamental concepts in the history of ideas.

7.2   THE THEORY OF COMPUTATION

The beginning of the 20th century was a foundation crisis for mathematics. Bertrand Russell had just published the paradox that bears his name. This devastating paradox shows the extreme difficulty to build mathematics upon solid foundations. In the absence of such foundations, mathematics looked like a sandcastle that could fall apart at the first wave. David Hilbert was aware of this issue. Consolidating the sandcastle became the priority of the greatest logicians and mathematicians. Frege, Cantor, Peano, Russell, Whitehead, Lebesgue, Zermelo, Fraenkel, and Tarski were some of the many great minds that tackled the problem. Decades of work accumulated into huge volumes that were more and more opaque to profane. But Hilbert did not despair. "We must know, we will know", he declared on radio.

But in 1931, a 25-year-old logician shattered Hilbert's hopes. This logician is often regarded as the greatest logician of all times. His name is Kurt Gödel. Gödel proved that all logicians' efforts would forever remain in vain: all mathematical foundations are bound to be no more than a sandcastle. More precisely, we will never be able to prove that a set of (sufficiently expressive) mathematical foundations is absolutely unwavering. This is Gödel's (second) incompleteness theorem.

But this is arguably not the most exciting aspect of Gödel's work. Gödel's work and the formal logic constructed by other logicians also stressed the importance of the rules of the manipulations of symbols. In its purest formal form, mathematics turned out to be a very precise language, with a very rigid syntax and grammar. The phrases of the language (assuming consistency) are then divided into four categories: provable, disprovable, syntactically invalid, and undecidable. What's more, determining to which category a given phrase belongs corresponds to wondering about the existence of a sequence of operations on symbols, which start from symbols composing phrases that are assumed to be true, so-called *axioms*, and end up constructing the given phrase (or its negation).

It may be this focus point on the operations on symbols that led Kurt Gödel, Alonzo Church, and Alan Turing to the independent discoveries of three distinct definitions of what a sequence of "physically valid" operations on symbols (like digits) is. Gödel defined the class of generally recursive functions, Church introduced λ-calculus, and Turing invented the computing machine that now bears his name. Surprisingly, Church and Turing found out that these three definitions were all equivalent!

This discovery seemed so profound that Church and Turing postulated that it was not a mere coincidence. They postulated the so-called *Church-Turing thesis*, which asserts that any notion of "sequence of physically valid operations", or "purely mechanical manipulations of symbols", or "computation that a machine can do" or "algorithm" was in fact equivalent to Gödel's, Church's, and Turing's. As Scott Aaronson suggests[1], "if you think enough you'll realise Turing machines are what you meant by computation all along".

Alan Turing even went a bit further and proved the fundamental theorem of the theory of computation. This theorem asserts the existence of so-called universal Turing machines. These universal Turing machines are able to simulate anything that any other Turing machine does, and can thus compute whatever Gödel's general recursive functions and Church's λ-calculus can compute. In other words, there exists a hardware which, as long as it is given the right software, can do anything that any other hardware can.

At first glance, it may seem that this notion of computation can only interest logicians, theorists, and perhaps the few scientists that run simulations. However, the Church-Turing thesis can be interpreted as a law about the physics of our universe. Indeed, it postulates the absence, in our universe, of computing machines able to solve problems that no Turing machine can solve. This postulate is about the whole universe. If the Church-Turing thesis were true, then the whole universe would not be able to do more than what a universal Turing machine can compute. In other words, all of our universe could be simulated by a universal

---

[1] *Remarks on the Physical Church-Turing Thesis*. FQXi. S Aaronson (2014).

Turing machine. In particular, if the Church-Turing thesis holds[2], then our brains too are no more than Turing machines.

The Church-Turing thesis will greatly influence the industry of new technologies. Since, in terms of computations, we know that we'll never be able to do more than a universal Turing machine, let's invest massively in the production of universal Turing machines! And such universal Turing machines have indeed invaded our daily life, under different names. We now call them laptops, tablets, and smartphones[3].

7.3   WHAT'S A *PATTERN*?

Turing machines have numerous consequences in mathematics, physics, and technology. But this is not why I have introduced them here. Perhaps most importantly, at least for us, Turing machines seem to be key in epistemology. Indeed, they seem to be the ideal tool to formalize the concept of *pattern*, which is so often informally discussed by mathematicians.

Consider the following sequence: 1, 2, 4, 8, 16. How would you complete it? You might have guessed that the next number should be 32. It's even very likely that you are particularly confident in your prediction. But why so? What makes the future of this sequence so predictable, even though I only gave you a small sample with 5 data points? And what about the credence you have in your prediction? Is it justified?

It's likely that the reasoning you have in mind is the following: to go from 1 to 2, multiply by 2. To go from 2 to 4, multiply by 2 as well. Do the same to go from 4 to 8, and then from 8 to 16. The next number of our sequence can be obtained by multiplying the last number of the sequence by 2. This is such a simple and regular *pattern* that the *pattern* seems doomed to carry on. Or, in other words, there exists a very simple computation rule, also known as an *algorithm*, that allows production of new elements of the sequence. And it seems that, because of Ockham's razor that we shall discuss later on, the *simplicity* of the algorithm is a nearly conclusive argument.

And yet, there is a totally different way to explain the sequence that was given. Take a disk, and locate two points on its boundary. Draw the

---

[2]Even quantum mechanics can be simulated by a Turing machine (especially once interpreted along the lines of Everett or De Broglie-Bohm). This simulation of quantum mechanics on a classical Turing machine might, however, require exponentially more time than a simulation on a quantum Turing machine.

[3]These are technically not actual Turing machines, as their memory is bounded.

line between the two points. The disk is now divided into 2 parts. Add a third point on the boundary of the disk, and now draw the two lines that connect this third point to the two previous ones. We now see an inside triangle and three pieces of disk outside the triangle. The disk is now made of 4 parts. We can carry on. Add a fourth boundary point and draw the three lines to the three other points, and you will have cut the disk into 8 parts. Adding a fifth point now yields 16 parts!

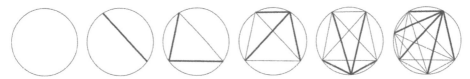

Figure 7.1   The addition of a point and of line segments connecting this point to previously added points produce 1, 2, 4, 8, 16 and 31 pieces.

The sequence 1, 2, 4, 8, 16 thus also corresponds to the number of parts of disk when a new boundary point is chosen, and when the disk is cut along the lines between the new point and old ones. Again, this is a nice *pattern*, which corresponds to a basic algorithm. We can thus be tempted to extend it by locating a sixth boundary point and count the number of pieces that the cutting process now leads to. Surprise! As depicted in Figure 7.1. The number of parts is now 31. Not 32. We have found another way to complete the sequence[4].

This seems to completely question the credence we should have on how to complete the sequence. Should it be completed by 31 or 32? Is there a right answer? And if not, what should be our credence in the different possible answers? Moreover, aren't there other possible "natural" completions?

You might feel that, although there is a good reason to complete the sequence by 31, and although there are probably still other possible completions, 32 seems nevertheless to be the most believable completion of the sequence. Could this intuition be formalized?

7.4   THE SOLOMONOFF COMPLEXITY*

In 1963, the mathematician Andrey Kolmogorov answered by the affirmative. Kolmogorov built upon the theory of computation of Gödel, Church, and Turing, to define a measure of complexity for numerical

---

[4] *A Curious Pattern Indeed.* 3Blue1Brown. G Sanderson (2015).

sequences like 1, 2, 4, 8, 16, 32 and 1, 2, 4, 8, 16, 31. This measure, now known as the *Kolmogorov complexity*, has since become a fundamental concept within the theory of computation.

Kolmogorov was not the first to think of this concept, though. Solomonoff preceded him, by publishing a 1960 preliminary research report. It would thus be fairer to call it the *Solomonoff complexity*! However, perhaps because the American Solomonoff combined the Kolmogorov complexity to a Bayesianism that was considered heretic in the United States, it was the Russian Kolmogorov that was more read and cited. Ironically, Solomonoff would receive the 2003 Kolmogorov award, for his discovery of Kolmogorov complexity!

Since, somehow, this chapter is a tribute to Solomonoff, I will shamefully allow myself to violate the commonly accepted terminology here. Instead of the Kolmogorov complexity, I'll be talking about the *Solomonoff complexity*.

So, roughly speaking, the Solomonoff complexity of a numerical sequence is the length of the shortest source code that, once run, generates the numerical sequence. However, as any programmer knows, the length of the shortest source code depends on the programming language that is used. A source code in Java will almost always be longer than a source code in MATLAB. The Solomonoff complexity is thus poorly defined. It depends on the language that is used. Or, if we consider source codes written directly in the language of the machine, it depends on the (universal) Turing machine that is used[5].

Fortunately, this dependency is not so huge. In fact, there exist computational programs, called *compilers*, whose job is to translate the source code of a programming language into its equivalent in the machine language. Some compilers are even able to translate codes of one programming language into codes of another programming language. Now, such compilers can be very long and complex. However, they have a finite length which, crucially, does not depend on the source codes to be translated.

Let's be more precise. Consider two Turing machines M and N. Call C the compiler from M to N, that is, the source code of machine N that can translate source codes of machine M into executable codes for

---

[5]It's worth mentioning that the minimal source code in Java will likely not resemble a "well-written" Java code. Typically, it will likely contain some decompression procedure. And the rest of the code will be some compressed file. In particular, such a source code will certainly be completely unreadable for a human.

machine N. This compiler C will be of a length compiler(N, M) that is independent from the code to be translated.

Now suppose that we have a source code S for machine M that successfully generates the numerical sequence. Then, we can obtain a source code for machine N that generates the numerical sequence. Indeed, to do so, we first write the compiler. We then append the source code S, and let the machine N run source code S, as interpreted by compiler C. The Turing machine N is now going to perform the same computations as the machine M, when M is run on source code S. Roughly speaking, we can write $M(S) = N(C, S)$. In particular, the machine N will then have generated the right numerical sequence.

Better yet, the length of the source code obtained for Turing machine N will not be much longer than that for Turing machine M: it will be the length of S plus the length of the compiler C. In particular, the Solomonoff complexity of the numerical sequence with respect to N will thus be at most[6] the Solomonoff complexity with respect to M plus a constant independent of the numerical sequence. We can write $K_N(sequence) \leq K_M(sequence) + $ compiler(N, M).

This all may seem obscure. Here's the takeaway. True, the Solomonoff complexity of a sequence is not an indisputably objective quantity. But it *almost* is. In any case, computer scientists have learned to live with that. Well, a bit further on, we shall see that the subjectivity of the Solomonoff complexity is precisely the cause of the subjectivity of the probabilities of Solomonoff's demon - and like computer scientists, Bayesians have learned to live with that.

Let's apply this to the sequence 1, 2, 4, 8, 16. Which of 31 or 32 is the most *obvious* or *believable* successor in this sequence? To find out, we can study the Solomonoff complexity of sequences 1, 2, 4, 8, 16, 31 and 1, 2, 4, 8, 16, 32. By Ockham's razor, the sequence whose Solomonoff complexity is smaller will be more believable. Of course, since the Solomonoff complexity is subjective, so is the question we raised! But any computer scientist will feel that the second sequence, the one completed by 32, is "easier to code" than the former, in any "reasonable" programming language[7].

---

[6]There is no equality because there may be a shorter source code for N that also generates the numerical sequence.

[7]For any given sequence, one can devise a language that generates the sequence with few instructions. However, if the sequence is complex, then the langugage that allows to do so will not seem "reasonable", nor "natural". It will be a language that is optimized for the given sequence.

In fact, our example here is too simplistic for the Solomonoff complexity to apply univocally. Indeed, in our case, it seems that, for most languages, the most succinct description of the sequence is an exhaustive description of it, term by term. For Solomonoff complexity to become relevant, we need to consider much longer numerical sequences.

Typically, instead of considering the first 5 powers of 2, suppose I now considered the sequence of the first 100 powers of 2. Then, the explicit listing of all the terms of the sequence would be tedious. In any "reasonable" language, computer scientists will rather write a short program that computes the successive powers of 2. We would have a case where the Solomonoff complexity of the sequence is much smaller than the length of the sequence, for any "reasonable" programming language. Any computer scientist would then reasonably conclude that the 101st element of the sequence is very likely to be 2^{100}.

7.5 THE MARRIAGE OF ALGORITHMICS AND PROBABILITIES

The Solomonoff complexity is a fantastic concept to study numerical sequences that follow from precise and unfaulty computations. However, as soon as we turn to empirical sciences, we are unavoidably flooded with imprecisions and errors. If the numerical sequence corresponds to physical measurements or social science data, we must expect it not to coincide with powers of 2. So let's suppose that the sequence that is given to us is actually: 0.9, 2, 4.1, 7.9, 15.8. It's then tempting to predict that the next value will be around 31 or 32, plus or minus an error of the order of 0.2.

The language of probability then becomes essential. In particular, instead of making a deterministic guess - like saying that the next element of the sequence will be precisely 32 - it seems more reasonable to make a probabilistic prediction. Typically, we might want to say that 32 is quite likely, but so are numbers like 31.9 and 32.2. In particular, for any possible completion of the sequence, a probabilistic prediction should be able to compute the probability of that completion.

This intuition led Solomonoff to combine the theories of computation of Church, Turing, and Kolmogorov with the probabilities of Bayes, Price, and Laplace into a theory of *algorithmic probability*. At the heart of this theory is a new proposal for the definition of what a theory is. This fantastic definition is the fruit of the labor and amazing genius of Solomonoff. Here it is. A *Solomonoff theory* is the source code of the Turing machine that, given a numerical sequence, computes the probability of the numerical sequence. In other words, a *Solomonoff theory* is a computable probability distribution over data[8]. Clearly, such probabilities are *subjective*; they are the opinions of the *Solomonoff theory*.

So far, for pedagogical reasons, I have been restricting myself to numerical sequences. However, especially given Shannon's information theory that we shall discuss in chapter 15, any sequence of data can be translated into a sequence of 0s and 1s. These are binary sequences, also known as *bit strings*. As a result, without loss of generality, according to computer science, all the data that our senses expose us to, images that we see, sounds that we hear, odors that we smell, or the sense of equilibrium that our inner ear measure, all such data can be described by some (very long) bit string[9]. A *Solomonoff theory* is then an algorithm that computes the probability of finite bit strings.

Such a definition of what a theory is may seem restrictive. Numerous theories commonly regarded as "scientific", such as the theory of evolution or Newton's laws, do not seem to be algorithms to compute the probability of bit strings. Indeed. In some sense, such theories may be regarded as too ambiguous, imprecise, and incomplete with respect to Solomonoff's standards.

However, conversely, scientists' brains seem to fit Solomonoff's definition. In some sense, these brains are what compute the probabilities of bit strings. Indeed, it's possible to tell a story of the evolution of life to these brains. And these brains will then be able to reply with an estimate of the probability of this story. *Yes, this seems very plausible*. Or, *no, I don't see this as likely*. Yet, according to the Church-Turing thesis, these brains are merely performing computations that a Turing machine with the appropriate source code could perform. Since the stories that were told can be encoded into a sequence of 0s and 1s (as is Wikipedia), it seems that scientists in general, and evolutionary biologists in particular, are actually not so distant from Solomonoff's formalism.

In particular, Darwinian evolution or Newton's laws would now correspond to a set of subprocedures that our *predictive algorithms* can

---

[8]Note that the fact that a theory is a probability distribution is actually a consequence of Bayes' rule, or, rather, of the equation of knowledge. Indeed, this equation involves *thought experiment terms* $\mathbb{P}[D|T]$. These quantities only makes sense if theory T *always* assigns a probability to *any* imaginably observable data.

[9]The throughput of information measured by our senses is estimated to be between 10^7 and 10^{10} bits per second, even if it's very quickly greatly compressed. See for instance: *Two views of brain function*. Trends in Cognitive Sciences. M Raichle (2010).

call. In fact, scientific theories are often some sets of equations. A computer scientist would model them by algorithmic structures. She would then combine these structures to the equation solver algorithms that we learn at school. This will allow her to construct libraries of codes. Such libraries will likely be useful to credible *predictive algorithms*.

We might then consider that the most widely used libraries correspond to some sort of fundamental laws of the universe. In the common language of scientists, it's indeed frequent to distinguish the theories that are invariant through time, from the additional data that are needed to make actual predictions. Typically, a physicist will distinguish the *laws* of physics from the physical *states* of the universe. The software developer will often distinguish the *instructions* of an algorithm from the *data* that the algorithm will be running with. And we tend to believe that some physical *law* may have a *truth* (or a validity) that is independent from any physical *state*.

However, back in 1936, Turing precisely proved that any distinction between *laws* and *states* could not be fundamental. After all, any sequence of instructions (or any physical law) can be encoded as data of a universal Turing machine. Like data, any instruction (or physical law) is just some piece of *information*. Besides, this was the fundamental principle that allowed John von Neumann to then conceive the modern computer architecture, and to guarantee that such modern computers were indeed universal Turing machines. In this architecture, *instructions* and *data* are all *information* stored in computer memory. And there is no fundamental difference between them[10].

Given what we have discussed, Solomonoff argued that the only kind of information that was worth regarding as *theories* were the descriptions of *predictive* algorithms. Insisting on (probabilistic) predictiveness then allows to clarify the notion of the *complexity* of a theory that is so crucial to the application of Ockham's razor. Indeed, a law of physics, alone, may seem very simple. The equation $\vec{F} = m\vec{a}$ contains only a handful of symbols. But so does the phrase "this is because of aliens". Why is invoking aliens usually considered not simple by scientists? Well, actually, both Newton's equation and invoking aliens, alone, are *not* predictive theories. They say very little, if anything about what can be expected. To make such theories *predictive*, they must be combined with

---

[10] *How do computers work? The Von Neumann Architecture.* Solid State Tech (2017).

more detailed descriptions of the physical state of the universe. But the level of such descriptions may differ from one theory to another.

Finally, we can distinguish the complexity of different laws. Laws that require extremely precise descriptions of the physical state of the universe to make predictions actually correspond to predictive algorithms of huge Solomonoff complexity. In algorithmic terms, the combination of such laws and of the description of the physical state will make up an extremely lengthy source code. By opposition, theories that manage to make predictions with a very partial description of the universe will be vastly simpler. This is typically the case of thermodynamics, as will be discussed in chapter 15. Such less data-expensive theories will be much more believable *a priori*.

So, in a nutshell, a *Solomonoff theory* must contain the (partial) description of the physical state of the universe, because it *must* be predictive. Ideally, it should also take into account its uncertainty about the actual physical state of the universe, as well as its uncertainty about its uncertainty. But importantly, *in the end*, it must make (probabilistic) predictions. Also, to be useful, it must also be *computable*. After all, according to the Church-Turing thesis, only such computable predictions can be made in our universe.

If we follow Solomonoff's footsteps, we must then believe that the set of theories worth studying is exactly the set of algorithms that compute the probabilities of finite bit strings. Just as Aaronson claims that "if you think enough you'll realise Turing machines are what you meant by computation all along", I would claim that, if you think enough, you'll conclude that any predictive theory is a *Solomonoff theory*.

7.6   THE SOLOMONOFF PRIOR*

Like any good Bayesian, Solomonoff then proposed to make all Solomonoff theories compete and to compute their credences. Of course, to do so, we first need to determine a *prior* on the set of algorithms to compute finite bit string probabilities. For this prejudice to obey Bayesianism, it must evidently be such that the sum of prior credences in all Solomonoff theories equals 1.

Of course, there are numerous ways to guarantee that. But to be concrete, given a programming language or a Turing machine, let's describe a quite canonical way to choose a prior probability on the set of Solomonoff theories[11]. Call $c_n \geq 0$ the number of source codes of length n that are Solomonoff theories. We can consider that each source code has a prior probability equal to[12] $1/(c_n 2^n)$. In other words, the prior probability of a theory T is here determined solely by the length $K(T)$ of its description in the considered language[13], via the equation[14] $\mathbb{P}[T] = 1/(c_{K(T)} 2^{K(T)})$.

In particular, the *Solomonoff prior* probability is thus intimately connected to the Solomonoff complexity, as it depends on the length of the descriptions of the Solomonoff theories. Yet, such lengths of descriptions depend on the choice of the programming language. They are therefore *subjective*. In other words, we have just unveiled the connection between the subjectivity of the Solomonoff prior and the subjectivity of the Solomonoff complexity. By universality of Turing machines and existence of compilers, we can thus conclude that the Solomonoff prior is subjective... but not that arbitrary!

As Solomonoff majestically explained it himself: "For quite some time I felt that the dependence of [my algorithmic theory of probability] on the reference machine was a serious flaw in the concept, and I tried to find some "objective" universal device, free from the arbitrariness of choosing a particular universal machine. When I thought I finally found a device of this sort, I realized that I really didn't want it - that I had no use for it at all! [...] It is possible to do prediction without data, but one cannot do prediction without a priori information."

7.7 BAYES TO THE RESCUE OF SOLOMONOFF'S DEMON*

Now that the Solomonoff prior has been constructed, all that is left is applying Bayes' rule to determine the most credible Solomonoff theories.

---

[11]Of course, this is not the *only* possible way. This choice is subjective. But the important argument is that, because the number c_n of theories of length n increases exponentially in n, according to any "reasonable" prior, the prior credence on a theory will decrease exponentially fast with the length of the description of the theory.

[12]To be more rigorous, one would need to treat specifically the cases $c_n = 0$.

[13]I am sweeping under the rug numerous technical difficulties, some of which shall be discussed later on.

[14]For so-called *prefix codes*, there is a more natural definition of the prior as proportional to $1/2^n$ for theories of length n. The Kraft-McMillan inequality guarantees that this will indeed define a valid prior probability.

Suppose we have observed so far the bit string[15] $d_1, d_2, \ldots, d_n$. The Bayesian credence in a Solomonoff theory T is computed by the following equation of knowledge:

$$\mathbb{P}[T|d_1, \ldots, d_n] = \frac{\mathbb{P}[d_1, \ldots, d_n|T]\mathbb{P}[T]}{\mathbb{P}[d_1, \ldots, d_n|T]\mathbb{P}[T] + \sum_{A \neq T} \mathbb{P}[d_1, \ldots, d_n|A]\mathbb{P}[A]},$$

where the Solomonoff theories A are alternatives to T.

However, this is not the goal of Solomonoff's demon. Her goal is prediction, not the computation of credences - although these are useful. To do so, we need to compute the prediction of any given Solomonoff theory T. According to T, the probability that the bit string $d_1, \ldots, d_n$ is followed by a 1 is

$$\mathbb{P}[d_{n+1} = 1|d_1, \ldots, d_n, T] = \frac{\mathbb{P}[d_1, \ldots, d_n, 1|T]}{\mathbb{P}[d_1, \ldots, d_n|T]}.$$

By combining this to the equations above, we can then derive the prediction of Solomonoff's demon. As we saw in chapter 3, it is obtained by taking the weighted average of all Solomonoff theories, where the weights are the Bayesian credences. After a few computation steps, which I invite you to do at home, we obtain the following prediction:

$$\mathbb{P}[d_{n+1} = 1 \mid d_1, \ldots, d_n] = \frac{\sum_T \mathbb{P}[d_1, \ldots, d_n, 1|T]\mathbb{P}[T]}{\sum_T \mathbb{P}[d_1, \ldots, d_n|T]\mathbb{P}[T]}.$$

This brilliant formula, called Solomonoff's induction, is what Solomonoff's demon computes day in day out to make predictions. What a spectacular formula! This magical formula is Bayes' rule in its purest and most fundamental form. In other words, to be a *pure Bayesian* is no more no less than to perform the above computation!

In particular, once the (universal) Turing machine or the (Turing-complete) programming language is chosen, all ambiguity vanishes. To be rational, to think, and to predict is no longer a blurry combination of numerous imprecise, arbitrary, and sometimes incompatible rules, that only a few misunderstood geniuses can metaphysically follow. Knowledge is just the result of a computation. And it's this computation that Solomonoff's demon performs.

---

[15]Note that contrary to frequentist statistics, we make no *iid* assumption, i.e., we do not assume that the bits are independent random variables, and definitely not that they are identically distributed.

7.8 SOLOMONOFF'S COMPLETENESS

Solomonoff's fundamental theorem is the completeness of his induction. Roughly, this completeness means that, if there is a computable *pattern* in the data, then Solomonoff's demon will find it. And what's impressive is that, no matter what pattern there is, Solomonoff's demon will find it with relatively few data. Indeed, the number of needed bits will be proportional to the Solomonoff complexity of the pattern[16].

More precisely, the more sophisticated the pattern is, the more data Solomonoff's demon will need to uncover the pattern. But Solomonoff's completeness shows that the amount of needed data will never exceed the sophistication of the data[17] - even when the pattern is noisy!

It seems to me that this fundamental theorem, and the fact that no other approach seems to perform as well, is an essentially conclusive argument in favor of *Bayesianism* - even though one needs to first convince oneself that any predictive theory is indeed a Solomonoff theory, and even though the ideal theorem would be a proof that any complete induction is necessarily a form of Solomonoff induction.

7.9 SOLOMONOFF'S INCOMPUTABILITY

Given all this enthusiasm, you could wonder if, as a result, Solomonoff's induction would not mean the end of our quest for a philosophy of knowledge. To a large extent, I would answer yes. However, there is a huge fatal flaw in Solomonoff's induction which will force me to write the rest of this book. Solomonoff's induction is in fact too difficult to be performed,

---

[16]More precisely, Solomonoff's completeness theorem says that if there is an underlying *Solomonoff theory* T^* to be found, then the expected cumulated sum of all prediction errors made by Solomonoff's demon will be bounded by the Solomonoff complexity of T^*. By measuring errors with the KL divergence which we shall discuss in chapter 15, we obtain

$$\mathbb{E}_a \left[\sum_{n=0}^{\infty} D_{KL} \left(\mathbb{P}[\cdot | a_{1:n}, T^*] \parallel \mathbb{P}[\cdot | a_{1:n}] \right) \middle| T^* \right] \leq 2K(T^*),$$

for a programming language with binary characters. Besides, we have here a justification of Hume's uniformity principle. This principle boils down to postulating the existence of T^*. Yet, this is precisely what's guaranteed by the Church-Turing thesis. On a side note, in episode 7 of the (French) podcast Axiome, I also claim and explain why Solomonoff's completeness solves the *grue* paradox.

[17]The word "sophistication" here corresponds to a very precise definition, which shall be introduced in chapter 18.

not only by ourselves and our limited cognitive abilities - recall that even Erdös failed to understand a basic version of Bayes' rule, but for computers as well.

Solomonoff's induction is *incomputable*. What does this mean? It means that no Turing machine can apply it rigorously. There is a simple reason for this, which is not really relevant, and a more subtle and devastating reason.

The simple reason is that, to be applied rigorously, Solomonoff's induction requires the simultaneous study of infinitely many Solomonoff theories. This is simply because there are infinitely many algorithms. However, no computer, nor any network of computers, can perform this infinite computation.

Having said this, it's worth pointing out that, by construction of Solomonoff's prior, theories with a large Solomonoff complexity are, in any case, associated to exponentially negligible probabilities. If we neglect the theories whose impacts on Solomonoff's prediction are provably small, could we compute a very good approximation of Solomonoff's induction? Unfortunately, no.

The real difficulty of Solomonoff's induction is not the number of Solomonoff theories to be considered, but the computations that the Solomonoff theories require. These days, the algorithms that run on your computers mostly run fast and well. This is because software developers did a great job! But in general, it's actually extremely difficult to know if an algorithm will be running fast. It's also difficult to know if an algorithm will be terminating at all, or if its computation will just keep getting more and more complex.

A striking (possible) example of such an algorithm is the *Syracuse conjecture*, also known as the Collatz, Ulam, Czech, or $3n+1$ conjecture. We start by giving a positive whole number to the algorithm. If the number is 1, the algorithm halts. If the number is even, the algorithm divides it by 2; otherwise it multiplies it by 3 and adds 1. Then, if it's not told to halt, the algorithm repeats its operations to its result. The Syracuse conjecture asks the following question: will the algorithm always halt, no matter which whole number we give it to start with[18]?

As weird as it may seem, we do not know the answer to this basic question. Worse, we have no idea how to approach this problem, nor if there even is a way to solve it. The great mathematician Paul Erdös once said: "Mathematics is not yet ready for such problems."

---

[18] *UNCRACKABLE? The Collatz Conjecture*. Numberphile. D Eisenbud (2016).

This case might be an instance of the undecidability discovered by Turing. Shortly after he defined the notion of computability by introducing the Turing machine, Turing asked whether one could foresee the halting of a computation before running it. Is there a way to determine if a computation halts? Or said differently, can we construct a Turing machine that predicts in finite time whether other Turing machines halt in finite time?

If you feel that Turing's question resembles a snake biting its tail, this is no accident. By mimicking Cantor's diagonal argument, Russell's paradox, and Gödel's theorem, Turing invoked self-reference to prove that the answer to his question was no. The halting of algorithms cannot always be predicted by a finite-time computation. The halting problem is said to be *incomputable* (or *undecidable*).

As it turns out, this is a huge problem for Solomonoff's induction. Indeed, to perform Solomonoff's induction, we need to compute the predictions $\mathbb{P}[d_1, \ldots, d_n | T]$ of different Solomonoff theories T. Ideally, only the predictions of halting theories should be taken into account. However, if we accept the Church-Turing thesis, then it's physically impossible to determine the halting of computations. In particular, the dramatic consequence of the reasoning is that, after a finite computation time, it's impossible to exclude the fact that the ongoing computations of Solomonoff theories will halt.

There is worse. It's also impossible to predict whether the ongoing computations that will halt will drastically change the result found so far. More generally, it will never be possible to measure the degree of validity of a partial result obtained after a finite amount of time. Solomonoff's induction is not only incomputable. Any approximation is incomputable too[19]!

7.10   SOLOMONOFF'S INCOMPLETENESS

Now this is embarassing. Granted, Solomonoff's demon, who transcends the laws of physics and the Church-Turing thesis, will be able to detect any pattern in a data set. But Solomonoff's demon has no physical

---

[19]Formally, this means that, for any tentative approximation algorithm ALG, there exists some past data $d_1, \ldots, d_n$ such that the difference of prediction for a new bit d_{n+1} by ALG and by Solomonoff's demon is very different, that is,

$$\forall \text{ALG} \exists n, d_1, \ldots, d_n, \quad |\mathbb{P}[d_{n+1} = 1 | d_1, \ldots, d_n] - \text{ALG}(d_1, \ldots, d_n)| > 1/3.$$

counterpart. We and our computers, who likely are constrained by the Church-Turing thesis, seem unable to perform any provable approximation of Solomonoff's induction. More generally, it will always be physically impossible to compute the adequate credence in our approximations of Bayes' rule. Why venerate this equation of knowledge?

Solomonoff's answer resides in another theorem of his. This theorem asserts that any *predictive algorithm* is necessarily *incomplete*. More precisely, any computable philosophy of knowledge will be unable to detect *all* patterns. Or in other words, all computable epistemologies have blind spots. This is the astonishing Solomonoff's incompleteness theorem - which I find even more fascinating than Gödel's.

Said differently, whatever your philosophy of knowledge is, as long as it is *computable*, there will be universes that can trick you. Such universes will make you *constantly* make very wrong predictions! *Computability* and *completeness* are incompatible[20].

In conclusion, the unfortunate incomputability of Solomonoff's induction is precisely what allows it to escape the incompleteness of computable epistemologies. For Solomonoff, this incomputability is not a pathology; it is a necessary property of any desirable philosophy of knowledge[21].

7.11   LET'S BE PRAGMATIC

I have been both mesmerized and dazed by Solomonoff. His theory was solidly grounded upon the most fundamental concepts of computer science and probability theory. Solomonoff's construction is wonderfully *natural*, in the sense that it's exactly what I started to develop when I started pondering pure Bayesianism. It's exactly what I would have come up with if I had thought long enough and if I had the necessary cognitive abilities. And yet, Solomonoff's conclusions are univocal, brutal, and unexpected - although, the more I meditated Bayes' rule, the more I saw them coming...

---

[20]The proof idea is to construct a pattern that exploits your computable epistemology to make your predictions fail.

[21]Note that Solomonoff's demon only performs predictions. Technically, it makes no decisions. But it's quite straightforward to turn it into an optimal decision-making demon, by introducing a reinforcement learning framework. Such a demon is known as AIXI. AIXI is sometimes regarded as the ultimate (but incomputable) artificial general intelligence. See *Universal Artificial Intelligence: Sequential Decisions Based on Algorithmic Probability*. Springer Science. M Hutter (2004).

Knowledge and rationality are forbidden by the laws of our universe. These laws force us to be content with approximations whose degrees of validity cannot even be measured. Worse, since the computational and time resources are bounded (including by the laws of physics), we have to restrict ourselves to an extremely heuristic variant of Solomonoff's induction.

This restriction will be all the more constraining as the amount of observed data increases, as is the case of *Big Data*. These days, both sense data and digital data are of the order of giga, tera, peta, exa, if not zettabytes. In other words, in practice, the sequence $d_1, \ldots, d_n$ contains billions, if not billions of billions, of elements! Such data cannot even be stored in memory. It's thus hopeless to hope to compute even a rough approximation of Solomonoff's induction. This seems like a call for greater humility and prudence.

Given this brutal observation, the rest of the book, like the world of science, statistics, and artificial intelligence, will have to focus on so-called *heuristic* philosophies of knowledge. Our knowledge will never be *accurate*. But perhaps it might be possible to make it *accurate enough*. To do so, given that we know what we should do to know *accurately*, in this book, we shall mimic Solomonoff's demon as much as possible, and aim at getting to conclusions that are similar to hers.

So let me now introduce a second (imprecise) philosophy, and a second fictional character. I shall call them, respectively, *pragmatic Bayesianism* and the *pragmatic Bayesian*. Contrary to the *pure Bayesian*, the *pragmatic Bayesian* will be constrained by limited computational power and memory space. As a result, it will be preferable to favor the computations of numerous fast and efficient algorithms, to the testing of a handful of algorithms that require unreasonable amounts of computations.

Now, this will demand an intimate knowledge of algorithms. In fact, *pragmatic Bayesianism*, more than its pure version, requires a highly sophisticated theory of computation and information, which is generally known as *theoretical computer science*. This theoretical foundation can then be complemented by computational experimentations that belong to the field of what might be known as *experimental*, or *empirical*, *computer science*. This modern field of science, initiated in the 1930s by Gödel, Church, and Turing, has become one of today's most fascinating, most promising, and most misunderstood fields of research.

Computer science is not only the art of creating and using modern technologies. In fact, for the competent Bayesian that I aspire to be, the concepts of computer science are actually the most important arsenal of thinking tools to determine some (near-)optimal pragmatic philosophy of knowledge - assuming we have just determined *the* right philosophy of knowledge. In his book *Quantum Computing since Democritus,* computer scientist Scott Aaronson goes as far as to propose to rename computer science as *quantitative epistemology.* He insists on the importance of *computational complexity theory* to any philosophy of knowledge[22].

However, I propose to leave aside this quest of a pragmatic philosophy of knowledge for now. We'll get back to it at length from chapters 14 to 19. Until then, I propose to first observe the ubiquity of Bayesian principles, in cryptography, sociology, biology, or in the emergence and meaning of the scientific consensus. In the next chapters, we'll distance ourselves from Bayesianism to explore areas that may seem completely unrelated to Bayes' rule at first. But we shall discover, again and again, that all these diverse phenomena are in fact relying on Bayesian principles.

FURTHER READING

Universal Artificial Intelligence: Sequential Decisions Based on Algorithmic Probability. Springer Science. M Hutter (2004).

Quantum Computing since Democritus. Cambridge University Press. S Aaronson (2013).

A Preliminary Report on a General Theory of Inductive Inference (Report ZTB-138). Zator Co. R Solomonoff (1960).

A Formal Theory of Inductive Inference. Part I. Information and control. R Solomonoff (1964).

A Formal Theory of Inductive Inference. Part II. Information and control. R Solomonoff (1964).

The Discovery of Algorithmic Probability. Journal of Computer and System Sciences. R Solomonoff (1997).

Algorithmic Probability: Theory and Applications. Information theory and statistical learning. R Solomonoff (2009).

---

[22] *Why Philosophers Should Care About Computational Complexity.* S Aaronson (2011).

Why Philosophers Should Care about Computational Complexity. S Aaronson (2011).

Two Views of Brain Function. Trends in Cognitive Sciences. M Raichle (2010).

Remarks on the Physical Church-Turing Thesis. FQXi. S Aaronson (2014).

Turing and the Halting Problem. Computerphile. M Jago (2014).

A Curious Pattern Indeed. 3Blue1Brown. G Sanderson (2015).

Circle Division Solution. 3Blue1Brown. G Sanderson (2015).

UNCRACKABLE? The Collatz Conjecture. Numberphile. D Eisenbud (2016).

The Universal Turing Machine. ZettaBytes, EPFL. R Guerraoui (2016).

Making a Computer Turing Complete. B. Eater (2018).

How Do Computers Work? The Von Neumann Architecture. Solid State Tech (2017).

II

Applied Bayesianism

CHAPTER 8

Can You Keep A Secret?

Arguing that you don't care about the right to privacy because you have nothing to hide is no different than saying you don't care about free speech because you have nothing to say.

Edward Snowden (1983-)

8.1   CLASSIFIED

During the Vietnam war, American generals wanted to determine the proportion of soldiers that were smoking marijuana. The trouble is that if a superior directly asks a soldier whether he smokes, then the soldier will likely answer no to avoid any punishment. The American army understood that it needed a survey that could not incriminate the repliants. It needed a survey that provably guaranteed the privacy of answers. Can you guess the American army's trick?

To keep information secret is a recurrent military challenge. It's often said that Julius Caesar encoded the messages he sent and received, by sliding the letters of the alphabet. Letter A was replaced by letter D, letter B by letter E, C by F, and so on. Later on, other military leaders used more sophisticated cryptographic encoding, by replacing A by any other letter of the alphabet, letter B by still another one, and so on. This is known as a *substitution code*, as each letter is systematically substituted by another letter.

The advantage of the substitution code is that, contrary to Caesar's sliding code, the number of possible encodings is now astronomical. Indeed, in the case of Caesar's code, the number of possible encodings is the number of possible slidings. There are 26 letters in Latin alphabet. Thus there are 26 possible slidings (including the one that consists of no sliding at all). The problem is then that a hacker will only need to test 26 possibilities to crack the code.

However, if we now allow any permutation of letters, the number of possible encodings becomes vastly larger. Indeed, we can replace A by any of the 26 letters, then B by any of the remaining 25 letters, then C by any of the remaining 24 letters, and so on. We see that the number of substitution encodings is then $26 \cdot 25 \cdot 24 \cdot \ldots \cdot 2 \cdot 1$, which is denoted 26! (and pronounced "26 factorial"). This number is gigantic. It's of the order of 10^{26}, which is roughly the number of stars in the universe! Even a computer will not run sufficiently fast to test all combinations before the collapse of our Sun.

World War II and technological progress led to the mechanization of cryptography. Some of the most important war machines used by the Nazis were not killing machines, but encoding and decoding machines. These machines were *Enigma* and the *Lorenz* machines, which we discussed in chapter 6. Here again, the number of configurations of machines was gigantic. It was hopeless to test them all.

8.2 TODAY'S CRYPTOGRAPHY

Since then, the technological context has evolved and we can now all afford highly interconnected supercomputers. We routinely send each other billions of messages through the Internet. To secure the privacy of our messages, cryptography has become essential.

We must receive our friends' emails, and prevent intruders from surveilling them. We must log into social networks, and no one else should be able to log into our accounts. We must ask our banks to perform financial transactions, and banks must be able to certify that these requests come from their clients; not from usurpers.

The solution to all of these everyday problems is *cryptography*. Thanks to cryptography, buyers and sellers have had access to new communication means to facilitate commercial trading. Without them, companies like Paypal, Amazon, Netflix, Uber, or Airbnb could not have been founded.

This cryptography essentially relies on two great discoveries. On the one hand, in 1976, Whitfield Diffie and Martin Hellman introduced the secret sharing protocol that bears their names. Using the Diffie-Hellman protocol, even if they never meet, Alice and Bob can create a shared secret by communicating out loud through the Internet. This shared secret can then be used by Alice and Bob to determine a common encryption scheme that only they will know. Thereby, Alice and Bob will be able to privately communicate without any physical contact.

The Diffie-Hellman protocol has been widely used since by many companies, including WhatsApp. This allows to provably guarantee to users that even these companies will be mathematically incapable to read their secured communications[1]. Today, we take it for granted. But prior to Diffie-Hellman's breakthrough, it was not clear that this could ever be the case[2].

On the other hand, in 1977, the next year, Ron Rivest, Adi Shamir, and Leonard Adleman (RSA) introduced *asymmetric cryptography*. To use RSA, Alice must create a pair of keys. She reveals her public key, but keeps her private key secret. Bob, like anyone else, can send encrypted messages to Alice. To do so, he uses Alice's public key to cypher his messages. Only Alice will be able to decrypt Bob's messages, as the private key is necessary to do so[3].

Even better, Alice can then sign a message that she sends to Bob, by encrypting (a hash of) her message with the private key. Once Bob has decrypted it with the public key, he will be guaranteed that the message is indeed Alice's. Indeed, the private key was essential to encode a message that the public key can decrypt. Thereby, by signing her messages, Alice can authenticate them. She can then prove to her bank or her social network that the messages that they receive are indeed hers; and not a usurper's[4].

In all the cryptographic examples that we have discussed thus far, security is guaranteed by the gigantic number of possible encodings, and the assumption that any hacker will have to test a major fraction of all possible encodings. For instance, if there are 10^{20} possibilities, even if

---

[1]In fact, this assertion relies on the unproved conjecture that the so-called "discrete logarithm" cannot be computed in polynomial time by classical Turing machines, and on the absence of sufficiently large quantum computers.

[2] *The Diffie-Hellman Protocol.* ZettaBytes, EPFL. S Vaudenay (2016).

[3] *Public Key Cryptography - RSA Encryption Algorithm.* Wandida, EPFL. EM El Mhamdi (2014).

[4]The actual signing algorithms are slightly more complicated than described here.

the hacker manages to reject 99% of the possibilities, then he will still have to test 10^{18} of them! This will take a huge amount of time, even with modern computers.

However, this reasoning is ignoring the Bayesian's two favorite tools: prejudices and Bayes' rule.

8.3   BAYES BREAKS CODES

On May 2, 1568, after numerous unfortunate adventures, Marie Stuart I, Queen of Scots, fled Scotland to stay safely with her cousin in England. However, since Catholics were considering that Mary was the legitimate inheritor of the throne in a newly Protestant country, Queen Elisabeth I of England treated her as a rival. She locked her away for the following 19 years.

It's in prison that Mary started to use cryptography. She knew that her messages were probably intercepted by the court of the Queen. Thus Mary used the substitution code to communicate secretly, especially with a certain Anthony Babington. However, it seems that her codes got decrypted, thereby revealing a plot to assassinate the queen. This plot, known as Babington's conspiracy, then led to the execution of Mary, on February 8, 1587.

But in the absence of computers, how could the substitution code be broken? As discussed earlier, the number of substitution encodings is larger than the number of stars in the universe. It's thus certainly *not* by testing them all that the court of Queen Elisabeth I managed to break Mary's code. The key was a prejudice. This prejudice is on what the English language is like.

Indeed, in the English language, the letters E, T, A, O, and I are more frequent than others. As a result, the most frequent letter of the encrypted message is likely to be the substitute of letter E. The second most frequent is probably the substitute of T, and so on. But this is not all! The words of the English language are also very rigid, since most combinations of letters do not make up a word. For instance, once we have decoded T*E and if we need to determine the missing letter *, we will guess that * is very probably an H, an I, or an O. Even better, if it's the first word of a sentence, then we can be very confident in the fact that it's an H.

As you see, the language of probability suddenly and naturally comes up in cryptography. In fact, behind the intuitive reasoning I have just presented is Bayes' rule. The problem is to determine our credences in the original message and the encoding that is used, given the encrypted message. Said differently, the problem is about determining the causes, given the consequences, which is precisely the problem that Bayes' rule solves.

It's for similar reasons that a randomly generated password is much more secure than the password "123456". Hackers have justified beliefs on users' passwords. They know that some passwords are more frequent than others. A clever hacker will thus launch an algorithm that first tests the most likely passwords. This was also a key principle that Turing relied on to break Nazi codes during World War II.

Of course, in the case of Mary and the substitution code, the set of causes is gigantic, which makes the exact Bayesian computation intractable. However, the fact that the code is done letter by letter allows to hack the intractable Bayesian computation. Indeed, instead of guessing the full substitution code, we can first ask what the substitute of E is. Then what that of T is. And so on.

This is something that Alan Turing could not do when he tried to break Enigma or the Lorenz machine. To break these more sophisticated codes, Turing had to further formalize the Bayesian computation. For instance, he introduced the approximation of the Bayesian computation on a logarithmic scale, which we shall get back to. Turing's computations would later be further formalized by Claude Shannon. Shannon would then go on to develop a full mathematical theory of communication and cryptography.

These days, computer scientists pay great attention to the statistical properties of encrypted messages to prevent hackable weaknesses. Amusingly, as Shannon would discover, a good way to do so is to first maximally compress the original message, and to then cypher this compression. Indeed, by performing this compression, we can destruct any rigidity, like the fact that, in the word T*E, the sign * can only be one of a few possible letters.

Having said this, these days, cryptographic systems derived from the mathematics of Shannon, Diffie-Hellman, and Rivest-Shamir-Adleman are considered secured, at least as long as scalable quantum computers are not out there. Other even more solid, so-called *postquantum*, cryptographic systems have been proposed. However, we have still to prove that no algorithm, classical or quantum, can break them - this problem is intimately related to the prestigious P versus NP problem.

8.4 RANDOMIZED SURVEY

However, all of this cryptography does not solve the survey problem of the American army on marijuana consumption. Granted, soldiers could cypher their answer. But if someone decyphers the answers, then the privacy will be violated. Soldiers would thus refuse to provide honest answers[5]. On the other hand, if no one decyphers the answers, then the American army will not make progress. It would have learned nothing.

In some sense, we want to let the army learn about its soliders, without learning anything about any of its soldiers.

The key trick is *randomization*. More precisely, before answering, each soldier will throw a coin. If the coin lands heads, then the soldier answers truthfully. If it lands tails, then the soldier answers yes. Crucially, when the military chief interviews the soldier, the chief cannot know, and will never know, if the coin landed heads or tails. Thereby, if the soldier replies yes, the chief cannot know if the soldier said yes because he smokes, or if it's because the coin landed tails. In other words, the soldier has *plausible deniability*. Nevertheless, by compiling all soldiers' responses, the American army could determine the fraction of marijuana-smoking soldiers.

Indeed, suppose we received 200 answers, including 160 "yes" and 40 "no" answers. We know that around half of the answers correspond to tails. Thus, around 100 of the "yes" answers were caused by the coin. The remaining 100 answers are thus honest answers. They are 60 "yes" and 40 "no" answers. As a result, we conclude that roughly 60% of American soldiers smoke marijuana[6]. We have successfully surveyed the soldiers without violating any soldier's privacy[7]!

Actually, not exactly. If we reason in Bayesian terms, we realize that our prejudice on the probability that a random soldier smokes marijuana has changed. Before the survey, the chief might be guessing that 20%

---

[5]We could imagine an urn in which soldiers would cast anonymized answers. However, as we shall see, anonymization (in general) does not guarantee privacy. Typically, we could imagine that the Army uses a fake urn that only contains "no" answers, except for the soldier's ballot. The army could then determine the answer of the soldier based on the presence or absence of "yes" answers in the urn.

[6]Actually, to be good Bayesians, we should apply Laplace's succession rule which we discussed in chapter 6. This gives a proportion of 61/102. Better, we should start with an informed prior on what the fraction of smokers is likely to be. Moreover, we should not forget to compute the uncertainty on our estimated proportion of smokers!

[7]*An Embarrassing Survey - Randomized Response*. Singingbanana. J Grime (2010).

of soldiers smoke. The survey would then have drastically modified the chief's prejudice on marijuana consumption. The chief used to guess that a random soldier has a 20% probability to be a smoker. Now, his guess is roughly 60%. But evidently, this change in the chief's belief is not surprising. It was, after all, the purpose of the survey.

However, the more problematic issue is the chief's suspicion regarding a surveyed soldier that answered "yes". Indeed, given that the soldier answered yes rather than no, it's more likely that he is a marijuana smoker than if he had answered no. And also more likely than if he had not been surveyed. To determine the (Bayesian) chief's credence on the soldier's marijuana consumption, again, we need to apply Bayes' rule. Let's write it:

$$\mathbb{P}\left[\ast|yes\right] = \frac{\mathbb{P}\left[yes|\ast\right]\mathbb{P}\left[\ast\right]}{\mathbb{P}\left[yes\right]}.$$

I spare you the details (but invite you to do the computations at home!). Given our data, the military chief should conclude with credence of 75% that a soldier who replied "yes" does smoke marijuana. This result is to be contrasted to the credence of 60% on an unsurveyed soldier. In other words, by accepting to be part of the survey, the soldier slightly incriminated himself. His privacy has not been entirely spared, even though it has not been entirely violated either.

On the contrary, though, there is a case where the soldier's privacy has been entirely violated: the case of a "no" answer. Indeed, as a result, assuming that the soldier did not lie, the chief can know for sure that the soldier does not smoke. Yet, this can actually be a big problem. Indeed, it's possible that in the future, for some mysterious reason, research discovers that the American soldiers sent to Vietnam who did not smoke marijuana are very likely to develop a colon cancer. Insurance companies might then want to increase the fees of the non-smoking soldiers.

To violate nobody's privacy, we must modify the survey mechanism. Here's one way. When the coin lands tails, the soldier will now throw a second coin that will determine the soldier's response. In other words, in this variant, any soldier has a $1/2$-probability to be truthful, a $1/4$-probability to answer yes because of the coin tosses, and a $1/4$-probability to answer no because of the coin tosses. As earlier, we see that this still allows the chief to estimate the proportion of smokers.

Moreover, by applying the same Bayes' rule as earlier, the military chief will have an *a posteriori* credence of 82% in the fact that a soldier who did reply "yes" to the survey is indeed a smoker. This time, though, the military chief will assign a nonzero credence to the fact that a soldier who replied no actually smokes. This credence will be of 33%. These quantities of 82% and 33% are to be compared to the 60% probability for an unsurveyed soldier. In particular, it has become common to measure privacy by the ratios 82/60 and 60/33. In this case, the loss of privacy never exceeds a factor[8] 2.

8.5 THE PRIVACY OF THE RANDOMIZED SURVEY

However, the ratios that we computed depend on the 60% proportion of smokers. Yet, military chiefs could not anticipate it prior to the survey. Given this, theoretical computer scientist Cynthia Dwork introduced a new theory to mathematically study privacy. At the heart of this theory lies the concept of *differential privacy*.

Contrary to the analysis we did, differential privacy aims at guaranteeing a level of privacy prior to the survey. To do so, it determines the worst possible loss of privacy that military chiefs might have caused, assuming any imaginable proportion of smokers[9].

Assume that 1% of soldiers smoke marijuana. And consider a soldier who answered yes. The posterior credence in the fact that he smokes is computed by Bayes' rule:

$$\mathbb{P}\left[\ast|yes\right] = \frac{\mathbb{P}\left[yes|\ast\right]\mathbb{P}\left[\ast\right]}{\mathbb{P}\left[yes|\ast\right]\mathbb{P}\left[\ast\right] + \mathbb{P}\left[yes|\text{not }\ast\right]\mathbb{P}\left[\text{not }\ast\right]}$$

$$= \frac{3/4 \cdot 0.01}{3/4 \cdot 0.01 + 1/4 \cdot 0.99} \approx 0.029.$$

This credence is almost three times larger than for unsurveyed soldiers! In fact, in the limit where the proportion of smokers goes to 0, the multiplicative factor will be exactly 3. This is also the worst case. Randomized survey with two coins is said to be (ln 3)-differentially private.

This rigorous privacy guarantee is to be contrasted with many naive methods of *pseudonymizations* of data. Such methods are unfortunately still widely used, especially in epidemiology, and consists of simply erasing the names of the surveyed persons. However, once we know the age, sex, address, social-economical level, food consumption, and level of

---

[8]Technically, we should also be looking at the other probability ratios of not being a smoker, which correspond to 40/18 and 67/40, one of which does slightly exceed 2.

[9] *What is Privacy?* Wandida, EPFL. LN Hoang (2017).

study of a person, it's generally quite easy to guess the identity of that person. *Pseudonymization offers no privacy guarantee.*

8.6   THE DEFINITION OF DIFFERENTIAL PRIVACY*

Cynthia Dwork's differential privacy is a very general criterion to identify the methods that offer provable privacy guarantees. As in the case of the randomized survey, suppose you wish to extract useful information from individuals' data. Thereby, your posterior, once the useful information is retrieved, will generally be different from your prior - this is, after all, the purpose of the extraction of useful information! However, intuitively, privacy requires that your posterior credences do not discriminate surveyed individuals, not between one another, nor between a surveyed individual and unsurveyed one.

This is precisely the intuition that is formalized by differential privacy. An information extraction mechanism will be said to be *differentially private* if, *a posteriori*, the credences on a surveyed individual are almost the same as for an unsurveyed one. By transitivity, this will also imply nondiscrimination between surveyed individuals. Indeed, assume that Alice and Charlie are surveyed, but that Bob is not. Then, by definition of differential privacy, the posterior credences on Alice and Bob are necessarily similar. So are those on Bob and Charlie. As a result, the posterior credences on Alice and Charlie will also be similar.

More formally and without loss of generality, Dwork supposed that surveyed individuals' data are located in some *database*. According to Dwork, it's the differential privacy of this database that needs to be guaranteed. The trick to do so is to forbid the reading of the database by mechanisms that are not proven to be differentially private.

The limit case is, of course, when no mechanism can access the database. This is evidently equivalent to the case where the mechanism always returns information that is independent from the database. In such cases, no information on the database content can be extracted. It's all as if the database did not even exist, or as if it was fully encrypted and no once could decrypt any of it. In such a case, it's clear that the database will be fully (differentially) private. But it will also be useless.

In fact, more generally, it's impossible to extract any useful information without partially violating differential privacy. Dwork's approach was then to quantify the tradeoff between useful information extraction and privacy guarantee. Roughly speaking, the privacy loss of a mechanism will then be measured by two parameters ε and δ.

A $(\varepsilon = 0, \delta = 0)$-differentially private mechanism will be fully private - but it will be extracting zero information.

But what are ε and δ? To answer this question, let's get back to the intuitive definition. We saw that a differentially private mechanism had to guarantee the absence of posterior discrimination between a surveyed Alice and an unsurveyed Bob. The trick to do so is to make sure that the information retrieved by the mechanism will be essentially the same as what the information would have been had Alice been removed from the database (in which case Alice would clearly be indistinguishable from Bob).

More precisely, call X the original database with Alice, and Y the one we obtain by removing Alice from X. Call R the response of the mechanism. A differentially private mechanism will return responses R similar for databases X and Y. And this similarity is measured by parameters ε and δ.

Still more formally, a mechanism is said to be (ε, δ)-differentially private if, for any database, with probability $1-\delta$, the posterior credences on any surveyed individual's characteristic is never e^ε times larger than the posterior credences on any unsurveyed individual's characteristic[10]. In other words, *a posteriori*, all should work out almost as if Alice was not in the database.

Let's now be purely formal. Differential privacy actually boils down to an easy-to-manipulate inequality for mathematicians. A mechanism is (ε, δ)-differentially private if, for any two databases X and Y that only differ by the addition or removal of a single individual's data, and for any possible response R, we have the inequality

$$\mathbb{P}[R|X] \leq e^\varepsilon \mathbb{P}[R|Y] + \delta.$$

In particular, we can then show that, to guarantee (nontrivial) differential privacy, the response R must be a random function of the database. In other words, a same request launched twice by a nontrivial differentially private mechanism on a single database must have a nonzero probability to return two different results.

8.7   THE LAPLACIAN MECHANISM

Differential privacy may seem quite cumbersome. However, for certain applications, it may actually not be an inconvenience. Typically, the

---

[10]In the case of the randomized survey, we had $\delta = 0$ and $\epsilon = \ln 3$.

so-called *Laplacian mechanism* allows us to perform differentially private surveys while providing useful information.

For instance, imagine that you are a hospital with patients' data. You would like to determine the proportion of lung cancers without violating patients' privacy. To do so, instead of revealing the exact number of patients with this cancer in the database, the Laplacian mechanism perturbs this exact number by adding a random quantity that obeys Laplace's law. The mechanism then returns the randomly perturbed result.

I won't detail Laplace's law here[11]. But note that it depends on a parameter that corresponds to the typical intensity of the random perturbation. For a $(\varepsilon, \delta = 0)$-differential privacy, the perturbation must be of the order of $1/\varepsilon$. Thereby, if the Laplacian mechanism says that there are 243 cases of lung cancer, then the actual number of lung cancers is probably not 243. But it will be close. It will be something like $243 \pm 1/\varepsilon$.

This may seem unsatisfactory. However, this is to be contrasted with the statistical fluctuations that accompany any survey. Suppose that we survey 500 persons at random. This sample can only be approximatively representative. In fact, if there is a fraction of $n/500$ lung cancers in the whole population, then, for each 500-person survey, we should expect roughly[12] $n \pm \sqrt{n}$ lung cancers. In particular, the uncertainty of the survey is of the order of $\sqrt{n}$.

As a result, if we expect roughly n cases, we may adjust the magnitude of the random perturbation required by differential privacy by setting $1/\varepsilon = \sqrt{n}$. Or, equivalently, $\varepsilon = 1/\sqrt{n}$. Thereby, we guarantee a $(1/\sqrt{n}, 0)$-differential privacy, essentially without deteriorating the accuracy of the survey. In particular, by surveying sufficiently many people, we may guarantee great privacy.

8.8 ROBUSTNESS TO COMPOSITION

In the last decade, differential privacy has become one of the most studied and exciting concepts of computer science. Beyond its intuitive appeal, this popularity can be traced back to two fundamental properties of differential privacy: robustness to subsequent compositions and addition of successive losses of privacy.

---

[11]Its density function is $f(x) = \frac{1}{2b} \exp(-b|x|)$, whose variance is $2b^2$.

[12]This is derived from the central limit theorem, or, better yet, from concentration inequalities like Chernoff bounds.

Let's start with robustness to composition. As discussed, the flaw of a mere pseudonymization is the possibility to cross metadata, like age, sex, or address, with other data sets to deanonymize data. Such techniques have been used, for instance, to identify the owners of malicious Bitcoin accounts, whose identifiers were pseudonyms.

Indeed, Bitcoin is a modern technology that allows financial transactions without the intervention of a central authority. Contrary to classical transfers that require banks' agreements, any computer on the Internet can validate Bitcoin transactions announced by Bitcoin accounts[13]. The possibility to quickly perform decentralized and pseudonymized transactions without control by any government was rapidly exploited by drug and arm dealers. The deanonymization of their Bitcoin accounts then became a matter of national security. In many cases, it was done successfully. The intelligence service's trick was to constantly cross the metadata of Bitcoin transactions (like the time of transaction and its recipient) with other data sets.

The takeaway of this story is that pseudonymization, even when done intelligently as was probably the case of drug and arm dealers, gives absolutely no guarantee of privacy. In particular, if this pseudonymization may seem sufficient at the time of the release of the data, it offers no guarantee of robustness with respect to composition with other kinds of information[14].

This weakness of pseudonymization methods is the strength of differential privacy. Whether it's at the time of the publication of results or two centuries later after analyzing plenty of other data sets, the privacy guarantee of differentially private mechanisms will remain the same.

This guarantee is preserved even in extreme cases. Imagine that Alice accepted giving her data to a study, but that all other individuals of the study are in fact well known. Despite the fact that all the database except Alice's entry is well known, as long as the only query to the database is (ε, δ)-differentially private, Alice's data will remain (ε, δ)-differentially private until the end of time!

In other words, with probability $1 - \delta$, the credence on Alice will never be multiplied by more than e^{ε}, between the case where Alice is in

---

[13]In fact, to have the power to validate transactions, it is necessary and sufficient to solve a very difficult math challenge, which is why, these days, the validation of Bitcoin transactions is performed by computing farms. These farms are directly remunerated in Bitcoin for their work.

[14]In Bayesian terms, this means that we can have both $\mathbb{P}[X|A] = \mathbb{P}[X|B]$ and $\mathbb{P}[X|A, Z] \neq \mathbb{P}[X|B, Z]$.

the database, and the case where she is not. And this holds both right after the response of the mechanism, or centuries later, after the analysis of many other data sets.

8.9   THE ADDITION OF PRIVACY LOSSES

Of course, this remark holds only if the database was only queried once by a (ε, δ)-differentially private mechanism. In practice, we might want to query a database numerous times, thanks to diverse differentially private mechanisms.

The other fundamental property of differential privacy is the fact that the losses of privacy will then add up. In other words, if we apply a first $(\varepsilon_1, \delta_1)$-differentially private mechanism, followed by a second $(\varepsilon_2, \delta_2)$-differentially private mechanism, then the total loss of differential privacy will be at most $(\varepsilon_1 + \varepsilon_2, \delta_1 + \delta_2)$.

This remarkable theorem can be derived from an (almost rigorous) reasoning on credences[15]. After the application of the first mechanism, with probability $1 - \delta_1$, the credence on a surveyed individual will never be multiplied by more than e^{ε_1} compared to the case where he was not surveyed. After applying the second mechanism, with probability now $(1 - \delta_1)(1 - \delta_2)$, his credence will not be multiplied by more than $e^{\varepsilon_1} e^{\varepsilon_2}$. Yet, a quick algebraic computation tells us that $(1 - \delta_1)(1 - \delta_2) \geq 1 - (\delta_1 + \delta_2)$ and $e^{\varepsilon_1} e^{\varepsilon_2} = e^{\varepsilon_1 + \varepsilon_2}$. By combining it all, we conclude that, after performing the two differentially private mechanisms, with probability at least $1 - (\delta_1 + \delta_2)$, the credence on an individual will not be multiplied by more than $e^{\varepsilon_1 + \varepsilon_2}$. This exactly corresponds to saying that the differential privacy of the two mechanisms is at most $(\varepsilon_1 + \varepsilon_2, \delta_1 + \delta_2)$. The loss of privacy of two mechanisms is the sum of the losses of the mechanisms!

This good news also raises a major difficulty posed by privacy. The more we query a database, the more we violate the privacy of its data. In fact, when we conceive a private system, it's important to control its entire life cycle, and to anticipate its total deletion once some critical privacy loss is reached. Indeed, for the same amount of useful information, a system whose query results are not dependent on future queries will be less differentially private than a system that was optimized for the set of all queries to be answered.

---

[15] *Interpretation of ϵ and δ's of Differential Privacy (Proof)*. Wandida, EPFL. LN Hoang (2017).

In the former case, we talk about *online* queries. In the latter, we talk about *offline* queries. For optimal privacy-accuracy tradeoff, a system should address all *offline* queries at once and delete the database once the *offline* queries are answered. Unfortunately, though, in practice, the designer of a private system often has to allow for *online* queries, since at the time of conception, he often does not yet know which queries will be made. This will typically be the case when queries are made by users, but not by the system designer.

As you can imagine, there would be much more to say about differential privacy. The concept was introduced in 2006, and research in this area is still very active[16].

8.10 IN PRACTICE, IT'S NOT GOING WELL!

However, there are multiple and complex gaps to overcome to apply the theory in practice. Much more time will probably be needed for hospitals and institutions to construct differentially private systems. But some works are in progress[17]. In particular, the designers and users of these systems all need to better understand privacy in general, and the limits of pseudonymization in particular.

Having said this, it's not yet clear that differential privacy is the ideal concept to guarantee differential privacy. Differential privacy might be too constraining in many cases. In particular, it must guarantee privacy, no matter which data are in the database, and no matter what prior knowledge hackers have. This is a lot to ask! Furthermore, it's worth pondering the fact that your personal loss of privacy in your life is the sum of all losses of privacy of all mechanisms that you are a part of.

A more radical approach to privacy is to forbid the aggregation of information and to make every individual the only owner of his data. Twenty-five years ago, before the advent of the web, this could be imaginable, as everyone's data were physically separated from others' data. At that time, each had a computer (or floppy disks), and our digital data were physically confined to our houses (although there were also printed and archived data in public institutions).

---

[16] *Differential Privacy and Alternative Definitions.* ZettaBytes, EPFL. C Troncoso (2018).

[17] *The Big Data Setup of the Human Brain Project.* ZettaBytes, EPFL. A Ailamaki (2017).

However, these days, private data travel throughout the Internet. None of us even has a clue about the geographical location of our personal data. Worse, much of these data are stored in gigantic data centers that are owned by only a few giants of the web, like Google, Apple, Facebook, Amazon, and Microsoft. Even most of the processing of these data, from our Facebook profile to the reservation of plane tickets, as well as photo editing, are now done in these data centers, to the point where Facebook is now likely to better predict which next articles or videos you will enjoy than your partner could.

In terms of data, the big companies of the Silicon Valley clearly are a step ahead.

8.11 HOMOMORPHIC ENCRYPTION

However, all is not lost. In particular, *homomorphic encryption* could soon revolutionize our relation to data and data centers. The principle of this encryption is to let data centers process our data, but in such a way that these data centers will be unable to read or understand the data they are processing.

Homomorphic encryption has already been applied to secure and guarantee the privacy of electronic voting, with online prototypes like *Helios* and *Belenios*. The principle is roughly the following one. Each voter has a private key. Using this private key, he can encode and sign his vote, but no one can decode it. Encoded votes are then combined by a publicly verifiable operation, and lead to a final encrypted result. The voters' private keys are then combined to form a sort of masterkey, with which the final encrypted result, and only this encrypted result, can be decoded. Thereby, we can mathematically verify the final result, without violating the privacy of any voter's ballot[18].

If, in principle, cryptography tricks seem to have solved the problem of secure electronic voting design, electronic voting does not exactly guarantee all the nice properties provided by the classical voting booth. In particular, in the voting booth, the voter is alone and not monitored. On the contrary, when the voter votes on his phone, he could be under the threat of a malicious blackmailer. Worse, there would remain the risk that the voter's device was attacked by some virus or hacker. He may

---

[18] Although this privacy guarantee is not as strong as the privacy provided by differential privacy.

then think he voted one way, while the virus or hacker actually made him vote the other way.

Nevertheless, it seems that the question should not be about the perfectness of electronic voting; the relevant question seems to rather be if electronic voting is preferable to today's voting. Yet, today's voting also has many flaws. For instance, it's quite time-wasting. But let's not dwell too much on this issue. This question is partly a moral question, and thus steps outside the realm of a philosophy of knowledge.

Let's get back to homomorphic encryption. In the setting of electronic voting, the encryption allowed combining encrypted votes to reach a final encrypted result. And this encrypted result corresponds to an actual computation of the actual votes. This is the magic of homomorphic encryption. It allows the computer to perform valid data processing, even though the computer is unable to know anything about the nature of processed data!

From a mathematical standpoint, the homomorphic encryption used by electronic voting is quite simple: add the "yes" ballots, and add the "no" ballots. The underlying mathematical operation is just addition[19]. However, the Holy Grail of modern cryptographic research would be to allow for more sophisticated operations than addition. Ideally, a computer would even be able to perform any computation on encrypted data, without the need to decrypt them. If this could be done, from your home, on your phone or laptop, you could ask a data center at the other end of the world to manipulate your encrypted data, to compute a result from these data, and to only send you the result. You would decrypt the result with your phone, in which your secret keys would be stored. You would then be able to read emails, watch holiday photos, or listen to your music. And the data center who sent you encrypted versions of these data would be unable to know the actual content of these data!

In fact, fully homomorphic encryption already exists. Unfortunately, though, it's still too inefficient. To perform the required operations, data centers would need to perform all kinds of complex operations. With today's fully homomorphic encryption schemes, data centers would need to spend vastly more time, memory space, and electrical energy than if these operations were done directly on decrypted data. But research is in progress. And it's progressing fast....

---

[19]Remarkably, this is sufficient for gradient-based distributed machine learning algorithms!

FURTHER READING

The Algorithmic Foundations of Differential Privacy. Foundations and Trends® in Theoretical Computer Science. C Dwork & A Roth (2014).

Differential Privacy. Automata, languages and programming. C Dwork (2006).

Differential Privacy. Encyclopedia of Cryptography and Security. Springer US. C Dwork (2011).

Maths from the talk "Alan Turing and the Enigma Machine". J Grime (2013).

An Embarrassing Survey - Randomized Response. singingbanana. J Grime (2010).

Mathematics of Codes and Code-Breaking. singingbanana. J Grime (2012).

Diffie-Hellman Key Exchange. Wandida, EPFL. J Goubault-Larrecq (2014).

The Diffie-Hellman Protocol. ZettaBytes, EPFL. S Vaudenay (2016).

The Big Data Setup of the Human Brain Project. ZettaBytes, EPFL. A Ailamaki (2017).

Why Privacy Matters. ZettaBytes, EPFL. C Troncoso (2018).

GDPR and the Future of Privacy. ZettaBytes, EPFL. C Troncoso (2018).

Differential Privacy and Alternative Definitions. ZettaBytes, EPFL. C Troncoso (2018).

Public Key Cryptography - RSA Encryption Algorithm. Wandida, EPFL. EM El Mhamdi (2014).

Differential Privacy (playlist). Wandida, EPFL. LN Hoang (2017).

CHAPTER 9

Game, Set and Math

A bet is a tax on bullshit.

Alex Tabarrok (1966-)

What the game is, defines what the players do. Our problem today isn't just that people are losing trust, it's that our environment acts against the evolution of trust.

Nicky Case

9.1 THE *MAGOUILLEUSE*

The *École Polytechnique* is one of the most prestigious French *Grandes Écoles* for science and engineering. Founded in 1794, it was then militarized in 1804 by Napoléon I, who used it as a recruitment pool to lead his armies. These days, the *École Polytechnique* is still managed by the Defense ministry. This is why, when they enter the *École Polytechnique*, all French students must undergo a three-week initial military training. Then, they are faced with an assignment problem.

There are 400 students. Roughly, 130 must go to the Army, 60 to the Navy, 60 to the Air Force, 60 to the military police, and the 90 remaining students must be assigned to other areas, like police, fire forces, or humanitarian forces. Rather than choosing random assignments, the *École Polytechnique* took advantage of having would-be engineers. It let students determine themselves how to assign their fellow mates. This is how a software called *magouilleuse* got developed. Each student enters his medical aptitude and his assignment preferences into the software. The software mixes it all, then applies some obscure algorithm, and determines the assignments of students.

Studying the properties of such software is a fascinating topic that occupied my thoughts for years. In fact, I found this problem so intriguing that I turned it into the topic of my PhD thesis. Unfortunately, after years of research, while it did help me better grasp the problem posed by the *magouilleuse*, my thesis failed to solve this vast problem.

One difficulty simply consists of determining the purpose of the *magouilleuse*. If I believe what I was told, the *magouilleuse*, as implemented in my days, aimed at minimizing a sum of quadratic losses. In other words, the *magouilleuse* gave a 1-point penalty to a first choice, 4 to a second choice, 9 to a third choice, and, more generally, a n^2-penalty to the n-th choice of a student. The *magouilleuse* would then minimize the sum of penalties. This is quite an arbitrary and questionable choice - which I discussed in length in my thesis[1].

But let's leave this major and nontrivial difficulty aside for now. What I want to insist on now is a poorly kept secret that became a recurrent discussion topic among Polytechnique students. Namely, your best strategy was not to reveal your true preferences. The trick was to put your favorite assignment as a first choice, and to then list highly demanded assignments like the Navy. Indeed, since such assignments are in high demand, they will be given to first choices. As a result, for the *magouilleuse*, the alternative to your first choice will then be a fourth or fifth choice. Yet, such alternatives are associated with large penalties. Thus, the *magouilleuse* will favor those who followed such strategies and disfavor those who entered their preferences truthfully. The *magouilleuse* incentivizes dishonesty.

Unfortunately, this fact is not specific to the *magouilleuse*. Preelectoral periods often lead to endless debates between heart and reason. During the 2002 presidential election in France, which occurs in two rounds, numerous leftist candidates ran. And all gathered a large number of votes in the first round. But this also meant that the ballots were spread, at the expense of Lionel Jospin, the main leftist candidate. Jospin got eliminated from the presidential race at the first round. Instead of a tight second round between Chirac and Jospin, France witnessed a landslide victory of Chirac over Le Pen. Chirac got elected with an indisputable (but misleading) legitimacy. And the millions of leftist French voters who did not vote for Jospin regretted their truthful voting.

Incentives induced by decision mechanisms like the *magouilleuse* or the two-round uninominal voting system are the objects of study of the

---

[1] *Measuring Unfairness Feeling in Allocation Problems*. Omega. LN Hoang, F Soumis & G Zaccour (2016).

theory of *mechanism design*. The Holy Grail of this theory is to determine interaction rules between agents (like Polytechnique students or French citizens) that incentivize them to reveal their true preferences (or, more generally, to behave ethically) and that lead to relatively desirable outcomes (like a fair assignment of students or the election of the candidate that best matches the will of the people). As we shall see, Bayesian philosophy has much to contribute to this effort - and it has already been awarded numerous Nobel prizes.

But before getting there, let's first discuss the *game theory* that mechanism design builds upon. And to do so, let's go to Britain.

9.2 *SPLIT OR STEAL?*

The TV game *Golden Balls* is almost over. Sarah and Steven collected a 100,150£ jackpot. They now enter the last stage of the game, called *Split or Steal*. The two candidates are face to face. Each has two balls. One is *Split*. The other is *Steal*. Each candidate must choose one of the two balls. If both play *Split*, then they will share the jackpot. If one chooses *Split* while the other chooses *Steal*, then the stealer will win the whole jackpot. Last but not least, if both play *Steal*, then both will go home with empty pockets.

Before choosing, the two candidates have some time to discuss. Sarah begs Steven to play *Split*. She is almost crying. Steven tries to reassure Sarah, and promises that he will play *Split*. The discussion is now over. Each candidate secretly chooses his or her ball. The suspense is at its peak. The presenter asks the candidates to reveal their choices. Shocking moment! While Steven did play *Split*, Sarah actually played *Steal*. Sarah won it all! Steven is completely devastated. Sarah is embarassed. She does not know where to look. And yet, she has just won 100,150£!

In the *Split or Steal* game as in our everyday life, uncertainty is ubiquitous. Our fate greatly depends on others' decisions, and our power of influence is often very limited. In this context, rather than forcing others to act in a certain way, it's often more reasonable to anticipate their actions and to act accordingly - although activism can also have an impact. The conceptual difficulty that then arises is that the way others will act depend on how we act. But we've just said that it depends on how they act, which depends on how we act, which depends on how they act... and so on *ad infinitum*.

In 1951, in a 28-page PhD thesis that only contains 2 references, the future Nobel prize winner John Nash proposed an equilibrium concept to shortcut this infinite reasoning. The now-called *Nash equilibrium* corresponds to a setting where each player is playing an action that is optimal with respect to other players' actions. Thereby, given what others are doing, each player will want to keep playing his or her strategy. It's in this sense that the Nash equilibrium is indeed a concept of equilibrium: once players play it, we should expect them to keep playing it.

Weirdly, in first approximation, Sarah and Steven's strategies form a Nash equilibrium. Indeed, given that Steven chooses *Split*, Sarah will want to play *Steal*, as she thereby doubles her gains. Meanwhile, if Steven knows that Sarah will be playing *Steal*, he has nothing to gain by switching from *Split* to *Steal*. Indeed, in both cases, Steven goes home with nothing.

This setting is (almost) that of the infamous *prisoner's dilemma*. This dilemma was imagined by Merrill Flood and Melvin Dresher in 1950, and was then formalized by Albert Tucker. In this dilemma, two bandits are arrested and interrogated separately. If one bandit is snitched on by the other, then his sentence will be greatly increased. However, the police promises a small sentence reduction to each snitching bandit, whether he is snitched on or not. In other words, each bandit is better off snitching, no matter what the other bandit does. As a result, the case where both bandits snitch is a Nash equilibrium. It is, in fact, the only Nash equilibrium.

But then, the two bandits are being snitched on. They both obtain harsh sentences, which they could have avoided by both remaining silent. The moral of this story is that individual incentives are not always aligned with the overall interest of the group. In first approximation, stealing at *Split or Steal* is similar to snitching at the prisoner's dilemma. In both cases, it's an individually optimal strategy that leads to global suboptimality.

9.3 BAYESIAN PERSUASION

However, these examples are certainly much too simplified to really represent Steven's, Sarah's, and the two bandits' thought processes. Indeed, besides financial gains or sentences, there is a huge psychological cost to publicly steal or to snitch on a friend. In particular, it's not unreasonable to imagine that Steven actually prefers to play *Split* than *Steal*. To win the whole jackpot and endure the disapproving look of the other candidate, of the viewers, and of one's family can be very painful. As a result, Steven might want to play *Split*, no matter what Sarah does.

Now, things get even more interesting if we now assume that Steven prefers to see Sarah losing the jackpot if she plays *Steal*. In other words, let's now imagine that Steven prefers the following outcomes in this order: (*Split*, *Split*), (*Steal*, *Steal*), (*Steal*, *Split*), (*Split*, *Steal*). The first action of the pair is Steven's choice, the second is Sarah's. In particular, now, Steven's optimal strategy depends on Sarah's choice. If Sarah plays *Split*, then Steven will want to play *Split*. But if Sarah plays *Steal*, then Steven will want to play *Steal*.

It's at this point that the prior discussion between Steven and Sarah - as well as Bayesian philosophy - will be playing a very important role. The discussion may affect Steven's credences in Sarah's upcoming choice. Evidently, in our case, Sarah managed to make him believe that she would play *Split*. She did it very well. Steven got quickly convinced that this would be the case. He even seemed to have dismissed the alternative, which explains his great disappointment.

Such communications can be formalized by a Bayesian formalism. After all, to communicate is to reveal information that the other will be using to update his credences. In fact, in 2011, economists Kamenica and Gentzkow wondered how a prosecutor could best incriminate a defendant in front of a Bayesian judge. Surprisingly, they showed that a good prosecutor could convince the judge to sentence more defendants than the number of guilty defendants the judge believes there are.

Let's give details. A Bayesian judge necessarily has a prior on a defendant's guilt. Let's assume that this prior is $\mathbb{P}[\text{😠}] = 0.3$. Moreover, we shall assume that the Bayesian judge will sentence a defendant if the judge has more credences on the defendant's guilt than on the defendant's innocence. To convince the judge, the prosecutor will then propose to launch an investigation that's a bit peculiar. When the defendant is guilty, the investigation will show it. But when the defendant is innocent, the investigation will purposely be sometimes mistaken. In fact, the investigation will incriminate an innocent 3 times out of 7. The prosecutor knows it. And so does the judge.

Of course, as a result, if the investigation incriminates the defendant, it might simply be because the investigation is mistaken. However, this can only increase the judge's suspicions. A Bayesian computation determines precisely this posterior suspicion:

$$\mathbb{P}[😈|🚩] = \frac{\mathbb{P}[🚩|😈]\mathbb{P}[😈]}{\mathbb{P}[🚩|😈]\mathbb{P}[😈] + \mathbb{P}[🚩|😇]\mathbb{P}[😇]} = \frac{1 \cdot 0.3}{1 \cdot 0.3 + {}^{3}\!/\!{}_{7} \cdot 0.7} = 0.5.$$

In other words, *a posteriori*, once he learns that the investigation incriminates the defendant, the Bayesian judge will have as much credence in the defendant's guilt than in the defendant's innocence: the judge will decide to convict him[2]. As a result, any incriminated defendant will be convicted. Yet, the probability of being incriminated is given by the law of total probabilities:

$$\mathbb{P}[🚩] = \mathbb{P}[🚩|😈]\mathbb{P}[😈] + \mathbb{P}[🚩|😇]\mathbb{P}[😇] = 1 \cdot 0.3 + {}^{3}\!/\!{}_{7} \cdot 0.7 = 0.6.$$

In other words, *a priori*, the judge will now have a 60% probability to convict the defendant, even though his prior credence in the defendant's guilt is 30%. The judge will necessarily convict too many defendants!

Just to be clear, however, this would absolutely not be a flaw of the judge's Bayesianism. If he had not used Bayes' rule and had restricted himself to his prior, then the judge would have exonerated the defendant. This would have led to the same probability of wrong sentence. More generally, the judge's goal is not to make the number of convictions equal the number of guilty defendants. The judge's goal is to minimize the error rate[3].

But then, would it be possible to trick the Bayesian judge and increase his error rate? The answer is no. In fact, Bayesian inference has a remarkable property that many other inductions do not possess: as long as the judge correctly interprets additional information and applies Bayes' rule, the judge's expected error rate will not decrease[4]. In other

---

[2]By replacing $3/7$ by $3/7 - \varepsilon$, we obtain a posterior slightly larger than 0.5, which definitely leads to the defendant's conviction.

[3]Note too that we have considered a setting where convicting an innocent and exonerating a guilty person are equally wrong for the judge. The setting can easily be modified to address different moral philosophies of conviction under uncertainty.

[4]More generally, consider a Bayesian with a utility function u that satisfies the von Neumann-Morgenstern axioms. This Bayesian will want to make decisions a despite not knowing the value of x. Without additional information y, the Bayesian would solve $\sup_a \mathbb{E}_x[u(a,x)]$. Suppose she now learns y. She would then maximize $\sup_a \mathbb{E}_x[u(a,x)|y]$. Her prior expected gain is then $\mathbb{E}_y[\sup_a \mathbb{E}_x[u(a,x)|y]]$. As long as $\mathbb{P}[x|y]$ is computed through Bayes' rule, the Bayesian will then have gained by learning y, i.e., $\mathbb{E}_y[\sup_a \mathbb{E}_x[u(a,x)|y]] \geq \sup_a \mathbb{E}_x[u(a,x)]$.

words, in (subjective) expectation, the Bayesian always wins by gaining information[5].

9.4   SCHELLING'S POINTS

Let's go back to *Split or Steal*. The game becomes even more interesting if we now imagine that Sarah's preferences are in fact the same as Steven's. In this case, both will want to share. It would suffice that they mutually persuade one another that they will share so that all goes right.

However, any suspicion could quickly degenerate. If Steven clumsily gave the impression that he might not play *Split* after all, Sarah could then increase her credence in Steven's playing *Steal*. There could then be a point where this credence is so great that Sarah would prefer playing *Steal* to *Split*, as she would rather avoid the humiliation of seeing Steven taking it all away from her. But if she then let her hesitations show, then Steven could then anticipate Sarah's *Steal* strategy, and would then gain comfort in playing *Steal* too. Worse, Steven may then want Sarah to play *Steal*, to avoid publicly stealing what she wanted to split.

This is known as a *coordination* problem. There are two symmetric Nash equilibria[6]. In one of these equilibria, Sarah and Steven both play *Split* and go home happy. In the other equilibrium, they both play *Steal* and go home sad. But if they don't coordinate, both will go home even sadder, either because their money gets stolen away, or because of the disapproving look of the audience that witnesses their steal. The outcome of the game would then greatly depend on the respective credences on what the other will do - and flawed credences could be catastrophic for both.

Such coordination problems transcend by far the *Split or Steal* setting. There is, of course, the Hollywood example of two love birds who hesitate to reveal their loves, because they doubt the other's interest in them. Bad credences can ruin what would otherwise be a beautiful love story - and the tormented learning of the right credences is often what makes the movie entertaining!

It is to address such coordination problems that we have traditions, conventions, or protocols. These elements that structure our social

---

[5]Of course, the notion of expectation is crucial, as the judge is now going to be sometimes misled into convicting innocents that he would have otherwise exonerated.

[6]In fact, there is even a third (unstable) Nash equilibrium where Sarah and Steven both play a certain random combination of *Split* and *Steal*.

relationships are known as *Schelling points*, after 2005 Nobel prize winner Thomas Schelling. In Bayesian terms, Schelling points correspond to prior credences on one another's behaviors in society.

Common sense often requires a good estimate of these Schelling points. It then plays an important role in individuals' behaviors. Thereby, we can imagine that, in a society where trust and honesty are two reliable Schelling points, Steven and Sarah would eventually play *Split*. On the contrary, in societies where the norm is not mutual trust and where distrust is advised, Steven and Sarah might be more likely to play *Steal*.

9.5   MIXED EQUILIBRIUM

There is a last variant of *Split or Steal* that will allow us to explore other subtleties in agent interactions. Now suppose that Steven and Sarah are both venal, and that they promise a 10,000£ to the other if one wins the whole jackpot. As a result, the preference order is now: (*Steal, Split*) with a 90k£ gain, (*Split, Split*) with a 50k£ gain, (*Split, Steal*) with a 10k£ gain, and (*Steal, Steal*) with no gain.

Oddly enough, if Sarah now asserts that she will play *Steal*, then Steven will have incentives to play *Split*, which would further comfort Sarah in her choice. On the other hand, if Steven claims he will play *Steal*, then the roles would be inversed. This game has two antisymmetric Nash equilibria. The asymmetry of the two equilibria then raises an intriguing problem: each player will want to convince the other to play the Nash equilibrium that best suits him or her. In practice, this will incentivize each player to claim he or she will be stealing the jackpot, and to try to be more convincing than the other.

Such strategies may sound foolish. Yet, surprisingly, when Nick and Abraham were faced at *Split or Steal* with a 13k£ jackpot, Nick asserted: "Abraham, I want you to trust me. 100%, I am going to pick the *Steal* ball." Abraham felt distraught and powerless. He asked: "Where is your brain coming from?" And then, "You're an idiot!"

But Nick was no idiot. He just wanted to make sure that Abraham had all the incentives in the world to play *Split* - and since he was ready for several tricks, Nick actually played *Split* too! What's most stunning is that, in a follow-up interview on Radio Lab[7], Abraham acknowledged that he first intended to play *Steal*. Nick's strategy was perfect.

---

[7] *The Golden Rule*. Radio Lab (2017).

Now imagine that, inspired by Nick, Steven and Sarah both claim they will play *Steal*. But none of them would agree to play *Split*. We then have a profoundly Bayesian problem. Each would be uncertain of what the other will be doing. Each must then reason probabilistically, and use prior credences.

Suppose for instance that Steven assigns a 1/2-probability that Sarah plays *Split* and a 1/2-probability that she plays *Steal*. If Steven plays *Steal*, then he has a 1/2-probability to win the gain of (*Steal*, *Split*), which corresponds to 90k£ and a 1/2-probability to win that of (*Steal*, *Steal*), which corresponds to 0k£. The expected gain of Steven would then be 45k£.

By opposition, if Steven plays *Split*, then he will win the 50k£ gain of (*Split*, *Split*) half of the time, and the 10k£ gain of (*Split*, *Steal*) the other half of the time. His expected gain would be 30k£. Steven's computation would thus lead him to *Steal* rather than *Split*, since the expected gain in the former case is larger[8].

Since this book deals with the subjective probabilities of Bayesianism, it's worth insisting on the fact that the probabilities at play here for Steven and Sarah are absolutely not frequentist. Despite the frequentist language I used for pedagogical reasons, Steven and Sarah actually do not play *Split or Steal* every day. What's more, the potential gains represent unique and isolated cases in their respective lives. The expected gain here corresponds to the gains estimated by their subjective probabilities.

Let's get back to Steven and Sarah. Suppose now that both are very clever and know that the other is just as clever. After prior discussions, Steven and Sarah both really gave the impression of playing *Steal*. They are in front of a wall. To avoid a dead end, Steven suddenly asserts that he will play *Steal* with probability 4/5, and he will play *Split* otherwise. He then adds that he invites Sarah to do the same.

Sarah can then perform expectation computations. Supposing that Steven will do what he claims, by playing *Steal*, Sarah would win 0£ four times out of five, and 90k£ the remaining times, which gives her an expected gain of[9] 18k£. On the other hand, by playing *Split*, she would win 10k£ four times out of five, and 50k£ one time out of five, for an expected gain of[10] 18k£ too. She then concludes that her choice will not affect her expected gain. What's more, she realizes that by

---

[8]We assume here that the players are venal and risk-averse, i.e., the utility $u(x)$ of winning x£ is $u(x) = x$.

[9]This is the computation $\frac{4}{5} \cdot 0 + \frac{1}{5} \cdot 90 = 18$.

[10]This is the computation $\frac{4}{5} \cdot 10 + \frac{1}{5} \cdot 50 = 18$.

following Steven's advice, she will be giving incentives to Steven to follow his announced strategy. Thus Sarah accepts. She even adds that she will play *Steal* with probability 4/5 as well. In technical terms, the two players have agreed to play a so-called *mixed* Nash equilibrium, that is, an equilibrium where players use randomized strategies.

If Steven and Sarah were maximizing their gains and refused asymmetry, then Nash's theory asserts that they would play this mixed equilibrium[11]. However, this Nash equilibrium, like many other Nash equilibria, is suboptimal. Indeed, Steven and Sarah would both be better off if they simply played *Split*.

This is where 2005 Nobel prize winner Robert Aumann intervened. In 1974, Aumann proposed to introduce a signal, a sort of traffic light, which would coordinate the players' decision making.

For instance, this signal could be a heads or tails throw. If the coin lands heads, Steven will play *Steal* and Sarah *Split*. If it lands tails, then it will be the inverse. The genius of the outside signal is that, given what the coin says, Steven and Sarah will both have incentives to do what the coin tells them to do. In other words, Steven and Sarah's optimal strategies, given the signal and given how they are supposed to react to the signal, will be to follow the instructions given by the signal. This almost corresponds to a Nash equilibrium. We talk about a *correlated equilibrium*. Any Nash equilibrium can in fact be interpreted as a correlated equilibrium with no signal.

However, the addition of a signal can greatly improve Nash equilibria. In the case of Steven and Sarah, Steven and Sarah's expected gains before the signal are both 50k£. In other words, the signal allows them to do as well as if there were no individual incentives to deviate from[12] (*Split, Split*)!

9.6 BAYESIAN GAMES

The combination of game theory and subjective probabilities of Bayesian philosophy constitutes a powerful arsenal to tackle complex decision problems under uncertainty. This is the case of poker. At the beginning

---

[11]Nash even proved that any symmetric game has a (possibly mixed) symmetric Nash equilibrium.

[12]It's not too hard to see that the convex hull of Nash equilibria is included in the set of correlated equilibria. However, the set of correlated equilibria is generally larger than this convex hull, which allows to sometimes determine correlated equilibria more optimal than any combination of Nash equilibria!

of a poker hand, each player knows his cards, but does not know the others'. However, as the game is being played, some players will give up their bets (they "fold"), while others will increase theirs.

Our *pure Bayesian* will then apply Bayes' rule to modify her credences on others' hands. Typically, if an opponent is more agressive than usual and bets large amounts, she will find it more believable that this opponent has good cards. Of course, this does not prove that his cards are indeed good. But to make optimal decisions, it's essential to update one's credences and to adjust one's strategies accordingly.

Without great originality, game theory under uncertainty is called *Bayesian game theory*. This theory was developed in 1967 in a series of three papers by John Harsanyi, who would then win the 1994 Nobel prize with Nash. By inserting uncertainties with respect to other players' cards, or to other players' preferences, Harsanyi made game theory more realistic and adapted it to numerous real-world problems where the relevant information is incomplete.

Harsanyi also used the Bayesian language to explain the relevancy of mixed Nash equilibria. Harsanyi thereby explains that the uncertainty on individuals' strategies could be derived from the uncertainty on their preferences. This is what is suggested by Harsanyi's *purification theorem*. More crucially, Harsanyi finally put Bayesian philosophy at the heart of economists' reasoning. 2007 Nobel prize winner Roger Myerson asserted in 2004: "The unity and scope of modern information economics was found in Harsanyi's framework."

The Bayesian variant of Nash equilibria is called a Bayes-Nash equilibrium. It's a setting where each player plays a best-response strategy to others' strategies. However, there is a subtlety to be found behind the Bayesian concept of "strategy". A strategy is now a way to react to private information, whether this information is individual preferences or cards in hand. For instance, in poker, such a strategy may consist of raising given pocket aces, and to fold otherwise - this strategy is probably not very good, though!

9.7 BAYESIAN MECHANISM DESIGN*

We are finally ready to go back to the *magouilleuse*. At first, instead of using a *magouilleuse*, one might seek a negotiation mechanism between students to obtain a global decision of assignments. However, this mechanism seems overwhelmingly complex. What's more, the theory of mechanism design gave us a nice theorem that greatly simplified the quest for a best-possible mechanism. The theorem shows that any such mechanism is equivalent to a so-called *direct mechanism* managed by a central authority, such as the *magouilleuse*. This authority needs to collect students' preferences, and then computes a global decision based on these inputs. The theorem that proves the equivalence between any mechanism and a direct mechanism is called the *revelation principle*[13].

To understand this principle, consider any process M. Suppose that each individual behaves in a way that maximizes his expected utility given his Bayesian credences. In other words, individuals play a Bayes-Nash equilibrium of mechanism M. Once the mechanism performed, we obtain a global decision x for the group, which may be assignments for students to military forces. The revelation principle allows us to obtain the same global decision x through a direct mechanism that is a sort of *super-magouilleuse*. This *super-magouilleuse* collects individuals' preferences and then simulates the individuals playing the Bayes-Nash equilibrium of mechanism M. In particular, the *super-magouilleuse* can then compute the results and derive the global decision x for the group. It's this global decision for the group that the *super-magouilleuse* selects.

What's great with the *super-magouilleuse*'s simulations is that, from the individuals' standpoints, all they need to do is to reveal their preferences. Our *super-magouilleuse* then derives the consequences of such inputs. Better, and contrary to uninominal voting systems or to the *magouilleuse*, each individual now has the incentives to reveal his preferences truthfully. Indeed, if an individual reveals preferences that are not his, then the *super-magouilleuse* will be simulating this individual's behavior, with preferences that are not his. The individual's simulated behavior will thus be suboptimal, which will lead to a global decision that is less advantageous for the individual.

This simulation trick is what the *attorney-client privilege* relies on. Indeed, what we would like is to determine a sentence for a defendant. Ideally, the verdict would take into account some information that only the defendant knows. However, this defendant usually has no incentives to reveal all that he knows - especially if he is guilty. Now, the idea of law courts is to set up an interaction system where the accuser and the defendant debate, and such that the final sentence is the outcome of this debating. But evidently, this puts the defendant in an awkward position. Such a debate would not be incentive-compatible, since the defendant will not have the incentive to debate truthfully.

---

[13]Mechanism Design and the Revelation Principle. Science4All. LN Hoang (2012).

The attorney-client privilege consists of adding intermediary agents that will simulate the debate between the accuser and the defendant without the direct intervention of the accuser and the defendant. The agents who simulate the accuser and the defendant are the prosecutor and the defense lawyer. However, to simulate the accuser and the defendant well, the two lawyers need to behave optimally for their clients. This requires them to know all they can about their clients. And to do so, the clients need to have the incentives to reveal to their lawyers what they know. The attorney-client privilege aims at providing such incentives.

In game theory, the main use of the revelation principle is to justify the fact that we may focus, (almost) without loss of generality, on centralized direct mechanisms that consist of collecting private information to make a global decision. This principle permits discovering a very general mechanism that both guarantees the truthfulness of individuals' preference revelations and the maximization of social welfare[14]. This remarkable mechanism, which is thus *incentive-compatible* and *welfare-maximizing*, is known as the *VCG mechanism*, after 1996 Nobel prize winner William Vickrey, as well as Edward Clarke and Theodore Groves[15].

9.8 MYERSON'S AUCTION

In the case of a single-good auction, maximizing social welfare boils down to assigning the good to he who most values it. It's what the VCG mechanism guarantees by then making the buyer pay the good at the prize of the second buyer. However, *a priori*, this so-called *second-price auction*, while incentive-compatible, does not seem to maximize the seller's gains. For the seller, should the good not be sold simply through a first-price auction?

The surprising answer, given by 2007 Nobel prize winner Roger Myerson, is no. If the good is sold to the first buyer, then no matter how the price he pays is set, and so long as the buyers behave like utility-maximizing Bayesians, the seller's expected gain will always be the same! This is Myerson's surprising revenue-equivalence theorem[16].

---

[14]Social welfare is defined as the sum of individuals' utilities.

[15]*A Mathematical Guide to Selling.* Science4All. LN Hoang (2015).

[16]Like VCG, the theorem assumes that individuals' utilities are so-called *quasi-linear*. What's more, revenue-equivalence assumes that the buyers' valuations of the good are independent and identically distributed.

There is, however, a small technical detail that Myerson's theorem relies on, which may give an idea of the complexity of human interactions in practice. For Myerson's revenue-equivalence theorem to hold, it must be assumed that buyers and sellers all have a common prior on what each seller is willing to pay for the good. It's a common hypothesis in Bayesian games, as it greatly simplifies computations and avoids metaphysical questions. But it's false. "All models are wrong".

In the general case, we need to invoke *higher beliefs*. These are beliefs like what a first buyer believes regarding the credence of another buyer on what the first buyer is willing to pay, if not on what he believes that the other believes that he believes on the preferences of the other. And so on. It's interesting to note that the *pure Bayesian* takes into account such considerations. However, the study of *higher beliefs* is extremely hard and far transcends the scope of this book. So let's get back to Harsanyi and Myerson's setting, which is that of the revenue-equivalence theorem.

Would it then be possible for the seller to achieve a larger expected gain with another auction? Surprisingly enough, Myerson's answer is yes. In his wonderful 1981 paper, Myerson managed to determine the seller's optimal auction. The details of this auction are a bit technical[17], but the key idea is very simple. To maximize his gains, the seller must use his Bayesian prejudices. In particular, he must refuse to sell the good if the prices claimed by the buyers are far smaller than the seller's credences on what the buyers really are willing to pay. Of course, thereby, the seller has a nonzero probability to win any revenue out of the auction. But what matters is the *expected* revenue.

Myerson' auction explains why foreign tourists have such a hard time negotiating. It's simply because, in many developing countries, sellers believe that foreign tourists are willing to pay larger amounts than local people. Worse, if a foreign tourist and a local want to buy the same good, then Myerson's auction will advise the seller to articulate a negotiation at a distance. He may accept to sell to the foreign tourist only if the price he proposes is at least twice that of the local individual. It may be tempting to believe that the seller and the local buyer are colluding to increase the price of the good - and this may very well be the case. But it doesn't need to be so to justify the seller's selling strategy. The seller may simply be applying Myerson's auction, which discriminates buyers based on what the seller believes about what the buyers are willing to pay. It's not just a matter of negotiating skills! It's mostly a matter of Bayesian credences!

---

[17] *A Mathematical Guide to Selling.* Science4All. LN Hoang (2015).

9.9 THE SOCIAL CONSEQUENCES OF BAYESIANISM

A major consequence of Myerson's auction is that a *pure Bayesian's* optimal behavior makes her discriminate individuals based on prejudices. If she wants to maximize revenue, the *pure Bayesian* will not treat all individuals equally.

Now, this does not mean that we should reject Myerson's theory for ethical reasons. Myerson's theory is a mathematical theorem. It has no moral value. Similarly, to forbid Bayesian thinking on the ground that, in this case, it leads to morally undesirable consequences would be similar to forbidding rigorous thinking because it helps billionaires become richer.

Moral values could, however, appear in sellers' and buyers' preferences. In particular, ethically questionable preferences of buyers and sellers could be fought on moral grounds. But, interestingly, what Myerson's theory shows is that banal and socially accepted preferences can lead to behaviors that are rejected by our societies. Quite often, the problem is not that the discriminating agents aimed to do harm; the problem is that they were not sufficiently incentivized to do good. *The cause of the worst harms is usually not the willingness to cause harm; it's often rather the absence of willingness to avoid harm.*

Of course, the context of Myerson's auction is *a priori* not that controversial - though tourists often complain about the way they were treated, and some ethnies are stigmatized for their wealth. However, I invite you to list more touchy variants. Should prejudices be used when hiring a new employee, judging a suspect, or computing insurance prices?

For the *pure Bayesian*, this is a question out of the realm of the philosophy of knowledge. This is a matter of ethics, of morals, of values, of objective function, and of preferences. The *pure Bayesian* has no say on such matters.

Having said this, ignoring prejudices to make decisions seems to gain popularity, especially among intellectuals. As discussed in the previous chapter, a large part of computer science research consists of guaranteeing the privacy of private data, to prevent others from adjusting their Bayesian credences.

However, there are cases where the inabilities of some people prevent us from treating them like anyone else. This is the case of the assignment of Polytechnique students. For medical reasons, some students could not be assigned to some forces. Non-discrimination was thus illegal. As a result, rather than ignoring the differences among individuals, it may be more desirable to include in our moral philosophy a way to correctly deal with differences between individuals.

In fact, it's often useful to exploit prejudices to determine more ethical policies. Typically, numerous buses throughout the world ask passengers to yield seats to the elderly, pregnant women, and disabled people. In other words, these buses ask their passengers to use prejudices based on other passengers' physical appearances to favor those that will profit the most from sitting down.

More generally, as we shall see in the last chapter of the book, there is a branch of moral philosophy for which prejudices are morally desirable. This moral philosophy is *consequentialism*. According to it, only (likely) consequences of our actions matter. The *Bayesian consequentialist* will then use her whole Bayesian arsenal to optimize the goods of her actions. Yet, this arsenal includes prejudices. *Ignoring prejudices would be a moral sin.*

In particular, (adequate) prejudices help us help more quickly and more efficiently those that need it the most. It seems immoral to ignore them.

FURTHER READING

Equilibrium Points in N-Person Games. PNAS. J Nash (1950).

Non-Cooperative Games. Annals of mathematics. J Nash (1951).

Subjectivity and Correlation in Randomized Strategies. Journal of Mathematical Economics. R Aumann (1974).

Games with Randomly Disturbed Payoffs: A New Rationale for Mixed-Strategy Equilibrium Points. International Journal of Game Theory. J Harsanyi (1973).

Optimal Auction Design. Mathematics of Operations Research. R Myerson (1981).

Comments on "Games with Incomplete Information Played by 'Bayesian' Players, I–III Harsanyi's Games with Incoplete Information". Management Science. R Myerson (2004).

Bayesian Persuasion. The American Economic Review. E Kamenica & M Gentzkow (2011).

Measuring Unfairness Feeling in Allocation Problems. Omega. LN Hoang, F Soumis & G Zaccour (2016).

The Golden Rule. Radio Lab (2017).

The Evolution of Trust. N Case (2017).

Game Theory and the Nash Equilibrium. Science4All. LN Hoang (2012).

Bayesian Games: Math Models for Poker. Science4All. LN Hoang (2012).

Mechanism Design and the Revelation Principle. Science4All. LN Hoang (2012).

A Mathematical Guide to Selling. Science4All. LN Hoang (2015).

CHAPTER 10

Will Darwin Select Bayes?

There is grandeur in this view of life, with its several powers, having been originally breathed by the Creator into a few forms or into one; and that, whilst this planet has gone cycling on according to the fixed law of gravity, from so simple a beginning endless forms most beautiful and most wonderful have been and are being evolved.

Charles Darwin (1809-1882)

10.1 THE SURVIVOR BIAS

During World War II, the English Air Force hired statistician Abraham Wald to investigate the optimal armouring of aircrafts. The Air Force noted that the aircrafts which came back from battle were hit everywhere but in the front, where the engine was. The Air Force concluded that it would be a good idea to decrease the front armouring to reinforce the back. *Wrong*, asserted Wald! Au contraire, he replied, the fact that aircrafts were hit mostly on the back meant that it was the front of the airplane that needed more armouring.

Wald's remark is quite surprising. Yet it's essentially the same argument as Charles Darwin's to explain the emergence of complex life structures. In both cases, the subtle process that seems invisible to most of us is an *elimination* process - or, if we care about survivors, a *selection* process. In the case of Wald, the elimination is that of aircrafts whose front got hit. Such aircrafts had their engines destroyed, if not exploded. Thus they could not come back home. Similarly, Darwin asserted that animal species whose deficiencies prevented their reproduction have necessarily disappeared. As a result, those who remain today necessarily have remarkably few deficiencies.

While it's unanimously celebrated by the scientific community, Darwin's theory of evolution now undergoes numerous pseudo-scientific criticisms that favor the intelligent design argument. The critics' argument is the following. Imagine yourself in the desert. If you find a weird rock, you will not be surprised to learn it results from natural processes. However, if you now find a watch with complex cogs, then it seems stupid to claim that it could have emerged from purely natural processes. It seems that the sophistication of the watch can only be explained by the thorough work of an intelligent designer. Similarly, the remarkable sophistication of the human body, from the biomechanics of its bones and muscles to the organization of its immune system, including the ingeniosity of its eye and the incomprehensible complexity of the brain, can only be the result of an intelligent design - and the intelligent designer can only be God.

This argument may seem very convincing. Yet, beside the disputable amagalm between an intelligent designer and God, it actually greatly underestimates the elimination process that we discussed above - and that Darwin dubbed *natural selection*.

10.2   CALIFORNIA'S COLORED LIZARDS

Let's go to the Central Valley of California, where three different varieties of male lizards live. Roughly, there are orange, blue, and yellow male lizards. These male lizards are of the same species, and thus try to reproduce with the same females. But their attributes and reproduction strategies are very different. The orange lizards are brutes. They control a territory and reproduce with all females of the territory. The blue lizards are monogamous and jealous. They monitor each and every one of their partners' actions. Finally, yellow lizards are Don Juans. They go for any female they meet, and move on to the next female.

Darwinian evolution suggests that the lizards that are most fit to reproduce will be the ones that survive. However, what's amusing is that the male lizards' ability to reproduce depends on the current male lizard population.

For instance, suppose that most lizards are orange brutes. Then, each orange lizard will conquer a large harem that he will not be able to monitor. As a result, yellow Don Juans will easily copulate with unwatched females, so that the females will in fact be more often inseminated by yellow Don Juans than by orange brutes. Slowly but surely, we should then expect the yellow Don Juans to take over the land over orange brutes.

Now imagine that the yellow Don Juans are predominant. Jealous blues will then charm females and keep them for themselves, so that each female is eventually assigned to a jealous blue. But then, as no female will be left unwatched, the yellow Don Juans will no longer be able to reproduce. The jealous blues would thus cause the extinction of yellow Don Juans.

Let's finally postulate that male lizards are almost all jealous blues. Then, orange brutes would fight jealous blues. They would enlarge their harem, one female at a time. Jealous blues would all become single and would not be able to reproduce. They would eventually disappear.

Let's recapitulate. Roughly, yellow beats orange, blue beats yellow, and orange beats blue. This looks like the classical rock-paper-scissors game, where rock beats scissors, paper beats rock, and scissors beats paper. This game has a unique Nash equilibrium, which consists of playing randomly yellow, orange, and blue. Guess what? In practice, we observe that the three varieties of male lizards co-exist in Nature, as though they decided to play the Nash equilibrium of rock-paper-scissors! In other words, the Nash equilibrium concept, which is supposed to be only relevant for intelligent players, seems to perfectly apply to the outcome of Darwinian evolution[1]. As we shall see, this is no coincidence.

10.3   THE LOTKA-VOLTERRA DYNAMIC*

In 1972, biologist John Maynard Smith invented the concept of *evolutionary stable strategies*. Smith defined these strategies as a certain composition of a population that is robust to the invasion of a (small) population with a different composition (for instance the addition of 100 yellow males). In practice, this typically corresponds to the random variations of the population due to statistical fluctuations. Will such fluctuations cause profound modifications of the population? Or will Darwinian evolution bring the composition of the population back to what it was before the statistical fluctuations?

---

[1] *Rock, paper, lizards*. Numberphile. H Fry (2015).

To answer such questions, we are going to dig into the mathematical details of a simplified model of Darwinian evolution. "All models are wrong". But the one we shall discuss has been useful to many biologists.

Call $x(t)$ the number of individuals of a variety at time t. In the next time step $t + 1$, the population will increase because of births, and decrease because of deaths. The numbers of birth and death will be roughly proportional to the current population. The proportionality coefficients are called birth rate (noted %☺) and death rate (noted %☹). The population then becomes $x(t+1) = x(t) + (\%☺)x(t) - (\%☹)x(t) = x(t) + (\%☺ - \%☹)x(t)$. In other words, the variation of the population is proportional to the population $x(t)$. Let's call fitness $= \%☺ - \%☹$ the coefficient of proportionality. We then obtain the equation of the population dynamics: $\dot{x} = x(t + 1) - x(t) = \text{fitness} \cdot x$.

The above equation holds for individuals of a variety. If we now distinguish the different varieties with an index i, we obtain the *Lotka-Volterra equations* $\dot{x}_i = \text{fitness}_i \cdot x_i$. In fact, these equations are slightly more precise, as they also indicate how the *fitness* of different varieties varies as a function of the populations of the other varieties. As we saw, if the dominant population is that of orange brutes, then the *fitness* of yellow Don Juans will increase[2].

However, what will be of interest to us is not the numbers x_i of individuals of variety i, but the proportions z_i of the different varieties i within the whole population. After a few algebraic manipulations, which I shall leave as homework, we obtain the equation that governs the variations of proportions of the varieties in the population:

$$z_i(t + 1) = \frac{(1 + \text{fitness}_i)z_i(t)}{(1 + \text{fitness}_i)z_i(t) + \sum_{j \neq i}(1 + \text{fitness}_j)z_j(t)}.$$

Do you see it? The equation that governs evolution is actually Bayes' rule in disguise! This is stunning! The subjective probabilities correspond to the proportions z_i. To go from time t to time $t + 1$, the probabilities undergo a sort of Bayesian inference where the *thought experiment terms* are replaced by quantities $1 + \text{fitness}_i$. Finally, in the denominator, we find the partition function, which guarantees that the sum of z_i's will still equal 1 at time $t + 1$.

Now here's the mind-blowing conclusion of this analysis. So long as the *fitness* at time t of variety i can be identified with the ability

---

[2]In their classical forms, the Lotka-Volterra equations assume that the *fitnesses* are affine functions of populations.

of a theory i to explain data at time t, Darwinian evolution will be indistinguishable from a rational being!

This comparison may seem crazy. But it is corroborated by a remarkable (but trivial) theorem proved by biologist John Maynard Smith in 1973. The theorem asserts that the proportions of the population which Darwinian evolution leads to, also known as *evolutionary stable strategies*, are necessarily Nash equilibria. What's surprising is that Nash equilibria correspond to the strategies in games played by intelligent and rational agents. In other words, like that watch found in the desert, at first, the proportions that describe Nash equilibria seem to be obtainable only through intelligent design. This is what we could naively believe. But it would be wrong.

What seems to necessarily be the result of intelligent design may actually be an unavoidable consequence of Darwinian evolution. This is the stunning conclusion of Maynard Smith's theorem.

10.4 GENETIC ALGORITHMS

For many tasks, Darwinian evolution is even far better than a pale copy of human intelligence. In fact, it can easily create structures that a human intelligence could never think of by itself. The classical example is that of the human brain, which evolution successfully led to. Yet, even once equipped with supercomputers, neuroscientists still only have an extremely partial understanding of it.

The stunning sophistication that Darwinian evolution leads to is such that computer scientists and applied mathematicians often turn towards so-called *genetic algorithms*, to find solutions to problems they cannot solve otherwise. Such algorithms mimic natural selection, but also genetic crossings and mutations.

Suppose for instance that we aim to determine the order of visiting the 100 largest French cities that minimizes the total traveling time. This problem is known as the *traveling salesman problem*. Each order of visit is a solution of the problem. The goal is to determine the optimal solution. The difficulty of the problem lies in the gigantic number of solutions. There are $100! \approx 10^{157}$ of them. Even if we combined all the supercomputers on earth to list all the possible orders of visit, doing so would take more time than the age of the universe.

Genetic algorithms propose a remarkably efficient approach for this kind of problem. The principle of these algorithms is to keep alive a certain diverse population of promising but suboptimal solutions. At each time step, the algorithm selects two solutions of the population. It makes them reproduce. And it then adds (favorable) mutations to obtain a new solution to be added to the population. Every now and then, natural selection also removes the worst solutions of the population. Oddly, this Darwinian approach to optimization performs surprisingly well. In fact, for many problems, it is the state-of-the-art algorithm!

In particular, Darwinian evolution performs here much better than human intelligence. The sophistication of Nature is thus not a convincing argument against the theory of evolution. We shall defend this idea further in the next chapter.

10.5 MAKE UP YOUR OWN MIND?

The distinction between science and pseudo-science is the favorite theme of a thinking movement known as *skepticism*. This movement insists on recurrent sophisms and cognitive biases displayed by pseudo-science defenders. These reasoning errors are indeed often the foundations of numerous conspiracy theories, alternative medicines, and paranormal phenomena.

For some, the right reaction is to make up your own mind by yourself. However, the danger of such a reaction is that it unavoidably leads to a questioning of any opinion that requires enormous intellectual or empirical backgrounds to be made relevant. This is the case of Linda's problem, of the *p-value*, or of the concept of differential privacy. More importantly, this is also the case of the efficiency of vaccines, of the algorithms of Google and Facebook, and of man-made climate change. Unless you spend years carefully studying such questions, the opinion you will come to by yourself will not be well informed; it will thus not be relevant.

It's tempting to believe that, after spending a few hours, you might still be able to lean towards the right side. But this is far from a given. Like in the case of Linda's problem, our intuitions often have an error rate larger than that of a chimpanzee that would answer randomly. This is what statistician Hans Rosling enjoyed showing. On many questions like the number of years girls study, the number of deaths by natural catastrophes or poverty in the world, we are worse than ignorant[3]; we lean toward the wrong side!

---

[3] *How not to be ignorant about the world.* TED. H Rosling & O Rosling (2014).

Worse, it's often extremely hard to estimate the degree of confidence we should have in our intuition. Indeed, even after having read and thought a lot on a topic, it may be remarkably hard to understand how much we have understood. Worse, Derek Muller's PhD thesis[4] shows that watching perfectly correct physics explanations often increases the students' confidence in their misconceptions, even when these misconceptions are refuted by the videos that the students have just watched!

This recurrent overconfidence that we all suffer from is, as you have understood, the main cognitive bias I aim to fight in this book. This is what Bayes' rule, Erdös' difficulties with Monty Hall, and Solomonoff's incompleteness should force us to admit: we are constantly overconfident in so many of our opinions. As the great logician Bertrand Russell once said, "the fundamental cause of the trouble is that in the modern world the stupid are cocksure while the intelligent are full of doubt".

In this context, it seems clear that believing correctly something *all by yourself* is a monstrously difficult task full of obstacles. I strongly discourage it. If it were that easy to have correct opinions, higher educations would not last that long, and knowledge would not be as chunked into disjoint disciplines as it is. Given that we usually cannot afford the financial, temporal, and cognitive resources to work out complex questions in depth, we should instead rely on others' opinions. This is not a bad reflex. Our *pragmatic Bayesian* prefers to profit from the decades, if not centuries, of work by others to polish his belief about the world. Even the *pure Bayesian* knows that others have had access to numerous data that she did not see; they thus have something to teach her.

10.6   AARONSON'S BAYESIAN DEBATING

In 1976, the future Nobel laureate Robert Aumann wondered whether two truthful *pure Bayesians* with two very different life experiences could agree to disagree. It turns out to be easy to prove that, assuming they started their lives with a common prior, the answer is no. *Pure Bayesians* cannot agree to disagree. Indeed by simply telling each other about their life experiences, they would eventually converge to the same posterior which results from learning both of their life experiences[5].

---

[4] *Khan Academy and the Effectiveness of Science Videos*. Veritasium. D Muller (2011).

[5] Suppose the two *pure Bayesians* have observed data D_1 and D_2 during their lives. By sharing these data, both would eventually conclude with posterior $\mathbb{P}[T|D_1, D_2]$.

However, three decades later, the computer scientist Scott Aaronson made the point that such a discussion would be extremely lengthy. Assuming such *pure Bayesians* have lived decades and collected gigabytes of data every second, they would need to share exabytes of data! Assuming they can only receive gigabytes of data per second, their discussion would need to last decades. Clearly, this is theoretically interesting, but pragmatically impossible. But, then, the next question to be asked is the following. Could the two truthful *pure Bayesians* quickly nearly agree? Suppose they argue about the probability of human-level AI by 2050. Would the two truthful *pure Bayesians* necessarily essentially agree, say up to a 5% disagreement?

I was stunned to learn that Aaronson's answer to this question was positive. Two truthful *pure Bayesians* can quickly agree[6]. In fact, Aaronson proved that, essentially, to guarantee an agreement up to 5%, they would need to exchange only roughly $1/0.05^2 \approx 400$ 2-bit messages! It would probably take them less than a minute of discussion to essentially agree[7]. What's most puzzling is probably the fact that this discussion complexity is independent from the amounts of data collected by the two *pure Bayesians*!

Another astounding feature of Aaronson's theorem is the discussion protocol that Aaronson proposed. Essentially, to reach agreement, the two *pure Bayesians* only need to say what they think[8]! Essentially, a transcript of the discussion would go as follows[9].

B1: I'd say that the probability of human-level AI by 2050 is 10%.

B2: I see. But I'd say 80%.

B1: I see. Then I'd say 60%.

B2: I see. Then I'd say 76%.

B1: I see. Then I'd say 70%.

B2: I see. Then I'd say 75%.

B1: Agreement up to 5% achieved!

---

[6] *The Complexity of Agreement.* STOC. S Aaronson (2005).

[7] Technically, Aaronson only proved that to get ϵ-agreement with probability $1-\delta$, it was sufficient to have $O(1/\delta\epsilon^2)$ messages.

[8] Interestingly, Aaronson showed that this discussion protocol was sometimes significantly outperformed by some other "attenuated protocol".

[9] Technically, the 2-bit message protocol is slightly different.

What happens is that the information provided by each *pure Bayesian* updates the belief of the other *pure Bayesian*. More interestingly, the way a *pure Bayesian* updates her belief is a strong indication for how the other *pure Bayesian* should update hers. This is what Julia Galef calls *meta-updating*[10]. If a truthful (Bayesian) individual does not change her mind despite your numerous claims, then you might be the one who should change opinion.

In particular, a remarkable easier theorem is the validity of the argument of truthful Bayesian authority. This theorem says that if you know that a truthful *pure Bayesian* knows strictly more data than you[11], then you should believe everything she says[12]! In such a case, you do not need to bother with long computations.

10.7   SHOULD YOU TRUST A SCIENTIST?

The argument of authority is thus a powerful tool to efficiently understand our world. However, it only holds for truthful Bayesian authorities. But in practice, authorities are not always clearly truthful. And more importantly, they are clearly not Bayesians. At least not *pure Bayesians*. But then, in practice, what authorities are reliable? Does an argument by Einstein have more value than an argument by Shakespeare? Can we blindly trust scientists?

When faced with such questions, some skeptics and some scientists will put forward the *objectivity* of the scientific method. Thus, according to this reasoning, scientists have come to their conclusions through a perfectly rigorous, objective, and peer-reviewed reasoning. Their conclusions thus have a value superior to pseudoscientists, who supposedly do not follow this approach.

However, better skeptics warn against this rough and caricatural argument. On one hand, some pseudoscientists follow an approach that is more or less along the rough lines of the scientific method. Worse, Bayesianism rejects the *objectivity* of the scientific method - and even its soundness! But most importantly, scientists almost never exactly follow the scientific method.

---

[10]*How I use "Meta-Updating".* J Galef (2015).

[11]And assuming she had the same prior as you.

[12]Formally, this corresponds to the equation $\mathbb{E}_x\left[x|\mathbb{E}[x|D]\right] = \mathbb{E}_x[x|D]$, where $\mathbb{E}_x[x|D]$ is what the *pure Bayesian* told you to believe. Note also that this essentially holds if you only believe with very high probability that she knows more than you.

Take a paper at random in the scientific literature. Chances are that the authors did not first state a hypothesis, then identified a restricted protocol, then performed the experiment according to the protocol, then computed the *p-value*, and then completed the pre-rewritten paper. Both modern and past science is rather a series of trials and errors, models and simulations, parameter tweaking, and new questioning during the experiment. It's often only once the experimental results are obtained that the writing of the paper begins. At this point, the angle chosen by the authors often consists of ignoring almost all wrong paths attempted by the laboratory to better synthesize the findings and to wrap up with an intriguing conclusion.

Worse, scientists are unavoidably victims of the same cognitive biases, and even sophisms, as pseudoscientists. Indeed, as discussed in the two first chapters, even the best scientists are unable to address quite simple problems, as evidenced by Erdös' troubles with Monty Hall. There were times when the best scientists thought that the Earth was the center of the universe, that geometry was necessarily Euclidean, or that artificial neural networks were a dead end of artificial intelligence research - this was my reaction when I discovered their mathematical formalization in 2011!

Even the great Einstein, whose breakthroughs seem miraculous to many physicists, got it wrong on several occasions, by defending a wrong theory of general relativity in 1913 or by introducing a cosmological constant in his equations to make the universe stable and eternal - which he would call the "biggest blunder of [his] life". However smart they are, the greatest scientists only have, and will always only have, limited cognitive capabilities.

There is even worse. The academic system creates incentives that are not always compatible with a never-ending fight against cognitive biases. Indeed, the prestige of a scientist, or her mere ability to keep her job, depends on the originality of her ideas, her number of publications, and the respect of her peers. In this context, a scientist has incentives to outrageously defend her ideas, often beyond what Bayes' rule would allow for. Her incentives will be to never denigrate the past theories that gave her glory, even when these theories get rejected. Similarly, she has no incentive to take the time to verify the validity of concurrent theories, since journals do not publish such corroborations of existing theories.

Finally, there are the extreme cases of scientists whose fundings required they reach some predetermined conclusions. This was the case of Ronald Fisher who sold his soul to the tobacco industry. Yet, the existence of such unsound fundings can never be fully excluded.

These numerous arguments seem to question the authority and credibility of scientists. Besides, when I see the shortcuts used by some renowned scientists in public conferences, my credence in their speeches suffers a blow. But when I make videos on Science4All or ZettaBytes, EPFL, whose main goal is to promote mathematics and computer science, I, too, prefer to violently dodge technical difficulties to deliver a clear and convincing message. On numerous occasions, this has led me to lie to my audience - including in this book. Besides, other scientists that I profoundly admire preceded me in this intentional lie. But this is not surprising. She who presents Gödel's theorem without introducing first-order logic is necessarily the author of a white lie. The effort to reach out to a large audience forces us to prefer the fluidity of the storytelling to the perfect rigor of our explanations.

10.8   THE ARGUMENT OF AUTHORITY

Having said this, the opinions of some experts on some technical questions have a totally different value to me. This was typically the case of my first-year university mathematics professor. Like many other students, I was stunned by the relevancy of his remarks. Whenever my beliefs on a math problem conflicted with his, not only did I immediately violently question myself and rejected my belief, but I even started to believe his beliefs, and to seek their origin.

Similarly, whenever a renowned scientist, whose intelligence I admire, makes a surprising claim on a precise question of his area of expertise, I quickly greatly increase my credences on this claim.

This was, for instance, the case when a logician friend told me that, contrary to what a rough reasoning concludes and to what Wikipedia asserts, there exist mathematical models in which all real numbers are definable. Having often observed his superior expertise in mathematical logic, and despite my great credence in Wikipedia mathematical articles, I greatly questioned what I thought. I even quickly believed my friend. Even if I did not understand why he believed what he believed.

As weird as it may sound, my reaction was perfectly rational! Indeed, Bayes' rule forces us to accept the argument of authority here. Note 👍 the fact that the authority defends a thesis, and ✓ and ✗ the status of the thesis. Bayes' rule invites us to compute the posterior credence as follows:

$$\mathbb{P}[\checkmark | \clubsuit] = \frac{\mathbb{P}[\clubsuit | \checkmark]}{\mathbb{P}[\clubsuit | \checkmark]\mathbb{P}[\checkmark] + \mathbb{P}[\clubsuit | \times]\mathbb{P}[\times]}\mathbb{P}[\checkmark].$$

Suppose we had a negative prior. In other words, suppose $\mathbb{P}[\checkmark] \approx 0$ and $\mathbb{P}[\times] \approx 1$. Let's also assume that, if the thesis is true, then the authority will defend it[13], that is, $\mathbb{P}[\clubsuit | \checkmark] \approx 1$. As a result, the coefficient by which the prior credence is multiplied is approximately $1/\mathbb{P}[\clubsuit | \times]$. In other words, Bayes' rule tells us to accept the argument of authority if and only if it's highly unlikely that the authority defends the thesis, assuming that the thesis is false.

This is why the climate skeptics' authority is rejected by the *pure Bayesian*. Given the enormous economic incentives of the oil industry, it's absolutely not surprising that they successfully found individuals willing to defend their positions. Then, there is a huge selection bias. If a TV show wants to organize a one-against-one debate, then the probability that the person invited to defend climate skepticism will defend climate skepticism is equal to 1, whether the climate skepticism thesis is true or not.

Now, this argument also holds for the opposite side. The probability that an ecology activist defends the climate-change thesis is also 1, even if the climate-change thesis is false. What is true of the activist can even hold for the scientist, given all the cognitive biases that scientists suffer from and the selection bias. In a nutshell, when it comes to climate change, or any controversial and polarized topic with huge economic or ideological incentives, the argument of authority essentially has no validity.

However, on the contrary, in the case of my logician friend, the probability $\mathbb{P}[\clubsuit | \times]$ that he asserts what he asserted, assuming that his thesis is completely false, is essentially zero, especially given that my logician friend has read Wikipedia himself and that he perfectly understands the rough reasoning I had in mind. In fact, the probability $\mathbb{P}[\clubsuit | \times]$ is even weaker than my prior credences in a conjoint error of my reasoning and of Wikipedia, even if these credences are minuscule. This is why, after our discussion, and even if I did not understand my friend's argument, I was convinced he was right.

So, I have come to believe things I did not understand. Worse, if I believed them, it was solely based on an argument of authority. Some would argue that my behavior was irrational. But, even though I did not

---

[13]Our reasoning also holds as long as $\mathbb{P}[\clubsuit | \checkmark]$ is not very small.

know it back then, it was in fact the only rational posterior belief - at least according to Bayes' rule.

10.9 THE SCIENTIFIC CONSENSUS

Let's get back to climate change. We saw that no scientist could be regarded as an authority. Besides, rather than mentioning a single climate expert, ecologists often rather put forward the opinion of the climate expert community. Yet the opinion of the community is univocal. Almost all of this community believes in climate change and in its anthropic origin - a digit larger than 98% is often given[14]. But if each scientist is not credible, why would the scientific community be so?

The *pure Bayesian* has a remarkable answer to this question: the scientific community applies Bayes' rule better than its members. Imagine a simplified model in which the scientific community is a territory and imagine that theories T are animal species living in this territory. At each time t, the more credible theories will reproduce more. They will be adopted by more scientists. Let's call $p_T(t)$ the fraction of scientists that adopt theory T at time t. The Lotka-Volterra equations apply to the evolution of ideas:

$$p_T(t+1) = \frac{\text{fitness}(T, t)p_T(t)}{\text{fitness}(T, t)p_T(t) + \sum_{A \neq T} \text{fitness}(A, t)p_A(t)}.$$

Do you see it coming? The Darwinian evolution of theories in the scientific community is a Bayesian inference in disguise!

In other words, it's all as if the scientific community applied Bayes' rule to put forward the more credible theories. This is the reason why the community often deserves a credence that far transcends the opinions of the individuals. Or rather, as long as the *fitnesses* of the theories are somewhat correlated with the *likelihoods* of the data given the theories, the community applies Bayes' rule better than any of its members.

10.10 CLICKBAIT

By analogy, it could be tempting to think that the more widespread opinions in a population are also more believable. This argument is often invoked to defend the principle of democracy. However, it is fallacious.

---

[14]Besides, climate change would still be of huge concern even if there were only a 90% chance of it being true!

The reason is simple: theories that best propagate in a population are not necessarily the most believable theories; they are rather the most *viral* theories.

In a video[15] released shortly after the results of the 2016 American presidential election, the great Derek Muller of the Veritasium channel confessed his initial naive optimism. He thought that the Internet would more quickly share facts. He hoped that this would lead to a global convergence towards common values and (scientific) beliefs. However, as Derek Muller explains it himself, this has not been the case. The explanation of the divergence of opinions seems to be found in a video[16] of another excellent educational YouTuber known as CPG Grey. CGP Grey suggests that the propagation power of a theory on the Internet is more related to its ability to spark emotional reactions than its ability to explain observed data. In other words, the *fitness* of a theory on the Internet and within the general public seems more due to its *clickbait* effect than on the *thought experiment terms* of Bayes' rule.

Worse, based on a publication of Berger and Milkman[17], CGP Grey suggests that the theories that best propagate are those that cause anger. Moreover, such theories are all the more viral if they get opposed to adversarial theories that cause equally much anger. Two opposing theories are then like two animal species in a symbiotic relation. They mutually feed on one another. And they win over the animal territory together.

Put differently, the *fitness* of theories on the web favor ideological bipolarisation and hatred against defenders of the opposite sides, essentially independently from logical or empirical grounds. This is where the origin of the recent rise of ideological extremism is likely to be found. In particular, data suggest that the United States has not been that divided for a very long time[18].

More generally, the proliferation of brief, shocking, striking, annoying, seductive, critical, intriguing, partisan, saddening, inspiring, activist, defamatory, accusing, and unfounded information is comparable to a cancer tumor that grows and seriously endangers the health of the human body. At the scale of our societies, clickbait seems to favor extreme political positions with short-term agendas that are based on fears of insecurity and unjustified hope, rather than informed, thoughtful

---

[15] *Post-Truth: Why Facts Don't Matter Anymore.* Veritasium. D Muller (2016).

[16] *This Video Will Make You Angry.* CGP Grey (2015).

[17] *What Makes Online Content Viral?* Journal of Marketing Research. J Berger & K Milkman (2012).

[18] *Is America More Divided Than Ever?* The Good Stuff (2016).

and long-term visions. Most importantly, clickbait makes the democratic vote misinformed, biased, and irrational.

10.11 THE PREDICTIVE POWER OF MARKETS

In an audacious book where he says out loud what no one dares to admit in secret, economist Bryan Caplan presents an unflattering portrait of the median voter[19]. By relying on empirical data of American voters and on responses to surveys, Caplan concluded that the median voter is worse than ignorant. According to Caplan, the median voter is irrational. Caplan even justifies this conclusion based on a basic economics model. At the heart of the model is the remark that every vote has a near-zero probability to have any effect. But then, the pleasure to irrationally express one's convictions when voting far transcends the very unlikely effect of a rational and informed vote, especially given that coming up with an informed vote requires tremendous cognitive efforts. This is why, Caplan argues, the median voter is *rationally irrational*.

Bryan Caplan proposes an alternative to democratic opinion: capitalism and free market. Caplan even goes as far as to assert that the markets and lobbies have saved democracies from a mercantile, over-regulated, and overprotectionist chaos. According to this argument, the great defenders of the rights of immigrants in the United States are supposedly not the American citizens; they are Google, Facebook, and other companies whose economic power strongly depends on highly qualified immigration and on revenues highly influenced by the brand reputation abroad.

Caplan even defends the controversial *Policy Analysis Market* (PAM) set up by the Defense Advanced Research Projects Agency (DARPA). In this online market, users could, for instance, bet on the number of American victims in Iraq, or on a terrorist attack in Israel in the coming year. As you can imagine, this governmental initative was greatly criticized and quickly nicknamed the *terror market*. The project had to be immediately stopped.

Yet, the preliminary analyses were remarkably promising. Surprisingly, such an online betting market seems very efficient at making predictions. Just like bets on horse races, elections, or military invasions. What is better is that the bet is remarkably close to what eventually

---

[19] *The Myth of the Rational Voter: Why Democracies Choose Bad Policies*. Princeton University Press. B Caplan (2011).

happens. After all, unlike in democratic voting, since money is at stake here, betters will take the time to inform themselves and think it through before expressing themselves. Better yet, if they are not sufficiently confident in their predictions, contrary to voters, betters will not express themselves at all. Thus, they will not pollute data with misinformed, biased, and irrational convictions.

However, explaining the predictive power of markets by the betters' expertise only may be erroneous. In 1988, 4 employees of the *Wall Street Journal* tried to randomly bet in the stock market. Each month, they threw a dart to choose an action to buy. Month after month, the performances of dart players were then compared to the performances of 4 professional investors. One hundred months later, results were compiled: the dart players beat investors 39 times out of 100. In other words, the investors won by a small barely significant margin. Worse, many economists argued that the investors only won because their choices of action were announced in the *Wall Street Journal*, thereby creating an anouncement effect! Even worse, despite this bias, the investors only beat the market average (called the Dow Jones) 51 times out of 100.

Nobel prize winner Daniel Kahneman studied investors and markets in depth. His conclusions are even more severe. The correlation between successive rankings of traders month after month was near-zero, as if the traders' successes were independent and identically distributed random variables. Worse, Kahneman showed that the most successful traders were the least active ones, as if, to maximize one's gains, it sufficed to blindly trust the market rather than trying to beat it.

All these experiments seem to show, again and again, that the market is more competent than any of its traders - and even more so than any of us. How is that possible?

The answer I propose is the same as the one I used to justify the relevancy of the scientific consensus: the market applies Bayes' rule better than any of its traders[20]! To understand this, consider that each trader's belief about the marker is akin to predictive theory. As good Bayesians, we shall want to multiply the credence in traders whose past bets did well, and divide the credence in the traders whose past bets did poorly. Guess what! This is precisely what is done by the markets!

---

[20]Or rather, it implements a sort of hierarchical Bayesian model, that is, Bayes' rule over other learning models. Such hierarchical models are discussed more thoroughly in chapter 19.

To understand this, consider a trader T. Call fortune(T, t) the fortune of T at time t. Because of the multiplicative nature of the markets, his fortune in the next time step is given by fortune$(T, t + 1) =$ perf$(T, t) \cdot$ fortune(T, t). This is exactly the same equation as Lotka-Volterra's, where the variety i in a population is replaced by trader T, and where the fitness of a variety is replaced by the perfomance of the trader.

In particular, by now focusing on the market share of trader T, we now obtain the following evolutionary dynamics:

$$\text{share}(T, t + 1) = \frac{\text{perf}(T, t)\text{share}(T, t)}{\text{perf}(T, t)\text{share}(T, t) + \sum_{A \neq T} \text{perf}(A, t)\text{share}(A, t)},$$

where A corresponds to the traders other than T. Once again, we obtain a sort of Bayesian inference! Once again, Bayes' rule is better performed by the entire stock market than by any of its members. Like the scientific consensus, the market consensus seems more reliable than the opinion of any expert of the market!

There are, however, three caveats to mention about this analysis. The first is that the market prediction, contrary to Bayesian prediction, is not a weighted average of traders' predictions. In fact, the mechanism that transforms traders' predictions into a market prediction is rather a sort of weighted median than a weighted average. Indeed, the equilibrium market price will be such that it divides traders into two sets: those who believe that the price is underestimated will be exactly those who invested (and have thus led to the increase of that price). Still, the larger the share of the market a trader has, the more he influences the equilibrium price.

The two other caveats are bigger issues. On one hand, we need to consider the continuous flow of new traders whose fortunes were not won by bets in the market. One recent spectacular example of this is that of new betters on cryptocurrencies, like Bitcoin, whose investments were not money gained by winning bets on cryptocurrencies. The flow of new traders acts like a bonus given to theories with no track records. In particular, it serves as an erasing of past failures. And favors the short term.

On the other hand, there is the early departure of some traders, due to the fact that traders rarely plan to carry on their trading for a long time - it has become common for young traders to stay in their companies no more than 5 years! Typically, to obtain a promotion, traders need to shine within the coming years. This motivates them to take big risks in the short term.

These two effects, and probably others that I have not thought of, greatly harm the predictive capability of markets. They are typically the cause of financial bubbles.

10.12 FINANCIAL BUBBLES

Let's move to the Netherlands. In the beginning of the 17th century, the tulip was in high demand. The Dutch were fighting for tulip bulbs. There was an explosion in the demand without an increase in the supply, which led to a quick increase of the price of tulip bulbs. In fact, the increase of prices seemed inexorable, so that the good investors would buy huge quantities of tulip bulbs at a very large price, and could then sell them at an even larger price. Investors got rich and became the great actors of the financial market. In 1635, things got so crazy that a single tulip was worth as much as a manor. It was *tulipomania*.

But all of sudden, in 1637, the tulip price ceased to increase. The tulip bulb investors started to fear the decrease in the value of their goods. They aimed to dispose of their stocks, even if that meant selling at loss. The price of the tulip dropped. Yet, the more the price dropped, the more investors wanted to sell fast, and the more they dropped their selling price. Worse, the more the price dropped, the more the buyers expected the price to drop, and the harder it was for sellers to find buyers, and more sellers had to drop their price. This was the implosion of a speculative bubble.

This phenomenon is far from unique in the history of mankind. Lastly, in 2008, the so-called *subprime crisis* violently hurt the credit market in the United States with international repercussions. This crisis was triggered by the implosion of the speculative bubble of the American real estate market. Before the crisis, Americans did not hesitate to get mortgages to buy houses, in the hope that the house value increase would allow them to pay their mortgages. And the more the Americans and their banks believed in the increase of the real estate prices, the more they were willing to buy, and the more the real estate prices increased. And the more buyers wanted to get mortgages to buy houses.

However, like the price of the tulip, as soon as the real estate prices ceased to increase, the payment of mortgages was in trouble. More and more families went bankrupt, and had to sell their homes. But the more house sellers there were, the more the house prices dropped, and the more defaulting there was. This vicious cycle amplified the phenomenon.

Worse, bank loans were then complex financial derivatives. Banks had sold their financial products to Wall Street investors, who saw nothing coming. This is how numerous defaults of payments of American individuals turned into a financial abysse for large investment funds. Their collapse would then entail the fallout of numerous other companies. This was a worldwide disaster.

We reach here a fundamental limit to the predictive capability of markets. The frequent entry of new traders and the regular departure of old traders lead to a divergence between the dynamic of markets and Bayes' rule. In particular, the characteristics of markets favor the short term. This turns out to be incompatible with the slower time of political and speculative bubbles. In statistical terms, the accelerated time of markets makes them overfit their immediate past. This is why long-term predictions of markets surely don't deserve the same credence as the scientific consensus.

FURTHER READING

Designing Effective Multimedia for Physics Education. University of Sydney. PhD Thesis. D Muller (2008).

The Myth of the Rational Voter: Why Democracies Choose Bad Policies. Princeton University Press. B Caplan (2011).

Thinking Fast and Slow. Farrar, Straus and Giroux. D Kahneman (2013).

The Theory of Games and the Evolution of Animal Conflicts. Journal of Theoretical Biology. JM Smith (1974).

The Complexity of Agreement. STOC. S Aaronson (2005).

What Makes Online Content Viral? Journal of Marketing Research. J Berger & K Milkman (2012).

Experimental Evidence of Massive-Scale Emotional Contagion through Social Networks. PNAS. A Kramer, J Guillory & J Hancock (2014).

How not to be Ignorant about the World. TED. H Rosling & O Rosling (2014).

How I use "Meta-Updating". J Galef (2015).

Rock Paper Lizards. Numberphile. H Fry (2015).

Khan Academy and the Effectiveness of Science Videos. Veritasium. D Muller (2011).

This Video Will Make You Angry. CGP Grey (2015).

Post-Truth: Why Facts Don't Matter Anymore. Veritasium. D Muller (2016).

What Causes Economic Bubbles? Ted-Ed. P Singh (2015).

Is America More Divided Than Ever? The Good Stuff (2016).

That Time Tulips Crashed the Economy (Maybe). The Good Stuff (2018).

Partisanship and Political Animosity in 2016. US Politics & Policy. Pew Research Center (2016).

Evolutionary Game Theory. Science4All. LN Hoang (2012).

Numbers and Constructibiliy. Science4All. LN Hoang (2013).

CHAPTER 11

Exponentially Counterintuitive

> It is in the nature of exponential growth that events develop extremely slowly for extremely long periods of time, but as one glides through the knee of the curve, events erupt at an increasingly furious pace. And that is what we will experience as we enter the twenty-first century
>
> Ray Kurzweil (1948-)

11.1 SUPER LARGE NUMBERS

"A linear progression is 1, 2, 3. An exponential progression is 1, 2, 4. It doesn't sound that different [...]. But at the time they get at thirty, the linear progression, and that's our intuition, is at thirty. The exponential progression is at a billion," argues futurist Ray Kurzweil. "Our intuition about the future is linear. But the reality of information technology is exponential, and that makes a profound difference."

For Kurzweil, our misunderstanding of exponential growth leads us perhaps to slightly overestimate the influence of new technologies in the short term, but undoubtedly to deeply underestimate it in the long run – on the scale of 5 years or more. But before discussing this, it's good to first taste the immensity of very large numbers.

Let's start with a number that seems reasonable at first glance. A million is a number to which our daily lives confronts us so regularly that it's tempting to think that we understand its immensity. In February

2016, VSauce was the first science YouTuber to break the 10 million subscriber mark. The salaries of football players are in the millions. The populations of most countries are of a few millions.

However, we should not confuse our familiarity with this number with the understanding we have of it. A million is a lot. A man would have a hard time counting up to a million in a year, even at the unimaginable rate of almost a number per second. A million is a number that can be thought of. But it's humanly impossible to reach it from scratch.

And yet, one million is a grain of sand in the immensity of large numbers. A billion is already vastly more gigantic! The richest people earn billions per year. This represents a hundred thousand times my salary! To realize the immensity of these numbers, we can emphasize that a dollar for me corresponds, proportionally, to one hundred thousand dollars for these billionaires. So, in the same way that I sometimes do not bother to pick up a one-dollar bill in the street, the billionaire would not bother to acquire a property of one hundred thousand euros which is offered to him, if he has to lift a finger to do this!

We can see things differently. Some studies show that from 70,000 dollars per year onwards, money does not bring happiness[1]. Or rather, someone who earns more than 70,000 dollars a year seems statistically no happier than someone who earns 70,000 dollars. Imagine now that a billionaire learns this and decides to spend exactly 70,000 dollars per year. One billion dollars would allow him to live 14,000 years without working! Or put differently, if he earns one billion dollars a year, he can then afford to guarantee a sufficient financial contribution to maximum happiness for 14,000 people - himself included.

But a billion is still a ridiculously small number on the scales of physics. Our galaxy contains hundreds of billions of stars, our brains are composed of millions of billions of neural connections, our Earth has billions of billions of grains of sand, and a drop of water is a cluster of millions of billions of billions of molecules. For such large numbers, we prefer to use notations like 10^{25}. This is the number described by a one followed by 25 zeros! These numbers are literally astronomical.

However, it can be emphasized that these numbers are *only* astronomical. And if modern physics makes it possible to meet even larger numbers, the powers of 10 are enough to reach the limits of physics. For example, according to some modern theories of physics, time may be discrete. It may run one Planck's time at a time, that is, by chunks

---

[1] *Can Money Buy Happiness?* AsapSCIENCE. M Moffit & G Brown (2012).

of about 10^{-44} seconds. Thus, only 10^{60} elementary time steps seem to have passed since the Big Bang. On another note, one can only count about 10^{80} atoms in the observable universe.

11.2 THE GLASS CEILING OF COMPUTATION

The limits of physics then necessarily translate into the limits of our computing powers. In fact, nowadays, computer scientists often consider that an algorithm that requires more than 10^{70} computation steps will never end in the history of humanity. After all, this number of calculation steps is greater than the number of elementary time steps since the Big Bang.

This figure can be deduced from other physical assumptions. Landauer's principle, which is derived from Boltzmann's equations, postulates the existence of a physical limit to the minimum energy required to perform a bit of (irreversible) operation. This energy is 10^{-21} joules per information bit at room temperature[2]. Yet, the energy of the solar system is finite. We can even estimate it at 10^{17} joules. As a result, the number of (irreversible) operations that can be performed with all the energy in the solar system is only 10^{68}.

More convincingly still, mathematician and biophysicist Hans-Joachim Bremermann combined two fundamental physical laws to derive another glass ceiling of computation. More precisely, he combined Einstein's mass-energy equivalency and the Heisenberg uncertainty principle to derive the fact that there was a limit of $c^2/h \approx 10^{50}$ bits of computation per second and per kilogram. As a result, a computer of the size of the Earth would only be able to perform 10^{75} operations per second, and thus at most 10^{85} operations per century!

These fundamental principles are at the heart of cryptography, which assumes that the protocols used in practice require more than around 10^{70} calculation steps to be broken. If cryptologists are right, this would mean that the messages encoded today with their technologies will not only *not* be cracked tomorrow; they will even never be cracked by humanity, thus guaranteeing the confidentiality of encrypted data.

Of course, it might still be possible to hack these data by bribing the right people or their acquaintances - this technique called *social engineering* is often the main cause of computer insecurity. Worse still,

---

[2] *How many thoughts are contained in a Mars Bar?* Sixty Symbols. P Moriarty (2016).

the capabilities of algorithms are still poorly understood and one cannot exclude today the existence of efficient algorithms that bypass the 10^{70} operations that cryptologists think are necessary - this is the heart of the *P versus NP problem*, the most prestigious open problem of theoretical computing. In fact, since 1997 and the discovery by Peter Shor, we even know that if quantum computers with sufficiently large quantum memory spaces can be built, then it will be possible to use a quantum algorithm to hack many of the cryptographic protocols used today, including the RSA protocol.

Let's detail RSA. RSA is an asymmetric cryptography protocol. This means that an RSA user has two keys that work in symbiosis. The first is a public key e; the second is a private key d. Crucially, given the public key e, the user can secretly choose two prime numbers p and q to effectively compute a compatible private key[3] d. However, for others to send encrypted messages to him, the user must also make public the product $N = pq$ of his two prime numbers. So, if a hacker is able to decompose the number N into the product of two integers p and q, he can then efficiently calculate the private key d from the public key e. He will have broken the RSA protocol.

RSA security is thus based on the assumption that a user will not be able to break the integer N into a prime number product. This is the *factorization problem*. Today, we do not know if there is a fast (classical) factorization algorithm. And for the security of RSA, given the lack of evidence of non-existence, we can only pray that no one finds such an algorithm. Worse, Shor's quantum algorithm quickly solves the factorization problem. And it seems to be only a matter of time before quantum computers worthy of the name are out there....

Now, the difficulty of factorization may seem surprising. One might think that it's enough to try to divide N by all the numbers a which are inferior to it. It's even possible to prove that it suffices to test the division by all the numbers between 2 and $\sqrt{N}$. However, the number N used in practice is a gigantic number. It will typically be in the order of 10^{300}. But then, the number of divisions to test is $\sqrt{N}$, which is about 10^{150}. And as we have seen, we can consider that 10^{70} operations cannot be performed in our physical world.

---

[3]More precisely, he must determine d such that $ed = 1 \mod (p-1)(q-1)$, which can be done quickly thanks to Euclid's algorithm and the computation $gcd(e, (p-1)(q-1))$. I'm omitting a lot of fascinating details here.

11.3   EXPONENTIAL EXPLOSION

The glass ceiling of any calculation seems to be an unattainable limit. It is indeed physically unattainable for linear growth. However, in an extremely counterintuitive way, these orders of magnitude are in fact quickly reached by the astonishing exponential growth.

According to legend, King Belkib of India adored the game of chess that the sage Sissa had presented to him. He therefore proposed to the wise man to choose his reward himself. Sissa humbly replied that he would be happy with 1 grain of rice for the first box of the chessboard on the first day, 2 for the next on the second day, 4 for the next, and 8 after, and so on. The king, surprised by the modesty of the request, accepted. What a mistake! After 64 days, the king had a debt of several billion billion grains of rice, which is about a thousand times the annual world rice production today! In other words, King Belkib had an eternal debt toward the sage Sissa.

The exponential growth of King Belkib's debt is stunning. In the same way, it's enough to fold a sheet of paper 42 times so that its thickness is the height of the distance between Earth-Moon, and 103 times so that it makes the width of our observable universe! Yet, the observable universe is very, very big. It is almost a million billion billion kilometers wide! Nevertheless, in just 103 steps, the exponential growth has exceeded the limits of astrophysics!

We find this crazy growth in family trees as well. Indeed, if we go back in time, since each of us is the child of two (biological) parents, the number of our ancestors grows exponentially when we consider older generations. In fact, using computer simulations and taking into account the fact that humans mate more closely with geographically close individuals, Rohde, Olson, and Chang estimate[4] that all living humans today are descendants of a same ancestor who lived between 2000 and 5000 years ago[5]. Yes, we are all consanguineous cousins a few hundred degrees apart! Better still, almost any human who has lived before this most recent common ancestor is either an ancestor of all living humans today, or the ancestor of none of us!

---

[4] *Modelling the recent common ancestry of all living humans*. Nature. D Rohde, S Olson & J Chang (2004).

[5] Do not confuse this most recent common ancestor with other concepts like mitochondrial Eve, the mother of mothers. Indeed, the number of mothers of mothers does not grow exponentially - so the mitochondrial Eve is necessarily much older.

Biological trees have academic counterparts. The entry of a researcher into the scientific world is often equated with the defense of her doctoral thesis, which is supervised by one or more professors. Our thesis supervisors are often referred to as our academic mothers and fathers, and the academic parents of our academic parents are our academic grandparents. Better yet, we can trace back the academic genealogy and dig up our academic ancestors. I dug up mine, and found out that I am a descendant of Georg Danzig, Carl Friedrich Gauss, and Leonhard Euler. I even discovered I descend from Jerzy Neyman (we do not choose our ancestors!) and Pierre-Simon Laplace (with irrational pride!). However, the presence of big names in my genealogy is not surprising. On the one hand, the number of mathematicians many centuries ago was very small. On the other hand, because some have several academic parents, the number of ancestors grows exponentially.

If the exponential growth is staggering, so is its decline. By dividing a piece of sugar in half about sixty times, you end up having to divide the sugar molecules. In the same way, consider a homeopathic dilution that divides by about 100 the number of molecules of an active substance. Well, despite the astronomical number of initial molecules, 12 dilutions are enough to statistically guarantee the disappearance of all molecules of the active substance! And all that comes from the incredible exponential decay of the concentrations in molecules, dilution after dilution.

For many, the emergence of the complexity of life from (almost) nothing seems unrealistic. But our inability to make sense of the complexification of animal species by natural selection is undoubtedly due to our linear intuition of the world. By opposition, the living undergoes a multiplicative dynamic, as we saw when talking about the equations of Lotka-Volterra. During a mitosis, a cell divides into two cells, then each daughter cell divides into two others, and so on. At each step, the number of cells is multiplied by 2. Thus, in a few days or months, a single egg cell can be transformed into a complex living organism composed of thousands of billions of cells.

Certainly, the exponential growth of the complexification of the living is much "slower", especially compared to the growth of technology. However, it has continued over time scales that are impossible for us to imagine, like billions of years. Thus, while evolution produces only few perceptible changes at the scale of the century, it's noteworthy that tens of millions of centuries have passed since the appearance of the first living cells. The immensity of these quantities already exceeds our understanding. Exponential growth over such durations exceeds it even more!

Typically, at this moment, the human population increases each year by 11%. This growth is not insignificant, but it does not seem unreasonable. And yet. A quick calculation[6] shows that with such a growth, within 8604 years, the human population will be such that the number of particles that make up human individuals will exceed the total number of particles in the universe! Exponential growth can be imperceptible in the very short term, and yet invade the entire universe in a not-so-long term.

Conversely, if the number of children per woman is less than 2 (which is already the case in many developed countries), then the *total* number of humans in the history of the universe will be surprisingly limited! For instance, at 1.9 children per woman, the total population of humanity across all ages, past and future, will be only in the order of a hundred billion[7]. Overpopulation will then be a limited problem, even if biologists somehow make us immortal!

To improve one's intuition of the exponential growth, I highly recommend the *Universal Paperclips* online game[8]. Be warned, though! This game is terribly addictive. It illustrates Nick Bostrom's story of an artificial intelligence that maximizes the number of paperclips in the world. To achieve its goal, this artificial intelligence invests massively in scientific research, which makes the pace of its technological advance faster. The more technologically advanced it is, the faster it progresses. Its growth is exponential. And this growth is frantic. After only two days of play, starting with a basic production of individual trombones, we end up using technology to conquer and invade the entire universe.

11.4   THE MAGIC OF ARABIC NUMERALS

Let's go back to cryptography. We have seen that cryptologists daily used numbers $N \approx 10^{300}$ so large that $\sqrt{N}$ calculation steps were a process that transcends the limits of physics. How is it possible to speak of a number so large that it is beyond physics?

For a long time, the great empires of humanity, like the Egyptian and Roman empires, had an inefficient way of representing numbers.

---

[6] *How many particles in the universe?* Numberphile. T Padilla (2017).
[7] This is the conclusion of the computation:

$$7 + 7 \cdot 0.95 + 7 \cdot 0.95^2 + 7 \cdot 0.95^3 + \ldots = \frac{7}{1 - 0.95} = 7 \cdot 20 = 140.$$

[8] decisionproblem.com/paperclips/.

The number 1888 was written, in Roman numerals, MDCCCLXXXVIII. Worse still, the Romans were forced to invent new symbols for ever larger numbers, so they were unable to write numbers like a million - unless by aligning M thousand in a row!

Conversely, the Babylonians, followed by Chinese and Japanese, but also and especially by Indians and Arabs, had the brilliant idea of developing the system of positional notation. The major feature of this system is the fact that the position of a symbol determines its numerical value. Thus, as we all know, 12 is a number different from 21, even if the symbols used are the same.

I won't describe how the Arabic numeral system works. You should have learned it from an early age. However, what you may not have realized is the remarkable conciseness that it allows. Although it uses a small number of symbols (called *digits*), the number of digits required to represent a number is much smaller than the number itself. Indeed, 10^{100} is a number that transcends the limits of physics. However, our numbering system is so efficient that 10^{100} can be represented with only 101 digits - a "1" followed by a hundred "0".

Our numbering system is even optimal, in a sense. We cannot represent more succinctly all the numbers between 0 and $10^{100} - 1$. After all, all combinations of the 100 digits were used to represent all these numbers. In fact, the size of the numbers is exponential in the length of their representations. In other words, any integer x is roughly equal to $10^{\#\text{digits of } x}$. An equivalent way of saying this is to say that numbers have logarithmically-long representations. The logarithm of a number is approximately the number of digits needed to represent that number. So $\log_{10}(10^{100}) = 100 \approx 101$.

Unlike the exponential growth, the logarithmic growth is incredibly slow. For example, the time it takes to search for an element in a sorted array is logarithmic. This means that even if the array in question is the size of the universe, then it will be possible to find the element in a few hundred iterations - we would be limited only by the finiteness of the speed of light!

This discovery is also at the heart of one of the most fundamental notions of computer science, the notion of an *address*. Imagine you want to find information on the web. Surprisingly, while there are exabytes, even zettabytes of data on the web, armed with the url address of the information you are looking for, you will be able to find the information in question almost instantly!

What's more, the url is incredibly brief. Its size is logarithmic in the size of the whole web! In particular, the entire url can easily be kept in memory - contrary to the information to which this url refers. We will come back to this fundamental notion of address when we approach the management of memory by the *pragmatic Bayesian*.

Now, more precisely, the logarithm, like the exponential function, depends on a parameter called the base. The sequence 1, 2, 4 is an exponential growth of base 2, since at each iteration, the factor by which we multiply the elements of the continuation is 2. Or put differently, the $(n-1)$-th element from the continuation is written 2^n. Conversely, the logarithm of a number x in base 2 counts the number of times you need to multiply by 2 to reach the number x. Thus, $\log_2(2^n) = n$, since 2^n is obtained by multiplying n times the number 2. When x is not a power of 2, the logarithm in base 2 will find a natural and clever way to determine a real number y such that $2^y = x$.

11.5 BENFORD'S LAW

Open the Wikipedia page which lists the countries of the world ordered according to their populations. Note the first digits of these numbers of inhabitants. You should notice with amazement something troubling: these first figures of the numbers of inhabitants are more often "1" than "9"! This bluffing remark is not specific to the number of inhabitants of a country. Consider the lengths of rivers, the number of subscribers to YouTube channels, or the salaries of multi-millionaires. In all these cases, the most frequent digit will be "1", and it will be about 6 to 7 times more frequent than the "9"! This is the surprising *Benford law*.

The origin of Benford's law lies in a recurrent property of many systems. Like the Lotka-Volterra equation discussed in the previous chapter, many dynamical systems lead to chain reactions that characterize exponential growth. For example, growth in YouTube subscriber numbers is typically exponential. This means that it doubles, for example, every six months. But then, for six months, the number of subscribers will be between 1,000 and 2,000. Then, the following six months, it will be between 2,000 and 4,000. Then, the following six months, between 4,000 and 8,000, and six months later between 8,000 and 16,000. Here we see the origin of Benford's law. The number of subscribers will remain between 1,000 and 2,000 for 6 months, while it will be less than a month between 9,000 and 10,000. This suggests that the number of subscribers starts around six times more often with a "1" than with a "9".

To better understand these exponential growths, it's useful to change the scale we are studying. Rather than studying the number of subscribers, one might want to study the logarithm in base 2 of the number of subscribers. Recall that $1024 = 2^{10}$ and $2048 = 2^{11}$. So, in six months, the log base 2 of the number of subscribers goes from 10 to 11. Six months later, it goes to 12, then to 13, and so on. Do you see what's going on? Every six months, the logarithm in base 2 of the number of subscribers simply increases by one unit.

But then, on the logarithmic scale that we've just built, the logarithm of the number of subscribers spends as much time between 10 and 11 as between 11 and 12 and between 12 and 13. Suppose now that we observe the numbers of subscribers of different channels taken at different times of their growth. The logarithms of these subscriber numbers can then be expected to be just as often between 10 and 11 as between 13 and 14. In technical terms, the logarithmic distribution of subscriber numbers is roughly uniform[9]. This is the technical condition to the validity of Benford's law: if the natural scale of a quantity is the logarithmic scale (and if it extends over several orders of magnitude), then the first figures of this quantity will be six times more often a "1" than[10] "9".

11.6 LOGARITHMIC SCALES

Physics and chemistry love these logarithmic scales. It's almost impossible to draw the solar system to scale, because the sizes of the planets are ridiculously small compared to the distances that separate them, which are themselves almost nil compared to the size of our galaxy, which is itself non-existent compared to the intergalactic distances, which are nothing in front of the immensity of our observable universe! On the other hand, microscopic, nanoscopic, or even subatomic scales are spread over so many orders of magnitude that it's impossible to represent both the constituents of the protons of the nuclei of atoms and the molecules

---

[9]In fact, David Louapre's study shows that the logarithms of the subscriber numbers of a selection of (French) scientific YouTube channels are roughly distributed according to a normal law. *De quoi dépend le succès d'une chaîne YouTube?* Science Étonnante. D Louapre (2017).

[10]Rigorous computations yield $\mathbb{P}[\text{1st digit} = d] = \log_{10}(d + 1) - \log_{10}(d)$. The probabilities that this figure is 1,2,3,4,5,6,7,8,9 are about 30%, 18%, 13%, 10%, 8%, 7%, 6%, 5%, 5%.

that combine them. To think all these scales at once, the logarithmic scale is unavoidable.

Similarly, sound volumes, seismic magnitudes, and solution acidity are usually measured on logarithmic scales such as decibels, Richter scale, and pH. These usual units are in fact (sometimes up to a negative sign) the logarithms of the amplitudes of the pressure variations, the energy of the seismic waves, and the ion concentration of H^+.

These logarithmic scales reveal the multiplicative nature of the objects of study. This multiplicative interpretation of measurement differences is in fact very familiar to us in many cases. For example, it's tempting to say that the difference between a channel with 200,000 subscribers and a channel with 1 million subscribers is less than that between the 200,000-subscriber channel and a 50-subscriber channel. This sentence makes little sense on an additive scale, since the difference in the first case is 800,000 subscribers, while in the second case, the difference is "only" 199,950 subscribers.

However, our intuition perfectly fits the multiplicative scale. Indeed, to go from 200,000 to 1 million, it's enough to multiply the number of subscribers by 5. However, to go from 50 to 200,000 requires to multiply the number of subscribers by 4,000. What is argued by the multiplicative scale is equivalent to what reveals the logarithmic scale. The difference between 200,000 and 1 million on the logarithmic scale is $\log_2(1 million) - \log_2(200,000) \approx 2.3$, while the difference between 50 and 200,000 on this logarithmic scale is approximately 12. This is why these logarithmic scales are so often used to talk about objects whose variations are multiplicative rather than additive.

Curiously, young children and primitive tribes seem to have a multiplicative intuition of numbers rather than an additive one. When asked to place the numbers from 1 to 10 on a scale, these people space out the first numbers, and bring the last ones more closely. This is what the logarithmic scale looks like - even though the scale of these people is actually not quite logarithmic. Conversely, any person formatted by mathematical education will space numbers in a regular way, thereby constructing an additive scale. Our intuition of numbers is strongly influenced by what we have learned. School taught us to think in the additive mode. Yet, the additive scale is no more natural than the multiplicative (or logarithmic) scale.

11.7 LOGARITHMS

As Mickaël Launay explains it very well on his YouTube channel[11], the additive and multiplicative scales seem to belong to two distinct worlds. The first is that of additions, subtractions, arithmetic averages, and integrals. It's the linear world. The second is that of multiplications, divisions, geometric means, and asymptotic equivalents. It's the world of YouTube subscriber numbers, earthquake intensities, and Darwinian evolution. For Ray Kurzweil, it's also the world of technological progress. And for us, it's above all the world of Bayes' rule!

From a mathematical point of view, however, the two worlds are not unrelated. In fact, there are even two intermediaries, two kinds of translators, who make the connection between these two worlds. These intermediaries have already been mentioned. These are the logarithmic and exponential functions. The exponential translates the objects of the additive world into objects of the multiplicative world. In particular, the addition between numbers will then become a multiplication between their exponentials. Conversely, the logarithm translates the objects of the multiplicative world into objects of the additive world. In other words, the logarithm brings objects and operations that are foreign to us to the world we are most familiar with.

Because adding is simpler than multiplying, these translators played a central role in numerical computation before the advent of calculators. To multiply a and b a few decades ago, students and scientists began by using logarithmic tables (or rulers) to determine the logarithms of a and b. Then, they added logarithms. Finally, they used the exponential to translate the result of the addition to the multiplicative world, and thereby obtained the expected result. This approach may seem convoluted. Yet, this was the best way to perform sophisticated multiplication quickly and with few errors.

In the same way, because adding is simpler than multiplying, Alan Turing used logarithms to perform his Bayesian calculations in wartime. The additive unit obtained by translating the Bayesian probabilities on the additive scale was (roughly) what Turing called the *banburismus*[12]

---

[11]*Addition contre multiplication*. MicMaths. M Launay (2014).

[12]Turing was actually interested only in relative probabilities, also known as *odds*. Thus Turing's *banburismus* were the units of *log-odds* of the form $\log_{10}(\mathbb{P}[T_1|D]/\mathbb{P}[T_2|D]) = \log_{10}\mathbb{P}[T_1|D] - \log_{10}\mathbb{P}[T_2|D]$ (which amounts to a difference of *bits* of Shannon, up to a multiplicative factor, since *bits* use base 2). This allowed Turing to bypass the computation of the partition function.

and, as we will see later, they are intimately related to the famous *bits* of Shannon - and thus to notions like entropy and the KL divergence. For now, I suggest to simply write Bayes' rule on the logarithmic scale:

$$\log \mathbb{P}[T|D] = \log \mathbb{P}[D|T] + \log \mathbb{P}[T] - \log \mathbb{P}[D].$$

It's often this version of Bayes' rule that artificial intelligence researchers prefer to study; and it also has applications in statistical physics and cognitive science!

11.8   BAYES WINS A GÖDEL PRIZE

It's only recently that computer scientists have understood how to exploit the madness of exponential growth. In 2012, Arora, Hazan, and Kale published a remarkable article that synthesized and unified many disparate but similar ideas into an algorithm of disconcerting simplicity and efficiency. This algorithm is the *multiplicative weights update* algorithm. The important word here is, of course, the word *multiplicative*. The trick of this algorithm is to use multiplicative scales to guide the choices, as opposed to the additive scale used to measure its performance. Staggeringly, this simple trick allowed the three researchers to effectively solve many of the problems that previous generations had been stumped by.

The *multiplicative weights update* is so effective that Arora, Hazan, and Kale suggested that it's one of the most important ideas in computing. At the beginning of the article, they proposed to include the teaching of their algorithm in introductory computer courses, along with other well-known methods like *divide and conquer*. The effectiveness of the *multiplicative weights update* algorithm is a testament to the major difference between the additive world and the multiplicative world. The fact that it was only recently discovered and understood is evidence of the fact that this distinction is very counterintuitive. It corroborates Kurzweil's postulate. Our intuition of exponential growth is very wrong.

One of the most remarkable successes of the *multiplicative weights update* is the technique of *boosting* in *machine learning*. But before explaining *boosting* and the *multiplicative weights update*, let's make a detour into the enlightenment century in France and a remarkable reflection of the Marquis de Condorcet.

At a time when many French philosophers defended the decentralization of power, Condorcet came to wonder whether it was preferable for judicial decisions to be made by a competent judge or a jury composed of several less competent citizens. Condorcet proposed a simple model where each citizen had a probability p strictly greater than $1/2$ to make a good decision. In this context, the probability of the jury making the wrong decision exponentially decreases with the size of the jury[13]. Condorcet deduced that a sufficiently large jury is much more reliable than a competent judge.

However, the case of Condorcet is too simple to be really relevant. In particular, the major underlying assumption is the independence of citizens' beliefs. However, especially if the members of the jury interact with each other, we should expect that the more loquacious ones will influence others - especially given that behavioral psychology experiments like Solomon Asch's show that the group effect can quickly make individuals believe facts that are quite clearly false. Worse still, it's difficult to anticipate how individual convictions will correlate - and empirical studies[14] of jury proceedings are very disturbing, since it seems that deliberation pushes jurors to more extreme conclusions than the most extreme juror's opinion. Is it then possible to combine different, fairly accurate but very unreliable and possibly correlated opinions to deduce a remarkably reliable opinion? This is the question that computer scientists Michael Kearns and Leslie Valiant asked in 1988.

In 1997, Robert Schapire and Yoav Freund answered this question in the affirmative. Their solution was named *Adaboost*. It earned them the prestigious Gödel Award in 2003 and led to the first face detection algorithms by Viola and Jones. The amazing success of *Adaboost* might suggest that this is a very sophisticated algorithm. However, *Adaboost* only exploits the disproportion between exponential growth and linear growth, between the logarithmic scale and the usual scale, between the multiplicative world and the additive world.

Better still, like the *multiplicative weights update* that generalizes it, *Adaboost* is nothing but a disguised Lotka-Volterra equation. In other words, *Adaboost* is nothing but an approximation of the Bayes' rule!

More specifically, suppose that we have experts with diverse and varied opinions. To start, for lack of data to distinguish the experts, we will

---

[13]Using Chernoff's inequality and calling n the size of the jury, it can be shown that the probability of error is upper-bounded by $\exp\left(-n^2 \frac{\left(p - \frac{1}{2}\right)^2}{2p(1-p)}\right)$.

[14] *The law of group polarization*. Journal of Political Philosophy. C Sunstein (2002).

start by considering them all indistinguishable from each other. Their aggregated opinion will be a simple average of their opinions. However, as experts' opinions are confronted with the data, the weight associated with the opinion of a given expert will be multiplied by the consistency of his opinions with the data. The aggregated opinion will still be an average of the opinions of experts. But this average will now be weighted by the weights of the different experts. More specifically, the credence in an expert will be the weight of her opinions relative to the weight of the opinions of other experts.

11.9   BAYES ON HOLIDAY

Let's take a simple case to better understand *Adaboost*. Suppose that, every year, you go on vacation to a distant country. Each year, to choose a destination, you ask each of your n friends to recommend a destination. To begin, not knowing what friend to listen to, you will write each friend's recommendation on a sheet of paper, and randomly draw a paper among these recommendations. That will be your destination.

This year, you drew Nigeria. Now is the time to confront the recommendation of your friend who suggested Nigeria with empirical data. And for that, you must unfortunately have a happy trip to the other side of the world! You come back tanned and happy. Your vacation was absolutely wonderful - and the positive Ebola test result does not worry you much.

Even if you are daydreaming, do not forget to update your credences in your friends. For that, you will multiply the opinion of the friend who suggested Nigeria by a number that represents your appreciation of the trip, on a scale of 0 to 1. You loved Nigeria? You can multiply the weight of the opinion of your friend by[15] 0.9.

However, to refine your credence in your other friends, it will also be necessary to imagine the holidays you would have had, if you had listened to their suggestions. Ideally, you should test all the vacations they suggest. But you lack time. Fortunately for you, your friends have themselves tested the vacations they suggested. And they love talking about their holidays. So you can easily get an idea of the appreciation you would have had of their trips. You then need to multiply the

---

[15]To make calculations neat, you should multiply the weight of the opinion of the friend by $(1 + \eta m)$ where $m \in [0, 1]$ is your appreciation of the trip.

weight of your friends' opinions by your estimate of your appreciation for their trips[16].

The next summer is coming, and you must choose your next holiday destination. To do this, you gather your friends, and collect their new recommendations. Again, you will randomly draw one of their recommendations. Except this time, you will make the probability of your friend's recommendation i proportional to the weight of friend i's opinion. In other words, the probability of following friend i's advice will be the credence in friend i. So, if friend i gave great advice in past years, then his recommendation will be more likely to be drawn.

This naive approach may seem convoluted and unconvincing at first glance. However, *Adaboost* and the *multiplicative weights update* mathematically guarantee that you will be doing almost as well as if you only listened to the opinion of the most reliable friend! In doing so, *Adaboost* guarantees that your successive decisions will be, in a sense that can be made rigorous, almost optimal[17].

An important point to emphasize is the negligible influence of the initial credence in the different friends. They had been assumed equal for all friends. However, because these credence have evolved exponentially, after a few iterations of the algorithm, the initial credence have completely disappeared.

More generally, the exponential evolution of Bayesian credence implies that, after a handful of Bayesian inferences, the role of the prejudices dissipates. It's for this reason that in the presence of enough data, Bayesian credence is in fact much less arbitrary than our linear intuition suggests. *Subjectivity need not be arbitrary.*

---

[16]Let's call w_i the weight of friend i's opinion. We then multiply w_i by $(1 + \eta m_i)$, where m_i is your estimate of your appreciation for his trip. The credence in the friend i is then given by the quasi-Bayesian formula which follows:

$$\mathsf{newCredence}(i) = \frac{(1 + \eta m_i)\mathsf{oldCredence}(i)}{(1 + \eta m_i) \cdot \mathsf{oldCredence}(i) + \sum_{j \neq i}(1 + \eta m_j) \cdot \mathsf{oldCredence}(j)}.$$

[17]More precisely, we can show that we then have

$$\mathbb{E}\left[\sum_t m(t)\right] \geq (1 - \eta)\left\{\max_{i \in [n]} \sum_t m_i(t)\right\} - \frac{\ln n}{\eta},$$

where $m_i(t)$ is your estimated appreciation of the trip suggested by friend i in year t, and $m(t)$ is your appreciation of the trip you actually made in year t.

11.10   THE SINGULARITY

Kurzweil goes further and blames the technological and academic world for not grasping the exponential growth of technologies, like Moore's Law which states (roughly) that the power of computers doubles every two years. In the same way, economists Brynjolfsson and McAfee suggest that too much importance is attached to various events of the past. If one concentrates on economic metrics, be it the number of inhabitants or the agricultural production, a major phenomenon is obvious and transcends all the historical facts like the fall of the Roman empire, the invention of printing, and the conquest of America. This major phenomenon is the inevitable exponential growth of economic metrics.

For Brynjolfsson, McAfee, Kurzweil, and many others, this exponential growth, driven by new technologies, heralds a very distinct future from the past. Artificial intelligence, 3D printers, nanotechnology, and advances in genetics herald a future in which human labor will no longer be needed to produce consumer goods in quantity and quality. Hunger in the world and diseases could then be eradicated. Our lifestyles would be upset.

In an excellent video entitled *Humans Need Not Apply*, CGP Grey even suggests that in the near future, human work will no longer be desired. For CGP Grey, the technologies may soon be such that (almost) any work done by a human can be done better and at a much lower cost by a machine. Brynjolfsson and McAfee share this belief, and announce the future extinction of jobs. I myself succumbed to such arguments, having predicted back in 2014 that the unemployment would be larger than 80% by 2034 - do not hesitate to remind me that I would have been wrong at that time!

McAfee is not a defeatist in this regard. On the contrary. In a TED talk[18], McAfee speaks of "the best economic news of our time". "Not that the competition is there," he adds. Abundance will be guaranteed. We will be able to rethink our societies, societies where there will be no need for the work of others - provided that the goods produced by the machines are properly redistributed. This has never happened in the history of humanity!

Nick Bostrom goes further. Bostrom suggests that technological growth may not be exponential, but super-exponential. To understand this, let's return to what characterizes exponential growth: at each time step, technology, or some more quantitative measure of it, is multiplied

---

[18] *What Will the Future of Jobs Look Like?* TED. A McAfee (2013).

by a constant. However, for Bostrom, it may be that the more the technologies develop, the faster the speed at which it develops gets large. In other words, at each time step, technology is multiplied by a number that increases with each iteration.

This phenomenon can be modeled using a differential equation. I'll allow myself a little mathematical jargon in this paragraph. Do not hesitate to skip it if you are not comfortable with this jargon. The exponential growth corresponds to the differential equation $\dot{x} = x$. The technological growth envisaged by Bostrom would be more like $\dot{x} = x^2$. But such an equation has a solution of the form $x(t) = 1/(1-t)$. In other words, this growth is so rapid that x reaches an infinite quantity after a finite time $t = 1$.

The conclusion of this small reflection is that there could be a singularity in technological evolution, a point where technologies would suddenly reach the limits of physics. This technological singularity is often interpreted as the moment when a superintelligence, that is to say an artificial intelligence superior to any human, would begin to improve its own intelligence. Being smarter than its designers, this superintelligence would find technological solutions out of our reach, and accelerate its self-improvement with a frantic pace. In a very short time, it would entirely change the whole world we live in. And its behavior would be fundamentally unpredictable - as it would result from an intelligence that is vastly superior to ours.

Bostrom does not venture to guess the precise date of this hypothetical singularity. However, he does not exclude its emergence within 50 years. Ray Kurweil goes further. He predicts that this technological singularity will take place in 2045. This prediction will upset more than one of you. For most of us, it seems even shamefully ridiculous. But for Kurzweil, it's because most of us are unable to extricate ourselves from our linear intuition.

I have my reservations about such predictions. However, I have even more reservations about my reservations, having repeatedly had the opportunity to see the limits of my intuition on exponential growth.

FURTHER READING

The Singularity is Near: When Humans Transcend Biology. Penguin. R Kurzweil (2005).

The Second Machine Age: Work, Progress, and Prosperity in a Time of Brilliant Technologies. W. W. Norton & Company. E Brynjolfsson & A McAfee (2005).

Superintelligence: Paths, Dangers, Strategies. Oxford University Press. N Bostrom (2014).

Modelling the Recent Common Ancestry of All Living Humans. Nature. D Rohde, S Olson & J Chang (2004).

The Law of Group Polarization. Journal of Political Philosophy. C Sunstein (2002).

The Multiplicative Weights Update Method: a Meta-Algorithm and Applications. Theory of Computing. S Arora, E Hazan & S Kale (2012).

Universal Paperclips. Decision Problem. F Lantz (2007).

Can Money Buy Happiness? AsapSCIENCE. M Moffit & G Brown (2012).

What Will the Future of Jobs Look Like? TED. A McAfee (2013).

Humans Need Not Apply. CGP Grey (2014).

The Accelerating Future. R Kurzweil (2016).

How Many Thoughts Are Contained in a Mars Bar? Sixty Symbols. P Moriarty (2016).

How Many Particles in the Universe? Numberphile. T Padilla (2017).

What is Singularity, Exactly? Up and Atom. J Tan-Holmes (2018).

DNA Encoding. ZettaBytes, EPFL. C Dessimoz (2018).

The Multiplicative Weights Update Algorithm. Wandida, EPFL. LN Hoang (2016).

Motivations and Applications of the Multiplicative Weights. Wandida, EPFL. LN Hoang (2016).

Theoretical Guarantee for the Multiplicative Weights Update. Wandida, EPFL. LN Hoang (2016).

CHAPTER 12

Ockham Cuts to the Chase

It is futile to do with more what can be done with fewer.

William Of Ockham (1285-1347)

Simplicity is the ultimate sophistication.

Leonardo da Vinci (1452-1519)

12.1 LAST THURSDAY

In 2002, in the US state of Ohio, Tonda Lynn Ansley was tried for the murder of her landlord. Ansley defended herself by saying that she believed she was living in the *Matrix*, in reference to the film trilogy of the same name. In this series of Hollywood movies, the *Matrix* is a computer simulation in which the overwhelming majority of humans live. Humans have been interacting with each other in this virtual world for so long that (almost) none of them distinguishes it from the real world. They take their simulated universe for reality.

However, *Matrix* is just a movie and believing in this movie is often interpreted as a sign of unreason. In fact, Ansley was found to be mentally ill and was acquitted. For many, *Matrix* is just fiction. You have to be mentally disturbed to believe it.

Yet, famous scientists like Stephen Hawking do not hesitate to seriously consider the *Matrix* hypothesis. Nick Bostrom even offers a very convincing argument in favor of this hypothesis: if technology allows it, it's reasonable to think that humans would prefer to ski down virtual ski slopes where the cold is not freezing and where avalanches will not endanger their physical integrity. Slowly but surely, the virtual universe could become humans' favorite universe. Thereby, the *Matrix* could be the future of any sufficiently advanced civilization. However, the most populated civilizations are those that are developed. As a result, one would expect most intelligent beings in the universe to live in a *Matrix*. But then, if we take any intelligent being from the universe at random - ourselves for example - the probability that this being is in a *Matrix* is very close to 1. Therefore, the *Matrix* hypothesis is not only worth considering, it's even very believable. In particular, there is nothing completely foolish about assigning a non-negligible credence to its truth!

It's even possible to go further into obscure metaphysical theories. An extreme theory is *Last-Thursdayism*. According to this theory, the whole universe was created last Thursday. All the universe, all our Earth, all our civilizations, all our monuments, all our books, but also all our memories. If you think you were on vacation in Nigeria last summer, it's only because, last Thursday, when everything was created, your brain contained memories of a vacation in Nigeria. Better still, *Last-Thursdayism* is unfalsifiable and perfectly consistent with the laws of physics. Regardless of future observation, it will be possible to determine causes from last Thursday[1].

For Karl Popper, however, the *Last-Thursdayism* and *Matrix* hypotheses have no value because they are precisely unfalsifiable. Such an answer may seem attractive. However, we have seen in chapter 4 that Popper's falsifiability has neither empirical nor theoretical solid grounds. I won't come back to it.

The traditional concept relevant to denigrate the *Last-Thursdayism* theory or the *Matrix* hypothesis is not Popper's philosophy. The relevant traditional concept is *Ockham's razor*, named after the philosopher William of Ockham. It is also known as the *principle of parsimony*, of *economy*, or of *simplicity*. In 1319, Ockham wrote: "*Pluralitas no is ponenda sine necessitate*". In English, "the multitude must not be advanced without necessity". Or, in other words, simpler theories are preferable.

However, it's not easy to see how *Last-Thursdayism* would be less simple than the alternative that the observable universe emerged 13 billion years ago, then generating all the complexity of galaxies, stars, planets, living things, and our human brains. Despite its apparent simplicity,

---

[1]*Is anything real?* VSauce. M Stevens (2013).

Ockham's simplicity principle is not simple! In particular, what may seem simple is not necessarily so - and what looks complicated is not necessarily complicated either!

In fact, a rigorous understanding of the simplicity of theories seems to require a theory of complexity like the theory of algorithmic complexity. Typically, Solomonoff's work seems to be an essential foundation for a good formalization of Ockham's razor.

12.2 IN FOOTBALL, YOU NEVER KNOW

For now, let's insist on the crucial importance of Ockham's razor, especially in order to determine a predictive theory. As data scientists learn early on, without Ockham's razor, the trap in which we keep falling into is that of *overfitting*. To see the damages of overfitting and the key role that Ockham's razor plays (or could play), let's take a detour to the world where overfitting is king: the world of sport.

Extra time has already begun. The image of Gignac's strike on the right post of the Portuguese goalkeeper still haunts the minds of French players and their supporters. This France-Portugal Euro 2016 final in Paris seemed promised to them. After all, France won the two major international football competitions they organized - there was one before the war that they did not, but it was such a different time. What's more, France had already won the Euro in 1984, then in 2000, as if an underlying rule seemed to promise the Euro to France every 16 years. Finally, the history of the France team shows that it only wins when it is led by an exceptional player. In 1984, it was Platini. For the 1998 World Cup and Euro 2000, it was Zidane. This year, Griezmann transcends the competition.

However, towards the end of extra time, Portugal scores the only goal of this final. Portugal becomes European champion, defying all the predictions and all the statistical rules that seemed to have been established. Statistics lied to us!

Or maybe not. The newspapers now assert that this Euro 2016 is the competition of all revenges. In the quarterfinals, Germany beat Italy for the first time in the history of international football tournaments. In semifinals, France beat Germany for the first time since a small final in 1958. And in the final, Portugal beat France for the first time, even though it remained on a series of 10 consecutive defeats in all competitions. This was the competition that halted all previous runs of defeats.

On another note, Griezmann seems to live a full and remarkable year, and his individual performances make him a favorite for the Golden Ball award, the equivalent of the Nobel Prize for football. However, Griezmann lost the Euro 2016 final against Cristiano Ronaldo's Portugal, after having won against Manuel Neuer's Germany. And a few months earlier, his club team, Atletico Madrid, lost the Champions League final against Cristiano Ronaldo's Real Madrid after beating Manuel Neuer's Bayern of Munich. A few months later, it's Cristiano Ronaldo who will be awarded the Golden Ball - Griezmann will be ranked third.

All the analyses I just mentioned are typical analyses of sports journals. Statistics are used to reveal intriguing, striking, and even troubling patterns. However, for the expert in *machine learning*, they probably have little value, because they most likely correspond to *overfitting*. Indeed, if we have a look at the history of football and if we torture the statistics of past matches, we will always find plenty of remarkable statistical regularities. With each new result, some of these regularities are destroyed - like the fact that France wins the Euro every 16 years - but the potential statistical regularities are numerous enough to ensure that all do not perish. On the contrary, the more data accumulate, the more ways there are to torture the data for apparent statistical patterns.

This is where *overfitting* kicks in. When the number of *ad hoc* explanations increases faster than the number of data, then, regardless of the data, a clever-looking explanation can be found for these data. This is typically what happens when sports commentators take the time to exanube all the information on a lot of players during a lot of games. This is why, every now and then, we discover that some player has just set some new record.

12.3 THE CURSE OF OVERFITTING

This phenomenon of *overfitting* is derided by the excellent Tyler Vigen on his site *Spurious Correlation*. Vigen examined numerous temporal data on the web and looked systematically for surprisingly significant correlations. What's particularly brilliant is that Vigen's correlations are so unlikely *a priori* that it's impossible to take them seriously.

Thereby, we discover that the years when Nicolas Cage appears in a lot of films are also the years when there are the most deaths by drowning in pools. Moreover, the years of high consumption of margarine are accompanied by a great rate of divorce in the state of Maine. Also, when Miss United States is old, the murders by steam and hot objects are numerous. Fortunately, even once these statistics are well known, politicians will neither seek to stop the cinematographic career of Nicolas Cage, nor to ban margarine, nor to put pressure on the jury of Miss United States....

The cases presented by Tyler Vigen are fascinating precisely because we tend to reject any causal link, despite very strong correlations. These are pedagogically excellent cases. They are an opportunity to remind that correlation is not causality, especially when the risk of overfitting is high - and in our case, it is, because the number of computed correlations far exceeds the number of data points in each temporal data series.

However, rejecting any causal link in spite of a strong correlation is not a reflex that most of us have, and the trap of overfitting is not restricted to the world of sport. News often takes overfitting correlations very seriously. And the consequences of this can be huge.

For educational purposes, the site FiveThirtyEight[2] offers you a web interface where you can easily play with US political data. After a few hacks, you can easily build a statistic that will suggest that your favorite party has a positive effect on the US economy. And what's most noteworthy is that, after only a few seconds, you will succeed in determining one that passes the threshold of the *p-value* required by the "scientific method"! In other words, your statistic will be significant enough to be published by a scientific journal. It will thus clearly be enough to be published in the *New York Times*!

The reason why FiveThirtyEight's approach can lead to any predefined conclusion is that the site offers a great multitude of ways to measure the effect of a political party on the economy. There are different metrics (unemployment, inflation, GDP, financial market), different representations of the party within the ruling authorities (president, governors, senators, representatives) with different ways of comparing the relative importance of these leaders, and even the possibility of taking into account economic recessions or not. As, moreover, it's possible to select combinations of these different parameters, for example both unemployment and GDP, the number of possible explanations offered by the website of the effect of a political camp on the economy reaches 2,048.

Now, remember, even in the absence of a real significant effect, the *p-value* method reports one in 20 cases as statistically significant! Therefore, in our case, we can expect that around 100 statistics will be scientifically publishable! What's even weirder is that, by playing with the data

---

[2] *Hack your way to scientific glory.* FiveThirtyEight (2015).

of the website, it's just as easy to obtain significant statistics in favor of the Democrats as in favor of the Republicans. In other words, by playing long enough with the data of FiveThirtyEight, you can easily publish an article with the clickbait title[3] "50 statistics that prove that x is bad for the economy", regardless of the value of $x \in \{Democrats, Republicans\}$!

And yet, the web interface of FiveThirtyEight is actually extremely limited. A journalist urged by his editorial board and sufficiently skillful with computer science - or having a friend that is skilled with computer science - can easily generate millions or even billions of possible explanations of the effect of a political camp on the economy. This would permit publishing a thousand significant statistics every day for the next hundred years. Such is the extent of the danger of overfitting. By exploring more and more plausible explanations, one is essentially guaranteed to find "statistically significant" statistics to defend any position - often without even realizing that the discovery of these statistics is in fact *not* miraculous. *While it's unlikely for each of the statistics to be meaningful, it's even more unlikely that none of the statistics is.*

This simple remark is the reason so why many articles contradict each other on clickbait topics, from politics to racism, through terrorism, food, and religion. After all, the more people are curious about a subject, the more time they spend researching on it, the more they will find sensational statistics that are likely to buzz, and the greater the interest in the topic will become. It's a vicious circle, which has the misfortune to create radical convictions. These beliefs are then almost exclusively based on overfitting except that this overfitting is invisible to most of us who only read the statistics that were carefully *cherry-picked* by journalists whose bosses sought the buzz. This selection bias seems to inevitably lead us to an uncontrollable profusion of misleading information.

There is even worse. According to psychologist Jonathan Haidt, social science experiments keep showing, again and again, that we humans first take position, and then search for arguments to justify our position. Reason is a tool that we mostly use to defend our pre-established convictions. We only involve our reason to cherry-pick the *ad hoc* arguments and data that will most confirm our intuitions. Unfortunately, given that *ad hoc* arguments are ubiquitous, we merely need to train our reason to convince ourselves of what we want to believe in.

---

[3]Our reasoning here is very approximate and not really valid. But it does give an idea of what can be learned from an analysis as proposed by FiveThirtyEight.

By favoring the *overfitting* on biased data, our reason constantly biases our worldview.

Overfitting is the trap in which we constantly fall. This is the core cause of superstitions and supernatural beliefs. This is also the big flaw of *Last-Thursdayism*.

For any new observation, there is a new explanation that makes the observation compatible with *Last-Thursdayism*. In fact, it's even likely that, to explain the world around us, a follower of *Last-Thursdayism* will eventually develop a model of the universe similar to that which scientists have developed. But then, the hypothesis of *Last-Thursdayism* will become superfluous. It will not be able to explain more than what the other elements of the theory already allow to explain. But as this hypothesis becomes superfluous, it should be removed by Ockham's razor.

12.4   THE COMPLEX QUEST OF SIMPLICITY

Rather than going back and forth between competing complex theories after each newly discovered piece of data, Ockham's razor suggests neglecting overly sophisticated theories, even if it means not fitting all the data. But that's fine! After all, the data are usually the result of so many causes that it's hopeless to explain them all perfectly.

When a die falls on a six, the position of each air particle can potentially have an impact on the outcome. However, it's illusory to follow the movement of each particle of air - especially because the number of these particles far exceeds the combined memory space of all computers created so far. And yet, a roll of a die is much, much, much simpler than the social issues that intrigue us so much. If it's impossible for us to fully explain the fall of a die, then it is completely hopeless to have the end word on explanations of politics, terrorism, and nutrition. We must accept and embrace the incompleteness of our models. "All models are wrong." And that's a good thing!

One of the first people to realize the need to avoid explaining everything perfectly was probably Galileo, the father of modern science. One of Galileo's greatest genius strokes was to challenge Aristotle's physics, as he argued that no, heavier objects do not fall intrinsically faster. This postulate of Galileo, known as the *law of fall*, is an experimental absurdity. Take a feather and a pebble, and let them fall. You'll see that Galileo was wrong.

However, Galileo's genius was to realize that the intrinsic fall of objects was only part of what made them move. The objects, and especially the feather, undergo air perturbations. This air brakes the movement of some objects more than that of others. It even allows birds to fly. Galileo then postulated that in the absence of air, there would be no air perturbation, and one would then observe the intrinsic falls of the objects. The falls would then be independent from the masses of the objects. In the void, Galileo argued, all objects fall at the same rate.

It's often said that Galileo climbed up the Pisa tower to test his law of falling bodies. But this story was very likely invented from scratch by Galileo's student. After all, if Galileo had made the experiment, he would have observed that the heavier object, less braked by air friction, falls faster. Empirical experiments did *not* make Galileo right. It's actually a thought experiment, which I won't detail here[4], which shows that the hypothesis that the mass of objects is the only variable to affect the falling of objects is self-contradictory - unless we assume that the mass of objects has no intrinsic effect on their fall.

Galileo's other revolutionary idea was the *principle of relativity*. This principle states that a man sitting in a windowless cabin of a ship would be unable to know if the ship is moving. "The movement is like nothing", Galileo wrote. Again, it's not clear that Galileo would be right in practice - you can imagine the boat would be more agitated when it's moving than if it were moored at the dock. However, the differences between theory and practice were sufficiently weak and random for Galileo to have full confidence in his relativity of movement. In fact, shortly afterwards, this credence in the principle of relativity would lead Galileo to put the Sun at the center of the universe.

In both cases, Galileo had the genius to prefer the simplicity and elegance of his principles to their adequacy with practice. This is a brilliant application of Ockham's razor to avoid the trap of overfitting in which others before him had fallen.

Besides, half a century later, it would be Isaac Newton's turn to postulate the fundamental principle of dynamics that is now summarized in four symbols: $\vec{F} = m\vec{a}$. Then, two centuries later, James Clerk Maxwell emphasized the simplicity and elegance of his equations to suggest that they could explain both electricity, magnetism, and light. All these brilliant theories rest on the same principle. They replace a multitude of disparate and *ad hoc* explanations by simple and universal principles - even if all the phenomena are not perfectly explained by the theory.

---

[4] *Don't heavier objects fall faster?* Relativity 12. Science4All. LN Hoang (2016).

12.5   NOT ALL IS SIMPLE

Having said this, it would be wrong to believe that the best theories are always simple. Models of meteorology are well known to be horribly sophisticated, while modern neuroscience strongly suggests that understanding the human brain will inevitably require horribly complex models - perhaps as complex as the brain itself!

In 2016, AlphaGo beat Lee Sedol in the game of Go. This artificial intelligence was so complicated that a whole computer was needed to represent it. The same holds for Cepheus and Libratus, those artificial intelligences that triumphed over human poker players.

In fact, when we talked about *Solomonoff's demon*, we already saw what was the complexity needed to study a phenomenon: it's the *Solomonoff complexity* (of the probabilistic law) of the data. While he did not know the formal version of this concept, Alan Turing understood this fact before anyone else. In his 1950 historical paper, Turing questioned the minimal complexity of a computer capable of speaking like a human. Drawing on results from neuroscience, Turing estimated that the size of the simplest algorithm able to communicate with humans was of the order of gigabytes. In other words, for Turing, the *Solomonoff complexity* of the spoken language may be of a few gigabytes. We shall discuss Turing's arguments at length in chapter 14.

In fact, the *Solomonoff complexity* of many biological, sociological, and economic phenomena may far exceed this amount. This would mean that these phenomena are out of reach of our human brains, whose memory space seems limited to a few petabytes. Any simple model would then be doomed to fail in biology, sociology, and economics.

However, the big models expose us to overfitting. What will then allow us to develop complexity without overfitting is the *buzzword* of data sciences: *Big Data*. The more data we have, the more complex our models can be made. There is even a rigorous formulation of this principle, known as the *fundamental theorem of statistical learning*[5]. Roughly, this theorem determines the number of data samples needed to adjust the parameters of a model. Or conversely, given data, this theorem suggests the appropriate complexity of models to consider.

The quantitative measure of complexity used in the fundamental theorem of statistical learning is the *VC dimension*, after computer scientists Vladimir Vapnik and Alexey Chervonenkis. The rigorous definition of

---

[5] *The fundamental theorem of statistical learning.* Wandida, EPFL. LN Hoang (2017).

this concept is a little too complicated for us[6]. Roughly, the VC dimension counts the number of allowed *ad hoc* explanations to explain the data. The rough rule that can be derived from the theorem is the following: the number of samples must be 100 times larger than the VC dimension of all the explanations considered[7,8].

12.6 CROSS VALIDATION

So far, I've been focusing on overfitting, because it's probably the trap we fall into most often. However, there is also the opposite counterpart called *underfitting*. Underfitting corresponds to not giving sufficient importance to the gap between theory and practice. This is typically what we are victims of, when we nonchalantly ignore data that contradict our beliefs - even if in our case and unlike learning algorithms, it's often more of a cognitive bias.

Striking a happy balance between underfitting and overfitting is a classic - and often considered unresolved - problem of data sciences. It is sometimes illustrated by the bias-variance dilemma. Imagine that we are trying to predict the y property of an x datum. To do this, we can collect many examples of (x_i, y_i) pairs. Let's call S the set of these pairs. We say that S is a *training set*. Then, we apply a certain approach to predict y from the data x and the *training set* S. Let's call $f(x, S)$ our prediction.

Suppose now that S is a random *training set*. This is typically what is assumed when S is obtained by sampling. The average squared error that one is led to do is then

$$\mathbb{E}_S\left[(f(x,S) - y)^2\right] = (\mathbb{E}_S[f(x,S)] - y)^2 + Var_S(f(x,S)).$$

---

[6] Here, a hypothesis is a function $X \to Y$. The VC dimension of a set of assumptions $\mathcal{H} \subset Y^X$ is the maximum size $|X_{max}|$ of a subset $X_{max} \subset X$ such that all functions $X_{max} \to Y$ can be obtained by restricting assumptions from $\mathcal{H}$ to X_{max}. See : *VC dimension*. Wandida, EPFL. LN Hoang.

[7] The formal formulation is based on the notion of PAC-learning. Roughly, the fundamental theorem of statistical learning says that one can determine an "ϵ - optimal" explanation within a set of hypotheses $\mathcal{H}$ with great probability $1 - \delta$ if the size of the learning sample is at least $\Omega\left(\frac{VCdim(\mathcal{H}) + \log(1/\delta)}{\epsilon^2}\right)$.

[8] Note however that the statistical learning theory only provides guarantees of non-overfitting for small enough VC dimension. It may still be the case that some models with large VC dimensions do not overfit. In fact, recent research suggests that highly overparametrized deep neural networks do not generally overfit, and instead often exhibit "double descent". Explaining this empirical observation is still an open research problem.

This equation is often rewritten as $error^2 = bias^2 + variance$. In other words, the error can be broken down into two pieces. On one hand, there is the error due to the fact that, on average, our algorithm predicts badly. This is the *bias*[9]. On the other hand, there is the error due to fluctuations in the prediction between one *training set* to another. It's the *variance*.

Underfitting then corresponds to using a learning algorithm that is too rigid. This rigidity prevents it from adapting to the data, which then causes bias in the prediction. To solve the underfitting, the simplest solution is often to increase the complexity of our learning algorithm. Typically, one can increase the number of its parameters.

However, we then risk overfitting. Overfitting means sticking to the data too much. Our parameters would thereby be too much influenced by the fluctuations caused by the sampling of the *training set*. To avoid this undesirable variance, we can reduce the number of parameters. The problem is that determining the correct *fitting* beforehand is tricky, since it seems to be an intrinsic property of the data.

In practice, data scientists usually use *cross validation*. The search for the best model is then divided into two phases. At first, we consider models that are simpler than a certain level of complexity K - typically models with at most K parameters. Among these models, we select the one that best explains the *training set*. Then, the performance of the selected model is calculated on another data set called the *test set*.

Cross validation then consists in optimizing the complexity degree K. Let's start with a very low value of K. We are currently in the regime of *underfitting*. Our set of models is too rigid to explain the data. When K goes up, the *test set* performance improves. This is not surprising, since we then typically allow more flexibility in our models. However, there comes a point where these performances stop increasing, as depicted in Figure 12.1. Here we come into the regime of overfitting. While the performance of the best model improves on the *training set*, its performance on the *test set* is now degrading. Finding the value of K where this transition takes place is one of the best ways to search for the right amount of *fitting*.

The quantity K in *cross validation* is what data scientists call a *hyperparameter*, as opposed to the parameters of the model that are optimized in the first phase of *cross validation*.

Cross validation, however, has its limits. In particular, it assumes that the *test set* is only used to test the hyperparameters of the model.

---

[9] It's sometimes referred to as *inductive bias*.

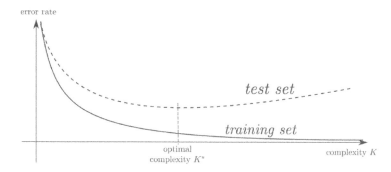

Figure 12.1 The solid curve represents the error rate at *training set*. The greater the complexity of the models, the lower the error rate. The dashed line represents the error rate in the *test set*. It represents the generalizability of the parameters calculated with the *training set*. We see that there is a compromise to be found. Too much complexity hurts generalizability.

However, it's often the case that the *test set* is used to test a very large number of different learning models, especially during competitions of machine learning like ImageNet, CIFAR, or other MNIST. The overuse of the *test set* means that it then becomes a kind of *training set*. We then risk overfitting on the *test set*.

12.7 TIBSCHIRANI'S REGULARIZATION

In 1996, statistician Robert Tibschirani had the idea to introduce another hyperparameter to adjust the *fitting* of *linear regressions*. Linear regressions are probably the most used techniques in science. By the end of the 18th century, Boscovich, Laplace, Legendre, and Gauss had defined and used such regressions to erase measurement errors of astronomical objects and to make predictions despite these errors.

In particular, linear regression can explain a variable of interest by p potential causes. Suppose we have a sample of n data. When n is much larger than p, we can serenely apply linear regression. However, in many problems as in genetics, conversely, the number p of potential causes is greater than the sample size n. Therefore, multidimensional linear regression is a very bad idea, since it will inevitably lead to a severe overfitting.

Tibschirani proposed to penalize some measure of the complexity of linear regression. Typically, to be retained, a linear regression that involves a lot of causes will have to explain the data far better than other linear regressions with few intervening causes. The formalization of this principle gave rise to the so-called LASSO regression[10]. The trick of the LASSO regression has since been generalized and reused in many problems of machine learning. This is called *regularization*.

A form of regularization might allow our cerebral cortex and its many neurons to avoid part of the trap of overfitting. After all, we live about 10^9 seconds, but our brain contains about 10^{14} neuronal connections. The risk of overfitting is enormous. However, regularization makes it possible to adjust the *fitting* of the models. In particular, regularization techniques have repeatedly demonstrated their usefulness in practice, and have become an essential tool in the analysis of data, whether for linear regression, linear classifications, or neural networks.

However, regularization seems mysterious. Why would regularization be a relevant guide to the best explanations? The fundamental theorem of statistical learning provides a very incomplete answer to this question. A better answer comes from *robust optimization*.

12.8   ROBUST OPTIMIZATION

The motivation of robust optimization comes from the following observation: data sets are full of inaccuracies, or even errors. In *machine learning*, we talk about *noise* in the data. Thus, any solution obtained after an optimization will be optimal only with respect to erroneous data. It could then be totally inadequate with respect to the real data.

To determine a powerful solution despite the noise of the data, robust optimization begins by identifying an uncertainty set[11]. This set is constructed so that, with very high probability, the actual data lie within this uncertainty set. Robust optimization then consists in choosing a solution that is suitable for all the data in the uncertainty set. Better still, it will choose the solution that is best suited, even for the worst case within the uncertainty set. It performs worst-case optimization.

Surprisingly, insisting on the inaccuracy of the measured data makes it possible to explain the *utility* of the recurrent dysfunction of neurons. Far from being a manufacturing flaw, the unreliability of neurons could

---

[10]The linear regression consists in decomposing the explanation of a variable y as a linear combination of causes $x_1, ..., x_p$ and of an error ϵ. Thus, we have $y = \beta_1 x_1 + ... + \beta_p x_p + \epsilon$. The classical approach is to determine the coefficients $\beta_1, ..., \beta_p$ that minimize the sum of the squares of the prediction errors ϵ^2 for the instances of the sample. Tibschirani had the brilliant idea of minimizing a combination of this sum of squares and the norm of the vector $\beta_1, ..., \beta_p$, typically the L1 norm, that is, the sum of the absolute values of the β_i.

[11]In dimension 1, these uncertainty sets correspond to confidence intervals.

actually be an asset. When a neuron is victim of a *bug*, it disrupts the signal, as if the data set was modified slightly to integrate the initial uncertainty concerning this data. By refining its model of the world over and over again, our brain explores and adjusts itself to a whole range of uncertainties - rather than adjusting to the raw data alone.

Interestingly, this trick is also used today by many practitioners of *deep learning*. Deep learning uses (deep) artificial neural networks as models to explain large data sets. Deep learning practitioners often randomly switch off a small fraction of the neurons from time to time, and test the functionality of their neural networks despite this. This technique is called *dropout*. It has been shown to be extremely effective in combatting overfitting.

Both regularization and robust optimization allow to combat overfitting. But what is the link between these two techniques? Well, they happen to be equivalent. In many problems, it can be shown that any solution obtained by regularization can be obtained by the choice of a certain uncertainty set and the application of robust optimization to this set. Conversely, for a given set of uncertainties, it's often possible to determine an equivalent regularization. In other words, the efficiency of regularization can be explained as a way of addressing noise in the data[12].

But there is better. Much better.

12.9 BAYES TO THE RESCUE OF OVERFITTING*

Regularization has a natural interpretation in Bayesian terms. Remember Bayes' rule translated into the additive world using logarithms. It's written as follows:

$$\log \mathbb{P}[T|D] = \log \mathbb{P}[D|T] + \log \mathbb{P}[T] - \log \mathbb{P}[D].$$

Machine learning and robust optimization usually consist in selecting the most credible theory T, given the data. This theory is called maximum a posteriori (MAP). It maximizes $\mathbb{P}[T|D]$, which is equivalent to maximizing $\log \mathbb{P}[T|D]$.

As a result, the quantity $- \log \mathbb{P}[D]$ is not relevant because it does not depend on T. Calculating the MAP therefore amounts to maximizing the

---

[12] *Robustness and regularization of support vector machines.* Journal of Machine Learning Research. H Xu, C Caramanis & S Mannor (2009).

sum $\log \mathbb{P}[D|T] + \log \mathbb{P}[T]$. The first of these two terms is log-likelihood. It measures the ability of the theory or model to explain the data. The second of these terms is the logarithm of the prior.

This prior is equivalent to a regularization term. Better still, by requiring that the sum of the prior probabilities be equal to 1, one is then tempted to distribute these parameters according to a law which exponentially decays towards 0, when the parameters take large values. This boils down to usual forms of regularization! In particular, regularization is a consequence of Bayes' rule[13]!

Better still, all the seemingly arbitrary hyperparameters (complexity levels K, regularizations, and uncertainty sets) are in fact evidence of the inevitability - or efficiency - of priors in the quest credible models! Regularization works because it forces us to introduce a prejudice. And as we have seen, prejudices form one of the pillars of rationality.

The *pure Bayesian*, however, sees a failure in the way that regularization and robust optimization are applied. Most *machine learning* algorithms conclude with a single model. They return a single choice of theory T. However, the methods of *ensembling* or *bagging* invite us to combine different algorithms of *machine learning*, especially when combined with techniques like *Adaboost*. Indeed, they show that the average of good theories often yields better results than the best of these theories. In particular, this is a great way to fight overfitting. *A forest of incompatible models is wiser than each of its trees.*

For example, when Netflix hosted a *machine learning* competition with 1 million dollars in play, the big winners took an average of 800 different models[14]! Yet, taking the average of the best theories is precisely what Bayes' rule requires!

Many researchers have realized this. In particular, in 2016, Yarin Gal published his dissertation *Uncertainty in Deep Learning*, in which he shows that many common techniques of *machine learning* can be reinterpreted in Bayesian terms. This is particularly the case of the *dropout* technique that we have just talked about! Indeed, each fault of a set of neurons corresponds to a model. The prediction of the neural network is then obtained by taking the average of the predictions of the different models, each of them being deduced from the faults of a subset of neurons.

---

[13]In particular, LASSO amounts to assuming an *a priori* distributed according to the Laplace law that we saw in chapter 8.

[14] *The Netflix Prize.* ZettaBytes, EPFL. AM Kermarrec (2017).

12.10   ONLY BAYESIAN INFERENCES ARE ADMISSIBLE*

There is even a theorem that insists on the importance of prejudices: the *no-free-lunch theorem*. Roughly, this theorem says that there is no best learning algorithm. More to the point, no matter how you choose a model, there will be problems for which your method will be surpassed by others. Such alternative methods will typically take advantage of more adequate priors.

The complementary theorem to this *no-free-lunch theorem* is the theorem of admissibility of Bayesian inferences in statistical decision theory. Imagine that there is a fundamental datum θ that you do not know. However, you receive some information x, which was partially caused[15] by θ. Now you need to make a decision whose optimality depends on θ. You might know x, but you still do not know θ. How can you use x to optimize your decision making?

The Bayesian approach is to first notice that you know $\mathbb{P}[x|\theta]$. However, you do not know θ. What should you do? Use prejudices of course!

The *pure Bayesian* will therefore consider a prejudice $\mathbb{P}[\theta]$. Then she will perform a Bayesian inference to determine $\mathbb{P}[\theta|x]$. Now that she knows the credible values of θ, she can optimize her decision making.

The *admissibility theorem* for Bayesian inferences then asserts that, whatever your decision-making mechanism, and regardless of the *pure Bayesian* prejudice, there will be a value θ for which the *pure Bayesian* will outperform you. No matter what previous data you had gathered, so long as all values of θ remain possible, you cannot be guaranteed to outperform the *pure Bayesian*, even if she is deeply misinformed[16]! We say that the *pure Bayesian*'s decision making is *admissible*. Of course, this does not mean she will *always* perform better than your approach; it all depends on the value of θ.

But this intriguing result is not the most intriguing aspect of the admissibility theorem. What's most remarkable is that, under some reasonable additional assumptions, regardless of your decision mechanism, there is a *pure Bayesian* with some prejudice $\mathbb{P}[\theta]$ whose decisions will *always* be as well or better than yours, no matter the value of θ! Or put differently, the set of admissible decision-making mechanisms is exactly the set of Bayesian approaches[17]. *Any non-Bayesian alternative will be inferior to a Bayesian method!*

---

[15]In fact, this is not a necessary assumption.
[16]Unless you always perform exactly as well as the Bayesian.
[17]*Admissibility and complete classes.* P Hoff (2013).

12.11 OCKHAM'S RAZOR AS A BAYESIAN THEOREM!

Let's finally get to one of my greatest moments of pure joy in my meditations of Bayes' rule. One day, I entered colleagues' office at the École Polytechnique Fédérale of Lausanne (EPFL) at lunchtime. Two of my colleagues were discussing the concept of *Ockham learning*, which is closely related to Ockham's razor. It got me thinking about the Bayesian interpretation of Ockham's razor. Could Bayes' rule be somewhat related to Ockham's razor?

Consider a language in which our theories will be described. This language can be English, formal logic, or a programming language. Each theory is then described by a (potentially very long) sentence in this language. Any such sentence is a finite sequence of language symbols, usually alphabetic letters. Let's call T_n the set of theories described by a sentence of n symbols. To be consistent with Bayesianism, the prior probability of these theories must be such that the sum of the credences $\mathbb{P}[T_n]$ in theories of n symbols equals 1. In other words, Bayesianism imposes the following condition:

$$\mathbb{P}[T_1] + \mathbb{P}[T_2] + \mathbb{P}[T_3] + \mathbb{P}[T_4] + \ldots = 1.$$

Now each quantity $\mathbb{P}[T_n]$ is nonnegative, and there is an infinite amount of such quantities. The theory of infinite sums then tells us that, if the infinite sum of these terms is finite, then the terms $\mathbb{P}[T_n]$ of the sum must become arbitrarily small for large values of n. As this thought suddenly crossed my mind, I threw myself at the board and wrote:

$$\sum_{n=1}^{\infty} \mathbb{P}[T_n] < \infty \quad \Longrightarrow \quad \lim_{n \to \infty} \mathbb{P}[T_n] = 0.$$

Yet this is exactly saying that theories that require more symbols to be described are less credible *a priori*. Unbelievable! Bayes' rule implies Ockham's razor!

Bayes' rule goes even further and tells us how much theories that are longer to describe are less credible. Indeed, the number of theories of n symbols is exponential in n. Thus the prior credence in a theory of n symbols exponentially decays with n. In other words, the more sophisticated theories are not just less credible; they are exponentially less credible!

I was deeply seized by this delightful discovery - especially since I had not yet met Solomonoff's demon at that time. Not only did this discovery further reinforce my credence on Bayes' rule, but it also shed light on Ockham's mysterious razor. For the *pure Bayesian*, Ockham's razor is not a philosophical principle that we must embrace without question; *Ockham's razor is a mathematical theorem of the Bayesian paradigm.*

FURTHER READING

The Righteous Mind: Why Good People Are Divided by Politics and Religion. J Haidt (2013).

Understanding Machine Learning: From Theory to Algorithms. Cambridge University Press. S Shalev-Shwartz & S Ben-David (2016).

Uncertainty in Deep Learning. PhD Thesis. University of Cambridge. Y Gal (2016).

Regression Shrinkage and Selection via the Lasso. Journal of the Royal Statistical Society. R Tibshirani (1996).

Robustness and Regularization of Support Vector Machines. Journal of Machine Learning Research. H Xu, C Caramanis & S Mannor (2009).

Spurious Correlations. Tyler Vigen (2020).

Hack Your Way to Scientific Glory. FiveThirtyEight (2015).

Admissibility and Complete Classes. P Hoff (2013).

Is Anything Real? VSauce. M Stevens (2013).

The Fundamental Theorem of Statistical Learning. Wandida, EPFL. LN Hoang (2017).

The Netflix Prize. ZettaBytes, EPFL. A.M. Kermarrec (2017).

CHAPTER 13

Facts Are Misleading

There are three kinds of lies: lies, damned lies, and statistics.

Benjamin Disraeli (1804-1881)

Politicians use statistics like drunkards use lampposts: not for illumination, but for support.

Hans Kuhn (1919-2012)

13.1 HOSPITAL OR CLINIC

You are seriously ill. You do your research and discover that for your disease, the hospital has a 50% survival rate, while the clinic has an 80% survival rate. No debate here. You definitely should go to the clinic rather than to the hospital... right?

Obviously!

Not so fast. After further research, you discover statistics that distinguish two types of patients: the slighly sick and the critically ill. In the clinic, the slightly sick patients have a 90% survival rate. Not bad. However, in the hospital, the survival rate for these same patients is 100%. In contrast, critically ill patients die in large numbers. The hospital nevertheless manages to save 40% of them. This is much better than the clinic that only saves 10% of them.

But think about it. Something extremely strange is happening. The hospital performs better for both slightly sick patients *and* critically ill patients. And yet, the clinic is doing better overall! How is it possible? How is it possible that each of the patients seems to be better off at the hospital than at the clinic, even though the overall survival rate at the clinic is higher than that of the hospital? Where to go for treatment? I invite you to stop reading and ponder this for a while.

If you are a little lost, know that it's perfectly normal. What I've just presented here, with fictitious data, is *Simpson's paradox*. This paradox is devastating. It shows better than any other paradox that statistics are surprisingly misleading, and that analyzing them adequately requires a tremendous amount of intellectual effort and expertise. Unfortunately, this expertise is extremely rare; and the average intellectual effort in interpreting statistics is almost nil.

In his book of statistics, Larry Wasserman writes that this paradox "is very confusing to many people including well-educated statisticians." In the introduction to his video on the subject, David Louapre[1] bets: "Once you've seen the video, I'm sure you will not look at the numbers the same way ever again."

If I do this well, this chapter should change the way you interpret statistical results. Carefulness, prudence, and humility should emerge as the key takeaways - and I hope that previous chapters have already encouraged such reflexes. In particular, what must always be on the top of your mind is that many apparently sound statistics are in fact virtually always very largely inconclusive. Much, much, much more inconclusive than what one might intuitively believe.

The key to understanding Simpson's paradox is the notion of *confounding variable*. In our case, the *confounding variable* is the health of the patients when they arrive at the healthcare center. In particular, if the clinic has a better survival rate than the hospital, it's simply because its patients are healthier when they arrive. The 80% survival rate at the clinic is therefore essentially the survival rate of slightly sick patients. Conversely, the reason why the hospital's 50% rate is so low is that it's essentially the survival rate of critically ill patients.

I have long felt that Simpson's paradox was not much of a paradox. I have even thought that it was a triviality. Once the data table is well filled, it's not difficult to check that, for each type of patient, the hospital is doing better, and to compute why the clinic still has better overall statistics. However, mathematically solving the problem once the table is completed is not the difficulty posed by Simpson's paradox. The real difficulty is that in practice, we often only have access to numbers like the 50% and 80% survival rates. And we are then so eager to conclude! Worse, even if one takes the time for reflection, it's often extremely

---

[1] *Le paradoxe de Simpson.* Science Étonnante. D Louapre (2015).

difficult to think of the right *confounding variables* to avoid the trap of Simpson's paradox.

In any case, the temptation to draw conclusions must be resisted. "We must hasten *not* to conclude," says Étienne Klein.

13.2   CORRELATION IS *NOT* CAUSALITY

In 2012, the journal *The New England Journal of Medicine* published a short article entitled *Chocolate Consumption, Cognitive Function, and Nobel Laureates*. The article suggested that chocolate consumption had beneficial effects on intellectual abilities. This staggering claim was based on a very clear correlation between chocolate consumption (per capita) in various countries and the number of Nobel Prizes (per capita) that these countries won. Quickly, the graph illustrating this correlation became viral and went around the Web. *Eat More Chocolate, Win More Nobel Prizes, Study Says*, said *Medical Daily*.

However, "we must hasten *not* to conclude". Correlation does not prove a causal link. In particular, while I cannot rule out the effect of chocolate on intellectual abilities, I am sure there are other far more credible explanations of the correlation between countries' chocolate consumption and their number of Nobel Prizes. And I invite you to think about it.

Statistics become particularly problematic when they are chosen by politicians, activists, or lawyers for their personal interests. Indeed, by playing with the confounding factors, it will usually be possible to find statistics that, at first glance, seem to defend this or that political position. As Winston Churchill would say, "when I call for statistics about the rate of infant mortality, what I want is proof that fewer babies died when I was Prime Minister than when anyone else was Prime Minister. That is a political statistic".

Here's a more concrete example. A strange and recurring phenomenon is the increase in crime figures right after the increase in police force, as if fighting crime inevitably favored crime. Such a correlation might suggest that investing in law enforcement is a bad idea. *Punishment does not work*, televisions then report. However, this conclusion is based on a misleading interpretation of statistics. In fact, there is another much simpler explanation of the correlation: the increase in the number of police officers increases the frequency of police checks. There are probably not more criminals. But there will certainly be more criminals arrested by the police. This is why crime figures are rising inescapably.

Similarly, healthcare investments usually improve medical diagnoses. As a result, the upgrade of medical means often leads to an increase in the number of patients to be taken care of! This is how a correlation between vaccinated children and autism can be explained. Since these vaccinated children receive more medical care, there is a good chance that, if they have autism, this autism will be diagnosed. Conversely, unvaccinated children with autism receive less medical care. Their autism is therefore likely to be undiagnosed.

These are cases of selection, survivor, or elimination bias - all are actually instances of Darwinian evolution. The numbers reveal how they were obtained rather than a cause-and-effect relationship. To avoid confusion, it's very important to properly interpret the figures that are presented to us. The number of arrested criminals is not the number of criminals. Rather, it's the number of criminals whose existence has been detected. Similarly, autism figures are the numbers of diagnosed cases. Not the number of autistic people. And this makes a huge difference when interpreting the figures!

That being said, the correlation between chocolate and the Nobel Prize does not seem to be a case of such a selection bias. Let's look for another plausible explanation of the correlation. When confronted with a correlation between A and B, it's tempting to think that A implies B. But, in fact, the notion of correlation between A and B is perfectly symmetrical. If A is correlated with B, then B is correlated with A. The explanation of a correlation can therefore simply consist in reversing the causal link that we wanted to make.

For example, top athletes are often very competitive. One might think that competing at a very high level creates a strong emulation and stimulates the desire to take on challenges - this is undoubtedly the case. However, a more straightforward explanation is that athletes with insufficient competitive spirit have not made sufficient efforts to reach the highest level. It's the spirit of competition that has allowed access to the highest level.

This effect is found in many situations. For example, politicians often have a thirst for power. Leading mathematicians have a deep appreciation for mathematical elegance. News headlines are particularly spectacular and dramatic. In all these cases, the explanation of the correlations lies in the way politicians, mathematicians, and headlines have been selected or eliminated in a systemic way.

Another example is the correlation between the large number of police officers at major events and the large number of incidents occurring there. It's not the presence of the police officers that causes the incidents; it's the risk of incidents that causes the presence of police officers. In all these examples, the correlation does not reveal so much the fact that A implies B, but perhaps more the fact that B implies A.

However, it's not entirely clear that the possession of Nobel Prizes increases the consumption of chocolate. To understand this correlation, let's look at another strange correlation: taking breaks at work outside reduces life expectancy. Although I have not done the empirical study, I am willing to bet that this correlation is pronounced. Why? I invite you to think about it for a few moments before continuing reading.

The explanation of this correlation comes in fact from a cause common to both variables involved: tobacco. Indeed, it's common for smokers to take many more outdoor breaks than non-smokers do. Yet tobacco is a major cause of lung cancer. Therefore, those who take breaks outside are also those who smoke. And they thus also are those with higher risks of lung cancer. Their life expectancy is hence smaller. We have explained the correlation between outside breaks and life expectancy by involving a common cause.

Could it be that the correlation between chocolate and the Nobel is the same? Is there anything that causes some countries to consume chocolate and also make these same countries win Nobel Prizes? Probably, yes. Indeed, countries consuming chocolate and receiving Nobel Prizes are all highly developed countries. The inhabitants of these countries enjoy a very high quality of life, a large consumption of luxury goods, and major universities. The correlation is explained here also by the existence of a common cause: wealth. We say that wealth is a *confounding factor* that explains the correlation between chocolate and Nobel Prizes as is represented by Figure 13.1.

Figure 13.1 The correlation between chocolate and the Nobel is explained by a *confounding variable*, wealth, which *causes* both more chocolate consumption and more Nobel Prizes. This graphical representation corresponds to a *Bayesian network*, which will be discussed in chapter 17.

Thinking of the right *confounding variables* is perhaps the most difficult task of statistics. So far, we have only discussed relatively simple cases. But *confounding variables* are sometimes much more subtle to find.

13.3 LET'S SEARCH FOR CONFOUNDING VARIABLES!

Here's a trickier example from an excellent TED-Ed video[2]. An English study discovered that, over a 20-year period, smokers had a better survival rate than nonsmokers. Such a study could be put forward, again and again, by politicians or lawyers funded by the tobacco industry. But evidently there is an interpretation problem. In particular, it should not be inferred that tobacco is healthy. Why? Because there is an important *confounding variable*. Do you see it?

Here's another example. While the debate over racism against blacks is raging in the US - as evidenced by the *Black Lives Matter* movement - a statistical study of the death penalty in Florida has been conducted. Are black suspects more likely to be sentenced to death? The statistics indicate no. This is a result that a political candidate denying racial inequality would likely quickly put forward.

However, there is an important *confounding variable*: the skin color of the victim. Since the culprits are very often of the same skin color as their victims, it turns out that black suspects are more likely to be tried for the murder of a black victim, while white suspects are more often judged for the murder of a white victim. However, given a same suspect's skin color, the judges are harsher when the victim is black. In contrast, for equal victim skin colors, black suspects are in fact significantly more often sentenced to death than white suspects. Such will be the statistics put forward by candidates who complain about racial inequality before the law.

Similarly, criminology researcher André Kuhn notes[3] that, in the vast majority of countries, the proportion of criminals among foreigners is higher than that among locals. However, he argues that this statistic is biased by Simpson's paradox. André Kuhn asserts that, given same age, sex, and socio-economic level, a foreigner is in fact just as likely to be criminal *a priori* as a local.

---

[2] *How Statistics Can Be Misleading.* TED-Ed. M Lidell (2016).

[3] *Comment s'explique la surreprésentation des étrangers dans la criminalité?* Vivre ensemble. A Kuhn (2012).

According to Kuhn, the cause of the statistical differences in all foreign and local populations does not come from a fundamental difference of kind between foreigners and locals, as some politicians often suggest. Kuhn's analysis shows that this difference actually mostly comes from the demographic difference between the two populations: foreigners are more often young men with little money, compared to locals who are comparatively more often wealthy older women. In light of this reflection, the statistics that seemed to incriminate foreigners completely loses its relevance!

Now that we have seen various examples, let's go back to the smokers. Did you determine the *confounding variable* that explains the better survival rate of smokers?

I invite you to think about this for a long time as soon as you pause in reading this book. Take the opportunity to feel the extent of your ignorance and to familiarize yourself with the extreme difficulty of interpreting the statistics that the Simpson paradox causes. And eventually, after much thinking, you may look up the answer in the TED-Ed video.

13.4   REGRESSION TO THE MEAN

Let's continue our quest for *confounding variables*. Should we reprimand or encourage? Instructors from the Israeli Air Force found that the pilots who had been reprimanded greatly progressed right afterwards. However, to their amazement, those who had received praise did not progress. Worse still, encouragement or congratulations seemed to make them regress, as if they had become complacent!

However, various scientific studies suggest that rewards are more effective than punishments for teaching. In particular, any teacher will undoubtedly consider that encouragement is *not* harmful to learning. What's happening here? Could the scientific community be wrong? Or could there be some *confounding variable* that invalidates the military instructors' conclusion?

The key ingredient to avoid the Simpson paradox trap is the *pure Bayesian*'s beloved prejudice[4]. To reason without prejudice, or without a model external to the data, is to jump altogether into the trap of the *confounding variables*! We indeed saw that there was an *a priori* difference between the hospital and the clinic (the health of patients

---

[4]In particular, to explain $\mathbb{P}[B|A] \neq \mathbb{P}[B|\text{no } A]$ we'll try to determine the characteristics Z to invoke such as $\mathbb{P}[Z|A]$ and $\mathbb{P}[Z|\text{no } A]$ differ greatly, and such that $\mathbb{P}[B|A, Z] \approx \mathbb{P}[B|\text{no } A, Z]$.

before being taken care of), between strangers and locals (age, sex, and socio-economic level), and between black and white suspects (the skin color of the victim). Could there be an *a priori* difference between the reprimanded pilots and the encouraged pilots?

Yes! The reprimanded pilots were reprimanded because they were particularly bad. Meanwhile, those who were congratulated were so because they were particularly good. However, a pilot who has made an unlikely mistake one day will probably not do it the next day, whether he has been reprimanded or not. In the same way, he who has achieved an exceptional feat one day will have a hard time doing as well the next day[5].

The phenomenon whose existence has just been revealed is therefore a special case of Simpson's paradox, known as *regression to the mean*. It can be summed up by the maxim "after the storm there is sunshine". Or rather, if the weather is particularly bad today, then the probability of tomorrow being even lousier is low. Not because tomorrow is likely to be better than usual. But because what tomorrow will be compared to is already exceptionally ugly. Or, conversely, Zinédine Zidane's sons will likely not outperform their father, because what they will be compared to is already an exceptional player. This phenomenon also explains why leaders whose mission is to rectify an abnormally bad situation have a good chance of being successful - whether they do something or not.

13.5 STEIN'S PARADOX

In 1955, Charles Stein discovered a mysterious solution to the *regression to the means*. Imagine that you have to estimate the levels of the pilots according to their performances. In an intuitive way, we risk overestimating the levels of the high-performance pilots and underestimating those of the pilots with clumsy performances. Can we avoid the trap of Simpson's paradox?

Yes, Stein answered. Instead of naively estimating the level of a pilot based solely on the pilot's performance, one can do strictly better by estimating it using the pilot's individual performance *and* the group's performance. This is called *Stein's paradox*[6]. This paradox is puzzling. How

---

[5] *Punishment or Reward More Effective?* Veritasium. D Muller (2013).

[6] Formally, for each $1 \leq i \leq n$, suppose that one draws independently $x_i \leftarrow N(\theta_i, 1)$ according to a normal law whose means θ_i is unknown. The naive (or the least square) estimator consists of estimating $\hat{\theta}_i^{naive} = x_i$. But we can do better, for

is it possible that by using the performance of others, we can improve the predictions concerning a given pilot?

In fact, Stein's paradox is a lot stranger than that. It also shows that estimates of a pilot's level, a country's chocolate consumption, and a hospital's survival rate can be strictly improved by using a combination of naive estimates of pilot levels, chocolate consumption, and survival rates, and the average of the three estimates! And what is very mysterious is that the improvement of the estimators will be guaranteed, even if the units of measurement of pilot level, chocolate consumption, and survival rate are incompatible!

This is extremely disturbing and counterintuitive. Even if there is absolutely no causal link between these three parameters, and even if the scales of these parameters have nothing to do with one another, Stein's paradox shows that, to estimate each parameter, we will *always* gain by invoking the other two parameters. In other words, in a rigorous sense, even though it improves the interpretability of models, *the fragmentation of knowledge is statistically inadmissible.*

As often, there is a Bayesian way of elucidating Stein's mysterious paradox. The trick is to add abstract explanatory concepts that will link the different quantities to estimate. However, these concepts are not really *confounding variables*; they are more sorts of *conciseness variables*. In particular, what Stein's paradox shows is that these *conciseness variables* are needed to identify the most credible models.

More generally, we tend to want to separate the fields of knowledge. We often consider that philosophy is philosophers' job, economics is economists', physics is physicists', and mathematics is mathematicians'. And that each expert should only restrict herself to her expertise. But according to Stein's paradox, the unification of all theories of all domains is actually a necessary step in the search for credible models[7]. This is why it's so crucial for any theory to be *universal*. As a result, the

---

example with the James-Stein estimator $\hat{\theta}_i^{JS} = \left(1 - \frac{n-2}{||x||_2^2}\right) x_i$. Indeed, for $n \geq 3$, for any value of the θ_i's, the expected quadratic error of θ^{JS} will be less than that of θ^{naive}, i.e.,

$$\forall \theta, \quad \mathbb{E}\left[||\theta^{JS} - \theta||_2^2\right] \leq \mathbb{E}\left[||\theta^{naive} - \theta||_2^2\right].$$

We say that the naive estimator is *inadmissible* because it is strictly dominated by another estimator. It turns out that the James-Stein estimator is also inadmissible, whereas all Bayesian estimators are admissible (see chapter 12).

[7]Although it's crucial for everyone to measure the extent of one's ignorance, especially in areas that are not one's field of expertise.

lack of universality of frequentist methods sounds like a strong argument in favor of Bayesianism.

In the case of pilots only, an example of a concision factor is the average level of any pilot. A possible Bayesian model then consists in assuming that the level of a pilot is equal to the average level plus some random fluctuation. The pilot's performance is then a second random fluctuation of the pilot's level. In fact, what we have just built here is a *Bayesian network*, the study of which is at the heart of much modern research in artificial intelligence. We'll come back to it.

It's noteworthy that this model has significantly fewer parameters than a model that would treat the cases of the different pilots separately. So it's *a priori* much more credible. But above all, by invoking a prejudice on the average level of a pilot, this model lends itself to Bayesian inference. In particular, by postulating some reasonable prior, this Bayesian inference leads us to something somewhat similar to Stein's statistical estimation. In other words, the strange Stein paradox disappears when one tries to make our models obey the principles of Bayesianism!

13.6 THE FAILURE OF ENDOGENOUS STRATIFICATION

These days, though, the Bayesian approach is not always favored by practitioners. Many still prefer so-called *stratification methods*. These methods consist of distinguishing comparable subpopulations. For example, it's better to compare foreigners and locals of the same age, sex, and socio-economic level, black and white suspects whose victims have the same skin color, and hospitals and clinics with comparable patients.

Around the 2010s, however, some statisticians emphasized the difficulties posed by the "hand-picked" choice of the strata considered. This choice often seems arbitrary. It often gives the impression of a lack of objectivity[8]. It can also be unjustified or insufficient, which would lead to wrong conclusions. Finally, it requires human intervention, and is therefore expensive in time and work. Would it not be possible to automate the choice of strata? Thus was born *endogenous stratification*.

In 2015, I had the opportunity to attend a statistics seminar at MIT. Many of the world's most recognized statisticians were in the room. The presenter, statistician Alberto Abadie of Harvard University, discussed an article[9] that presented the results of an endogenous stratification by

---

[8]Albeit the *pure Bayesian* does not attribute any value to this particular objection!

[9]*Research Student Aid Before You Reform.* Chronicle of Higher Education. A Kelly (2012).

Sara Goldrick-Gab and her collaborators[10], applied to Wisconsin students. The endogenous stratification divided the students into three categories: those whose chances of finishing high school seemed low when they entered the university, those whose chances were average, and those whose chances were high. She showed that, for the first group, receiving scholarships had a clear beneficial effect. Fellows who seemed unlikely to finish university fared much better than non-fellows who seemed unlikely to finish university.

So far, nothing is disturbing. However, this same endogenous stratification also showed that for students in the third group, those whose data at the time of entry to university predicted a high probability of success, receiving scholarships had a negative effect! It is as if, when we are good and are given more money, we become complacent....

Or not. To the surprise of several people in the room, Abadie showed that the conclusions of the *endogenous stratification* were unfounded. The automation of stratification had created its own *regression to the means*! The conclusion of the *endogenous stratification* was not due to the data; it was an artefact of *endogenous stratification*!

Unbelievable ! It was 2015, and the best statisticians on the planet were still discovering that relatively simple statistical models, models that were used nonchalantly by other leading statisticians, were fundamentally flawed! I was stunned.

A few weeks later, I took a few days off and visited a friend in the Silicon Valley. My friend was working for one of the web giants. I shared my nascent fascination for the tricky and subtle difficulties of statistics. A few months earlier, Ramesh Johari, a Stanford professor, had already astonished me when, in a seminary, he proved that the *p-value* method would always end up rejecting a hypothesis, provided that one collects enough data to conclude - we already talked about it in chapter 5.

But my friend was far more intrigued by Abadie's critique of *endogenous stratification* than by Johari's of the *p-value*. He made me repeat my explanations once. Then twice. And then a third time. Until, all of a sudden, he declared: "I think that's what we just did to test our latest product!"

Indeed, to do this, my friend had measured user clicks. He then compared them with clicks for the old product. He had not measured a statistically significant difference - Johari would say he just didn't wait long enough! However, *endogenous stratification* allowed him to conclude.

---

[10] *Need-based financial aid and college persistence: Experimental evidence from Wisconsin.* S Goldrick-Rab, D Harris, J Benson & R Kelchen (2012).

According to his analysis, the new product fared better than the old one for users whose click-through rate was initially low, and worse for those whose click-through rate was originally large. My friend had even found *ad hoc* explanations, including based on the geographical origins of the different users. This reinforced his belief in the fundamentally biased conclusion of endogenous stratification[11].

A few days later, I sent him Abadie's paper and he replied: "I analyzed one of the experiments, and I [saw] that we used an endogenous variable [...] to classify the results. The experiment was good, but the way to analyze the groups afterwards was incorrect [...]. The solution is to use an exogenous variable that does not modify the grouping. Thank you for making me think about that! I shared this analysis, so the next experiments will be [analyzed] correctly...."

What a pleasure for a theorist like me to see such immediate effects in practice! Unfortunately, I have never received any financial compensation from his company....

13.7 RANDOMIZE!

Simpson's paradox shows that any good data analysis must study plausible *confounding variables* that are not in these data. But if we only have data, how can we find these *confounding variables*? How to combat Simpson's paradox?

Even though I painted an evil portrait of Ronald Fisher, whose dogmatism and relentlessness against contrary opinions have certainly been damaging to the advancement of statistics, there is a very Fisherian phrase that must be emphasized, repeated, and celebrated: "Randomize."

Indeed, to test a product on a population, it's essential to compare the subpopulation exposed to the product to that which was not. The former subpopulation is called the *experimental group*, while the latter is the *control group*. However, to prevent any *confounding variable*, the assignment of individuals to groups should not be left to chance. Or rather, it should be! It must in fact be left to pure chance. Experiments designed

---

[11]This is more generally known as Hypothesizing After Results are Known, or HARKing. It corresponds to justifying only the validity of the results, as opposed to searching for caveats. Unfortunately, publishing incentives and cognitive biases may tend to overemphasize explanations in favor of the results, which then oversells the validity of these results. One proposed way to solve HARKing is by discussing explanations prior to finding out results, which is also known as expliciting prejudices. See chapter 5.

with randomized assignments to experimental and control groups are known as *randomized controlled trials*[12].

Designing randomized controlled trials is critical to avoid Simpson's paradox. Indeed, if the assignment is done systemically, that is, determined by a system chosen by the scientist or the environmental context, then we can be sure that there will be *confounding variables*. Even if these *confounding variables* play a negligible role, we can never be sure that this is not the case. And we should thus never have full confidence in the results of the experiment.

In the Fisherian tradition, the gold standard of medicine for testing new drugs is the *randomized double-blind test*. In this test, each patient is treated by a doctor with the new drug or fake drug (placebo). Crucially, the patient *and* the doctor must not know if the drug being administered is the new medicine or the placebo. This is a crucial point. Indeed, on one side, there is the *placebo effect*. If a patient believes he has had a medication while he has a placebo, then there will be positive physiological effects on his health. The patient will get better then if he knew he had a placebo.

But on the other side, it's also absolutely necessary that the doctor does not know if he has administered the new medicine or a fake. Indeed, experiments have shown that doctors who voluntarily administer a fake drug show less confidence and enthusiasm, so that the placebo effect on the patient is lessened. The randomized double-blind test controls such *confounding variables*, and thus aligns the results of the experiment with the phenomenon that we really want to study: the intrinsic effect of the new medicine[13].

More generally, this control of *confounding variables* is the reason why the good old scientific experiments like Galileo's matter so much. Ideally, these scientific experiments would replicate a large number of times two types of quasi-identical experiments. The only distinction between the two types of experiments would be the variable whose effect is to be determined. We are indeed studying the variable, "all things being equal".

---

[12] *How to take a scientific approach to charity.* Looking Glass Universe. M Yoganathan (2018).

[13] Technically, randomization is equivalent to comparing $\mathbb{E}_Z[\mathbb{P}[B|A, Z]]$ and $\mathbb{E}_Z[P[B|\text{not } A, Z]]$, as opposed to comparing $\mathbb{P}[B|A] = \mathbb{E}_Z[\mathbb{P}[B|A, Z]|A]$ and $\mathbb{P}[B|\text{not } A] = \mathbb{E}_Z[P[B|\text{not } A, Z]|\text{not } A]$. And contrary to the criticism that was raised in chapter 5, here A is actually taking a drug, not having taken it in the past. Nevertheless, it remains possible to conclude that A has no effect, while it may be hugely beneficial for some values of Z, and catastrophic for others.

It's interesting to note that this notion presupposes having specified all the things likely to vary. For instance, the fall of bodies depends both on the height of the object, but also on its potential energy. However, if we believe classical physics, it's impossible to test only the variation of heights, "all potential energies being equal". It seems that the notion of "all things being equal" presupposes the use of a model, and is, therefore, like every other aspect of knowledge, fundamentally subjective. In fact, even the results of randomized controlled trials should be taken with a grain of salt.

13.8　CAVEATS ABOUT RANDOMIZED CONTROLLED TRIALS

More generally, it seems important to note the limits of randomized controlled trials. While they allow for the collection of quality data, it should be noted that this data should still be analyzed with care. Perhaps more importantly, it should not be regarded as the only way to do science. In fact, quite often, it seems that it is not even the right way to do science.

Let's start with the caveats about the quality of data collected from randomized controlled trials. First, it should be said that the trials usually test a biased fraction of the population. Typically, psychology experiments often involve university students, sometimes dubbed Western, educated, industrialized, rich, and democratic, or WEIRD[14]. The data collected with such subjects do not seem straightforwardly generalizable to most humans[15]. Worse, the experimental conditions of randomized controlled trials are usually confined to some particular environment. It is far from clear that such results say much about what happens outside the laboratory[16]. In particular, even data from randomized controlled trials should *not* be trusted blindly[17,18].

---

[14]*Most people are not WEIRD.* Nature. J Henrich, SJ Heine & A Norenzayan (2010).

[15]Worse yet, any quantity $\mathbb{E}_Z[\mathbb{P}[B|A, Z]]$ will depend on the distribution of Z considered. But this so-called potential *distributional shift* does not seem to always be discussed as much as it should be.

[16]In other words, the experiment computes quantities $\mathbb{P}[B|A, Lab]$, which may be very different from $\mathbb{P}[B|A, IRL]$, where IRL means "in real life".

[17]An another important caveat is the lack of measures of long-term effects in randomized controlled trials. See *Can we study long-term effects?* Robustly Beneficial (2020).

[18]*Fixing poverty is hard, and science is complicated.* J Galef (2015).

But perhaps most importantly, designing randomized controlled tests is often extremely costly, both in terms of resources, time, and human labor. Yet, only a little data can be extracted from such tests. By opposition, *Big Data* usually consists of collecting all sorts of raw data, regardless of the randomization advocated by Ronald Fisher. While *Big Data* is often not made of *quality data*, it may still be preferable, simply because it is *big*, or because it was easy to collect. Yet, as we discussed in the previous chapter, obtaining huge amounts of data is critical to build up complex models without *overfitting*. In fact, most of the statistics we encounter in our daily lives are derived from such *Big Data* that weren't derived from randomized controlled trials.

But then, crucially, Simpson's paradox invites us to insist on the fact that such raw data will not be telling the full story. The context in which the data was collected is crucial to the analysis of the data. What is the age, sex, and socio-economic level of foreigners? Who are the victims of white criminals? Who are the patients of the clinic? Did the patient know that it was a placebo? Why was the pilot reprimanded?

Statistics collected without Fisherian randomization are *always* undermined by likely contextual pitfalls. *Without context, the wrong probabilities get tested.* We cannot trust statistics blindly. We must interpret with extreme caution. According to the *pure Bayesian*, it's essential to analyze them in the light of (several) *a priori* credible models. Such models will be key to suggest the appropriate *confounding variables* for the analysis of the statistics. And even then, one should not lose sight of the limited credibility of these models. "All models are wrong". In particular, "we must hasten *not* to conclude".

Having said this, I cannot insist enough on the fact that this skepticism towards statistics must not be interpreted as an acceptance of alternatives. On the contrary. If even statistics can so easily mislead us, given how malleable other kinds of data are, any conviction that cannot be statistically justified must be monitored even more closely! Especially if it is not empirically grounded....

13.9 THE RETURN OF THE SCOTTISH BLACK SHEEP

This allows us to come back to the story of the biologist, the physicist, and the mathematician traveling to Scotland and discovering a black sheep, which we introduced in chapter 4. Remember. The mathematician then mocked the excessive generalization of the physicist, because the physicist had concluded that the other half of the sheep, the one that we do not see, must be black.

Yet it seems that the physicist's generalization is not completely far-fetched. After all, there is a very clear correlation between the color of an animal on one side and its color on the other side. Few cats are exactly half white and half black. And even lower is the probability of seeing this cat at such an angle that the visible part is of a single color. Therefore, the generalization of the physicist does not seem unreasonably excessive.

However, the color of half of one animal does not seem to cause the color of the other half, and vice versa. How can we explain the fact that the colors of an animal on one side are almost always the same as the colors of the other side? Again, there is a common cause: the genes of the animal in question. In fact, if we go back in time, the animal was originally a single cell, which contained in particular a molecule called DNA. This cell then replicated, copying its DNA molecule identically. This is how the (almost) identical DNA is found in all the cells of the animal. It then determines, through the expression of its genes, the colors of the animal on both sides.

What's amusing, however, is that this explanation is a very modern explanation. After all, the structure of DNA was only discovered in 1953. For millennia, the common cause explanation that we advanced above was out of reach. However, it's hard to imagine anyone thinking that the correlation between the colors of animals on both sides could be explained by a causal link. How was this correlation explained? How to explain the fact that animals are dressed in a homogeneous color, without invoking the DNA molecule?

13.10 WHAT'S A CAT?

In 2012, Google made headlines with a strange and disturbing announcement: Google's artificial intelligence, headlines read, had discovered the concept of a *cat*[19]! Many probably found the news banal. But for me, it was a monumental and amazing breakthrough in machine learning, perhaps even more impressive and unexpected than AlphaGo's victory over Lee Sedol four years later.

To understand, it must be said that Google's artificial intelligence is a (virtual) artificial neural network, equipped with sensors that allow it to "see" digitized images. Google showed ten million images, without

---

[19] *Google's Artificial Brain Learns to Find Cat Videos.* Wired. L Clark (2012).

context, to its artificial neural network. Google measured the real-time neural activations of its artificial neural network, when this neural network was exposed to different images. This is akin to making it undergo a sort of magnetic resonance imaging (MRI), though all of Google's "measurements" are purely algorithmic. Google then realized that some of the neurons were activated, roughly, if and only if the image that was shown contained the image of a cat!

What's really remarkable is that it was not for this purpose that the artificial intelligence of Google had been designed. The goal of this artificial intelligence was to analyze, synthesize, and explain the content of images in the most relevant and effective way. The artificial intelligence had to design a model of the images it saw. To do this, it needed, among other things, to explain recurring correlations between the colors of certain pixels of the image. For example, when some of the pixels in the image were arranged in a certain way, there was very often a similar copy of these pixels slightly to the left or right. An image of an eye is often accompanied by another image of another eye. How to explain that an eye rarely comes alone?

What's amazing is that the artificial intelligence's explanation seems to be very much the same as the one most of us would give: animals often have two eyes. The fact that men, dogs, and cats (almost) all have two eyes is an assertion that seems so obvious to us that we hardly take the time to think about it. Our daily life is so accustomed to it that we do not even look for more explanations - yet it's a fascinating biological question that requires in particular the mathematical understanding of parallax!

But what's even more fascinating to me is that we can (globally) agree on what humans, dogs, and cats are to the point that we do not even realize that these concepts are both abstract and fuzzy! What is a cat? How to define the concept of a cat? And above all, does a cat *exist*?

To understand the strangeness of the concept of a cat, we can start by insisting on the vagueness of the definition. One could naively think that a cat is what you get after having two cats mating. Or said otherwise, the parents of a cat are cats. So far so good. This is probably even absurdly obvious. And yet....

Let's take a cat of today. Its parents are therefore cats. But the parents of its parents are cats too as well as parents of parents of parents. And so on. Except that there is no limit to this regression in time! According to this reasoning, one is forced to go back up the phylogenetic tree of life, until a time when there were no cats at all! Indeed, if we go back a few hundred million or even a few billion years, we get to a time when there was no mammal, nor vertebrate nor eukaryotic cell! To say that the parents of a cat are cats and to say that cats can only be the result of mating between cats are therefore logically inconsistent statements!

Let's be a bit less naive. You might now argue that the concept of DNA is the key to define the concept of a cat. However, this poses several problems. First, we have not sequenced all the genomes of all cats, and we will never sequence the genomes of all cats (since the cats of the future are not yet born!). It's therefore not easy to define a set of DNA codes that correspond to cats. Second, even if this were the case, having to sequence the genome of an animal to determine if it's a cat is a very inadequate solution in practice. Third, one can seriously question the fact that an isolated cell of a cat, for example a cell of its hairs, still is really a cat. Yet this isolated cell will be containing the DNA of a cat! Last but not least, the concept of DNA was completely absent half a century ago. At best, this means that Schrödinger, Darwin, and Aristotle did not know what they were talking about when they talked about cats.

In fact, there is arguably no satisfactory definition of the word "cat". If I asked you what a cat was, you could not provide a universal and indisputable definition. And it's not so surprising. After all, this is not how you learned to recognize and use the cat concept! If you know roughly what a cat is, it's because you've seen thousands, if not millions, of cat images, you've heard them, and you've read about them. You learned what a cat is by observing a very large amount of data. But you never had a formal definition of what it was! In fact, in the history of humanity, no one has ever known (or had to give) a formal definition of what a cat is.

Even better. There must have been a first human to think of the concept of a cat. As this human was the first to get there, no one could have taught him. This human then invented the concept of a cat himself. Why? And how did it happen? Where do the new concepts introduced by humans come from? Is it specific to human intelligence?

I find the discovery of Google's artificial intelligence fascinating, precisely because it gives us the answer to these questions. No, it's not a specificity of human intelligence, since Google's neural network got there. And it did it because it wanted to analyze, synthesize, and explain the correlations between the colors of the scanned image pixels. It wanted a model of what it sees. And the model it found naturally led to the creation of the cat concept!

13.11 POETIC NATURALISM

This leads us to the most fascinating question in this chapter. The cat is a concept of an abstract model. But then, does a cat *exist*? The question may seem stupid. You probably want to scream: *but, of course, since we see cats every day*! Yet, if we consider the most credible theory of physics to date - namely, the standard model of particle physics - the world is composed only of quantum fields whose excitations are quantized and form the electrons, quarks, photons, and other physical constituents. Nowhere in this model, and nowhere in physics, does one find the concept of a cat. Physical theories even reject the existence of objects that are not quantum fields. In particular, the standard model of physics refutes the existence of cats. At best, it's a pile of electrons, protons, and neutrons arranged in a certain way. But this is no more than saying that a digitized image of you is a bunch of bits.

In his excellent book *The Big Picture*, physicist Sean Carroll wonders if, when you see a cat running behind a mouse, it's scientifically correct to say that the cat *wants* to eat the mouse. Do we have the right to accept the existence of the cat, the mouse, and the intention that a cat could have?

For Sean Carroll, even if theoretical physics rejects the reality of all these notions, talking about cats, mice, and intentions is nevertheless the best way to talk about the situation we have just presented. Indeed, in doing so, we consider another model of reality, potentially in contradiction with the notion of theoretical physics, but which nevertheless remains compatible in terms of observable data. In particular, the so-called *emergence* phenomenon, which bridges the gap between the finer but more complex descriptions of reality and the coarser but more useful descriptions, is in fact a well-known phenomenon within physics. Notions like temperature and pressure emerge macroscopically, even if particle physics rejects their existence.

Sean Carroll thus defends a new epistemological position which he calls *poetic naturalism*. He asserts in particular that any theory of reality is a kind of poetry, which introduces its own concepts and allows its own predictions. Such useful concepts refer, according to Carroll, to some form of reality. This notion is similar to the concept of model-dependent realism of Stephen Hawking and Leonard Mlodinow, according to which any theory defines its own reality. The cat is then not real according to particle physics. However, the cat exists *in the model of reality* that is most familiar to us - the one that says that a cat running behind a mouse is very likely to want to eat it.

The positions of Carroll, Hawking, and Mlodinow seem at least partially Bayesian. The *pure Bayesian* does not care at all about the existence of anything outside of her models, except for the data measured by her sensors. Knowledge then only boils down to determining the most credible models, given these data - and language is only useful insofar as it describes the contents of these models. This is also how Google's artificial intelligence works. Abstract concepts, like the notion of cat, are components of credible models that explain the correlations in the data - or intermediate stages of computations necessary to make a prediction. But they need not have an existence beyond the models in which they are defined.

In fact, these abstract concepts are exactly the *confounding* (or *conciseness*) *variables* that make it possible to explain correlations without resorting to cause-and-effect links - and we'll see in chapters 17 and 18 that they play a key role in many models of *machine learning*. These abstract concepts are the key to explain why the hospital is better than the clinic, and why it's reasonable to think with great confidence that a black sheep on one side is also black on the other.

And in the end, the existence or realism of these concepts is of little importance to the *pure Bayesian*. "All models are wrong". What matters is that such concepts are *useful* to analyze the data that she is exposed to and to help her exploit these data to make relevant predictions of future observable data.

FURTHER READING

All of Statistics: A Concise Course in Statistical Inference. Springer Science & Business Media. L Wasserman (2013).

The Big Picture: On the Origin of Life, Meaning and the Universe Itself. Dutton. S Carroll (2016).

The Grand Design. Bantam Books. S Hawking & L Mlodinow (2010).

Inadmissibility of the Usual Estimator for the Mean of a Multivariate Distribution. Proceedings of the Third Berkeley Symposium on Mathematical Statistics and Probability. C Stein (1956).

HARKing: Hypothesizing after the results are known. Personality and Social Psychology Review. NL Kerr (1998).

Chocolate Consumption, Cognitive Function, and Nobel Laureates. The New England Journal of Medicine. F Messerli (2012).

Comment s'explique la Surreprésentation des Étrangers dans la Criminalité? Vivre ensemble. A Kuhn (2012).

Endogenous Stratification in Randomized Experiments. National Bureau of Economic Research. A Abadie, M Chingos & M West (2013).

Need-Based Financial Aid and College Persistence: Experimental Evidence from Wisconsin. S Goldrick-Rab, D Harris, J Benson & R Kelchen (2012).

Research Student Aid Before You Reform. Chronicle of Higher Education. A Kelly (2012).

Building High-Level Features using Large Scale Unsupervised Learning. QV Le, MA Ranzato, R Monga, M Devin, K Chen, G Corrado, J Dean& A Ng (2012).

Google's Artificial Brain Learns to Find Cat Videos. Wired. L Clark (2012).

Maths: Simpson's Paradox. singingbanana. J Grime (2010).

Is Punishment or Reward More Effective? Veritasium. D Muller (2013).

How Statistics can be Misleading. Ted-Ed. M Lidell (2016).

Simpson's Paradox. Minute Physics. H Reich (2017).

Are University Admissions Biased? Simpson's Paradox Part 2. Minute Physics. H Reich (2017).

How to Take a Scientific Approach to Charity. Looking Glass Universe. M Yoganathan (2018).

Fixing Poverty Is Hard, and Science Is Complicated. J Galef (2015).

III

Pragmatic Bayesianism

CHAPTER 14

Quick And Not Too Dirty

Far better an approximate answer to the right question, which is often vague, than an exact answer to the wrong question, which can always be made precise.

John Tukey (1915-2000)

Truth [...] is much too complicated to allow anything but approximations.

John von Neumann (1903-1957)

14.1 THE MYSTERY OF PRIMES

On March 11, 2016, Robert Lemke Oliver and Kannan Soundararajan experimentally discovered that the last digits of prime numbers do not behave very randomly. In particular, the latest figures tend not to repeat themselves. The prime number following a prime number that ends in a 3, like 23 or 43, will have a last digit that will be more often a 7 than a 3 too. This was a monumental surprise for the mathematical community. But for others, what was most surprising was perhaps rather the fact that this discovery could be surprising....

The prime numbers are whole numbers whose only divisors are 1 and themselves. The first prime numbers are 2, 3, 5, 7, 11, 13, 17, and so on. Their study has fascinated generations of mathematicians for millennia. More than 2,000 years ago, Euclid proved that there was an infinity of them. In 2002, Agrawal, Kayal, and Saxena found a polynomial-time algorithm to determine if a number is prime. And in 2012, Yitang Zhang proved that an infinity of consecutive prime numbers are separated by 70 million or less. Much has been discovered about these elementary bricks of the multiplicative structure of integers.

However, many basic questions remain open. Goldbach's conjecture postulates that every even number is the sum of two prime numbers. The twin prime conjecture postulates that there exists an infinity of prime numbers at distance 2, such as 3 and 5, 41 and 43, or 137 and 139. And the Riemann hypothesis postulates that the distribution of prime numbers can be deduced from some mathematical properties of a mysterious function called the Riemann zeta function. We'll come back to it.

One of the most difficult open problems is to determine a fast algorithm to compute the n-th prime number. An almost equivalent problem is to determine the number of prime numbers less than n with a fast algorithm. Whoever solves these problems will be covered with glory! However, it's not a given that these problems will one day be solved. To date and to the best of my knowledge, the best algorithm is that of Deléglise and Rivat, published in 1996. But the calculation time of this algorithm remains exponential in the number of digits of n.

This raises another problem with the problem of prediction. The case of prime numbers is neither fundamentally random, nor epistemically uncertain, nor even chaotic. If the prime numbers may nevertheless remain fundamentally unpredictable, it's because the amount of computation necessary to predict the googol-th[1] prime number could necessarily exceed what's physically possible, for our brains and our calculators!

In the same way, Ramsey's theorem raises problems with known methods of resolution. But these methods of resolution all require unreasonable calculations. For example, consider the problem of determining the minimum number of vertices that a complete graph must have to be sure that, regardless of the coloring of the edges in red or blue, the graph necessarily has at least one monochrome subgraph with n vertices. If you don't understand the problem, don't worry. The details don't matter.

In any case, it turns out that, for $n = 3$, the answer is 6. And the proof of this result is relatively easy. For $n = 4$, we know that the answer is 14. But "it's not quite so simple anymore", mathematician Paul Erdös adds. What happens for $n = 5$? "Well, nobody knows. It's between[2] 41 and 55."

"Suppose an evil spirit would tell mankind: 'Tell me the answer for 5 people, or I will exterminate the human race'. I said as a joke: It would be best in this case to try to compute it, both by mathematics and with

---

[1] A googol is equal to 10^{100}.

[2] In 2019, we now know that the answer is between 43 and 48.

a computer," Erdös says. "If he would ask for $[n =] 6$ [...], the best thing would be to destroy him before he destroys us, because we couldn't do it for $[n =] 6$". Some problems are impossible, not because we do not know how to tackle them, but because the computing power required to solve them far exceeds what physics allows us to calculate.

14.2 THE PRIME NUMBER THEOREM

Since they couldn't find a way to bypass the necessary computations, number theorists naturally turned to approximate calculations starting with Carl Friedrich Gauss. Around 1800, Gauss had fun studying the differences between successive primes. Between 3 and 5, the gap is 2. Between 7 and 11, the gap is 4. Gauss calculated that on average, the successive prime numbers less than 100 were spaced by 4. Those below 1000 were, on average, spaced by 6; those less than 10,000 by 8.1; those under 100,000 by 10.4. In other words, every time we multiply the prime numbers considered by 10, we add a little more than 2 to the average difference between successive prime numbers (2.3 to be a little more precise).

Does this ring a bell? This is a transformation of the multiplication (by 10) into an addition (by about 2.3). The average difference between successive prime numbers thus seems to be a logarithmic function of these prime numbers. Gauss came to the intuition that the distance between successive prime numbers less than n was well approximated by $\ln n$, where $\ln n$ is the *natural logarithm*[3]. This natural logarithm is the one whose base is Euler's constant $e \approx 2.718$. Equivalently, Gauss postulated that the number of prime numbers smaller than n, often denoted $\pi(n)$, seemed to be well approximated by $n/\ln n$.

More rigorously, Gauss conjectured the fact that the relative error of the approximation of $\pi(n)$ by $n/\ln n$ vanishes when n goes to infinity. We say that at infinity $\pi(n)$ is asymptotically equivalent to $n/\ln n$. This conjecture became the prime number theorem in 1896, when, independently, the mathematicians Jacques Hadamard and Charles Jean de la Vallée Poussin provided a proof of this remarkable approximate description of the exact distribution of prime numbers. Even though this approximation does not provide the exact location of prime numbers, this remarkable theorem has become one of the monuments of number theory[4]!

---

[3] We can see that $\ln(10) \approx 2.3$, as given by Gauss' observation.

[4] *Primes are like Weeds (PNT)*. Numberphile. J Grime (2013).

In 1854, Gauss' student, the brilliant Bernhard Riemann, went even further in the approximation of $\pi(n)$. He managed to find an exact formula for $\pi(n)$ using another equally mysterious function, now called the Riemann zeta function ζ. In particular, some numbers, called zeros of the zeta function, make it possible to calculate exactly $\pi(n)$.

Of course, there is a catch. There are even two. The first is that we do not know exactly where these zeros are (and the famous *Riemann hypothesis*[5] conjectures that non-trivial zeros are in fact aligned on a vertical line of the complex plane). The other problem is that there are infinitely many such zeros. The exact calculation of Riemann thus requires infinite calculation.

14.3 APPROXIMATING τ

That being said, mathematics is in fact full of infinite calculations. The most famous of these calculations is probably that of the famous circle constant τ (which is related to its historical "imposture" through[6] $\pi = \tau/2$). In the 14th century, the Indian genius Madhava discovered the stupendous equation $\tau = 8 - 8/3 + 8/5 - 8/7 + ...$ In other words, the constant of the circle τ is exactly 8 times the alternate sum of inverses of odd integers!

One might think that these infinite calculations are useless. Well, actually, they are ubiquitous in the preliminary models of applied mathematics, for example in the equations of fluid dynamics (a derivative or an integral is an infinite computation). Indeed, while these infinite calculations represent an incomputable ideal, they generally suggest an excellent way to perform approximate calculations. Madhava's equation indeed makes it possible to determine approximations of τ, and thus of the circumference of circles whose radii are known. Such approximations will be used by engineers in practice.

In fact, if you ask your favorite calculator or Google to give you the value of τ, it will lie to you and will probably only give an approximation with 13 decimals. This is not restricted to τ. Your calculator does not handle all the numbers whose binary writing is not finite[7], including irrational constants like e or $\sqrt{2}$, or rational numbers like $1/3$ or 0.2. In particular, because it only works with approximate numbers, your

---

[5] *Visualizing the Riemann zeta function and analytic continuation.* 3Blue1Brown. G Sanderson (2016).

[6] *Pi is (still) wrong.* Vi Hart (2011).

[7] *Why Computers are Bad at Algebra.* Infinite Series. K Houston Edwards (2017).

calculator may find erroneous results, like $1/3 \cdot 3 \neq 1$, or $(x+y) - x = 0$, even if y is strictly positive, and as long as x is much larger than y.

It often happens that mathematicians consider that computer calculations are approximations of mathematical theories. The position of Solomonoff's demon, however, is the opposite. For Solomonoff's demon, real numbers, for example, are meta-models that help structure and think better about algorithms. But they do not enable computable predictive theories. In particular, the model whose credences Solomonoff's demon seeks to measure is not the one whose parameters are real numbers; it's the one that will be implemented in a computer file with a truncation of real numbers of the mathematical model. Thus, unlike its mathematical idealization, the size of an artificial neural network is measured in bits[8].

14.4 LINEARIZATION

While approximating numbers like τ with great precision is of only limited interest, approximating curves, functions, and physical, biological, or mathematical behaviors has a very wide range of applications. This is the case of the approximation of the function that counts the prime numbers via the prime number theorem.

There is a very general tool that, given any model, calculates a much simpler approximate version of this model. This tool is *linearization*. Typically, linearization will approximate a small piece of circle by a line, or a small area on the surface of the Earth by a plane. In algebraic terms, this amounts to replacing so-called *non-linear* equations by affine equations of the form $y = ax + b$. For phenomena with reasonably small variations, these approximations will be perfectly fine. This explains why, in schools, we spend so much time studying these simple equations, and why we find them across all sciences[9].

---

[8]Moreover, the properties of neural networks strongly depend on this truncation of the mathematical model. Thus, a "mathematical" neural network has a VC dimension of at least $\Omega(|edges|^2)$, while any finite truncation of its actual weights yields a VC dimension of only $O(|edges|)$. More generally, the VC dimension is upper-bounded by the number of allowed bits to computationally describe models.

[9]Formally, the general case corresponds to the Taylor-Lagrange approximation. Consider a function $f : \mathbb{R} \to \mathbb{R}$ infinitely differentiable. The "linearization" of f at the point x_0 corresponds to the approximation of f around x_0 by

$$f(x) \approx f(x_0) + f'(x_0)(x - x_0).$$

Perhaps the most spectacular example of a linearization that is essential to physicists is the linearization of Einstein's general relativity equations, at low speed and weak gravity field. This linearization yields Newtonian physics! In other words, Newton's laws, especially the laws of gravity, are no more than an approximation of Einstein's, which is perfectly acceptable in many contexts.

Some defenders of science may persist in claiming that Newton's laws are "true" in their field of applicability. Our *pure Bayesian* does not agree, however. "All models are wrong." Put differently, in Bayesian terms, you should never put all your credence on a single model.

But that's not all. The *pure Bayesian* in fact attaches almost no credence to Newton's laws, since these laws explain strictly fewer phenomena than an adequate (even wobbly) combination of general relativity and quantum mechanics, without being much more succinct to describe.

14.5 THE CONSTRAINTS OF PRAGMATISM

The problem of the *pure Bayesian*, though, is that she is not pragmatic. Remember, in its purest form, which is Solomonoff's demon, she violates the laws of physics! In particular, the *pure Bayesian* is not bounded by computationally unavoidable processing time, and can instantly solve the equations of quantum field theory and space-time curvature, or determine the googol-th prime number.

In practice, however, computing powers are limited. We saw it. It's illusory to hope to make more than 10^{70} calculations in our solar system. This drastically reduces our ability to calculate Bayes' rule. If she wants to study all theories describable in 1,000 characters (two or three pages), the *pure Bayesian* will have to compute likelihood values $\mathbb{P}[D|T]$ for each of these theories! But the number of such theories is enormous. Assuming only the 26 letters of the alphabet are used, this corresponds to a corpus of 26^{1000} theories. As a result, the *pure Bayesian* will have to make at least 26^{1000} calculations. It's physically largely impossible.

Now, remember, this is the case where the *pure Bayesian* only considers theories of 1,000 characters. The human brain, in comparison, contains about 10^{15} synapses, which means that exactly describing the

---

Better approximations with higher order terms can be obtained as follows:

$$f(x) \approx \sum_{k=0}^{n} \frac{f^{(k)}(x_0)}{k!}(x - x_0)^k.$$

The Taylor-Lagrange theorem quantifies the smallness of approximation errors.

brain would require about 10^{15} bits of information. Studying all theories with so many characters would represent at least $2^{10^{15}}$ calculations! Applying Bayes' rule to these theories is therefore completely illusory, even with googols and googols of universe.

14.6 TURING'S *LEARNING MACHINES*

In 1950, Alan Turing marvelously transposed this reasoning on the inevitable complexity of algorithms to artificial intelligence. In an incredible article entitled *Computing Machinery and Intelligence*, and published in the journal *Mind, a Quarterly Review of Psychology and Philosophy*, Turing first asked a question very distant from that of the necessary length of algorithms. "Can machines think?" This is the question that Turing posed. However, the ambiguity of the word "think" forced him to clarify the problem. Instead, Turing wondered if a machine can act like a man.

More specifically, Turing proposed the following test: ask a human A to exchange written messages with two other entities X and Y. Among these two entities X and Y, there is another human B and a machine. The machine will then have passed the test if human A is unable to determine which of the entities X and Y is human B, and which is the machine. In other words, the machine will have passed the test, if it has been imitating a human so well that no human will be able to detect its inhumanity. This is what Turing called the *imitation game*, and is now called the *Turing Test*.

In sections 3 to 5 of his article, Turing reintroduced his 1936 work and rigorously defined what a machine is - which would lead to the invention of the computing machine, and then to the advent of the digital age! Then, in section 6, Turing refutes 9 classical arguments that seek to demonstrate that machines are incapable of thinking. But most importantly, in section 7, Turing anticipates the difficulty of his test and its resolution. While computers did not really exist yet, Turing already not only predicted their future existence, but also the fact that their performances would eventually be largely sufficient to win the imitation game. "It seems unlikely that [engineering advancements] will not be adequate for [passing the test]". For Turing, "the problem is mostly a programming problem".

In particular, in a visionary manner, Turing anticipated that the code of a program that passes the Turing test would necessarily be around 10^9 characters. In other words, by reusing the terminology that was introduced in chapter 7, Turing bet that the Turing test has a Solomonoff complexity that is of the order of gigabytes. To arrive at this estimate, Turing based himself on the only machine able to pass the Turing test that he knew of. I am talking about the human brain! Think about it. Who better than a human to imitate a human? Fortunately for Turing, neuroscience had already provided an estimate of the complexity of the human brain, advancing a figure between 10^{11} and 10^{15} synapses connecting neurons - recent estimates vary between 10^{14} and $5 \cdot 10^{14}$.

Turing postulated that a small fraction of these synapses was absolutely essential to anyone who would try to pass the Turing test, hence the figure of 10^9 characters. Turing adds: "At my present rate of working I produce about a thousand digits of programme a day, so that about sixty workers, working steadily through the fifty years might accomplish the job [of programming an algorithm that will pass the Turing test], if nothing went into the waste-paper basket. Some more expeditious method seems desirable."

This stroke of genius does not seem restricted to the Turing test. It's a safe bet that many tasks also require very long programs to be solved, starting with the mastery of natural language, the possession of a "common sense", or the art of empathy. Worse still, some tasks, particularly in biology and social sciences, may require algorithms that are longer than the size of the human brain. Therefore, not only the algorithms we write would fail, but our brains too would necessarily be unable to solve these tasks.

Predicting the next financial crisis could be beyond the cognitive abilities of our limited brains. "If people do not believe that mathematics is simple, it's because they do not realize how complicated life is," John von Neumann once noted. This assertion can even be made rigorous by measuring the complexity of the disciplines using the Solomonoff complexity[10] - or better again, the *Solomonoff sophistication* that will be introduced in chapter 18!

Let's go back to Turing. Turing then pointed out that the human brain is able to pass the Turing test. He then postulated that, to pass the Turing test as well, machines should imitate the thinking process that allowed the human brain to acquire its skills. Turing noted that the upbringing of the child is a crucial part of how his brain ended up the way it is. "Presumably, the child-brain is something like a note-book as

---

[10]Unfortunately, the *halting problem* implies the incomputability of Solomonoff's complexity.

one buys from the stationers", Turing wrote. "Rather little mechanisms, and lots of blank sheets[11]." Turing then proposed to help the machine fill its own blank pages by allowing it to learn from data. Thus was born the concept of *learning machines*, that is, machines capable of learning.

The idea of *learning machines* is to allow a machine to write itself a program of several billion characters - or more if necessary. In more algorithmic terms, this means that *machine learning* finally allows the study and exploration of algorithms whose description exceeds the gigabyte. And crucially, what guides this exploration is not the fingers of a programmer; the guides are the raw data. Like those that children exploit.

In particular, Turing's reasoning contradicts what many intellectuals, including some experts, repeat so often. "Machine learning works well - mathematicians do not know why," *Wired* wrote in 2015. But by 1950, Alan Turing had predicted the future success of machine learning, even specifying that it would be successful at the end of the twentieth century. Better still, Turing had pointed out the precise reason why *machine learning* would outperform man-made programs for many tasks: only *machine learning* is able to explore the space of algorithms whose description requires billions of bits.

Similarly, some experts often blame large neural networks for being impossible to interpret. It's not surprising, however, that a large network of neurons is not describable in a few characters. At least not in a complete perfect manner. Indeed, if it were, this description in a few characters would be a short algorithm able to generate another algorithm (the neural network), which would solve problems like the Turing test. As a result, the short algorithm would have solved the Turing test. However, it has been postulated that this test cannot be solved by short algorithms.

What is left is to specify how to explore the space of long algorithms. Alan Turing did not say *which method* to use. He simply said that the ability to learn was going to be key - although he did hint at what is now known as *reinforcement learning* and *genetic algorithms*. At the beginning of the 21st century, many approaches were proposed. We have already talked about several of them. These include linear regression, logistic regression, decision trees, decision forests, support vector machines, neural networks, Bayesian networks, or Markov fields. We'll

---

[11]We'll see in chapter 19 that this is now rejected by neuroscience, by virtue of Bayesian principles!

review at length these last three algorithms of *machine learning* in this chapter and the following ones.

For now, I want to recall what the *pure Bayesian* proposes. If she restricts herself to algorithms that are at most 10^9 characters long, the *pure Bayesian* will want to test and compare them all. This will require an amount of computation (vastly) larger than 2^{10^9}, which remains completely illusory, even with googols of universes within a time frame well longer than the age of these universes. In particular, if Turing was right about the Solomonoff complexity of the Turing test, then the purely Bayesian approach to solving the Turing test is doomed to fail in practice.

14.7 PRAGMATIC BAYESIANISM

To solve problems of large *Solomonoff complexity* like the Turing test, the *pragmatic Bayesian* is forced to give up Bayes' rule. He must prefer methods that do not require unrealistic computation times. This leads us to modify the meaning of *usefulness*. For the *pure Bayesian*, theories with great credence were useful. For the *pragmatic Bayesian*, theories with great credence that require unreasonable computation time are in fact useless. The *pragmatic Bayesian* will rather focus on the most credible theories, among the set of theories with low computation time.

This allows us to justify our pragmatic credences in the approximations given by the prime number theorem and Newton's laws. For the *pure Bayesian*, these two descriptions are completely useless, since they are both largely dominated by the exact calculation of prime numbers on one hand, and by Einstein's general relativity on the other. However, the *pragmatic Bayesian* will happily embrace these two approximations because their computational needs are much lower. Computing a good approximation of $n/\ln n$ can be done in logarithmic time in n, whereas the vector and differential computations of Newton's theory are much faster than those of the tensors of Einstein's theory. Provided that we place ourselves in a context where these two approximations are sufficiently acceptable (large values of n and low gravity), these two approximations will be, for the *pragmatic Bayesian*, much more *useful* than their exact counterparts!

This notion of utility relative to computing times could also be the primary explanation for the amazing success of neural networks. Indeed, unlike more sophisticated algorithms, neural networks, or at least those called *feedforward*, have necessarily limited and weak computation times.

In fact, if we consider a sufficiently broad definition of *feedforward* neural networks (depicted in Figure 14.1) then the set of these networks is exactly the set of fast algorithms, once these algorithms are parallelized. To perform *machine learning* on *feedforward neural networks* (with an appropriate Bayesian prior if possible) is therefore to try to explain the data as best as possible with the help of fast algorithms. It's therefore a first step towards *pragmatic Bayesianism*.

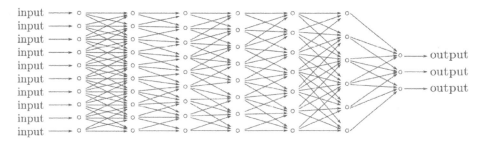

Figure 14.1 A neural network is a series of communications between neurons and elementary calculations by neurons. It is called *feedforward* if the communication is acyclic, that is, intuitively, if it only goes in one direction and never loops back.

Neural networks are not the only fast algorithms. Similarly, calculating the prediction of a linear regression is also a quick calculation. But probably too much so. In an article I co-authored with Rachid Guerraoui[12]: we even argue that the weakness of many approaches of *machine learning* is to restrict oneself to the study of algorithms whose parallelized computation time is composed of only a handful of steps. The problem is that we then ignore all the slower algorithms. However, for reasons that will be discussed in chapter 18, it's highly likely that many algorithms that are relevant for analyzing raw data, or for solving the problems that matter to us, are in fact necessarily relatively slow algorithms. This remark has led to a theoretical partial justification for the success of *deep learning*. We'll come back to it.

14.8 SUBLINEAR ALGORITHMS

So far, I have discussed candidate algorithms for machine learning. I also said that the accumulation of a large amount of data was essential to guide us in this quest for the best predictive algorithms.

---

[12] *Deep Learning Works in Practice. But Does it Work in Theory?* R Guerraoui & LN Hoang (2018).

However, not all data are equal. In particular, the continuous flow of a surveillance camera installed in the depths of a desert is arguably almost entirely uninteresting. In today's *Big Data* world, this raises a new kind of problem: reading uninteresting data to check that they are indeed uninteresting can take an unreasonable amount of time - that's what I tell myself to give me a good conscience when I do not read all the messages sent to me!

To solve this problem, theoretical computer scientists turned their attention to so-called *sublinear* algorithms. These algorithms have the particularity of extracting the "meat" of the data without taking the time to read all the data. In other words, these algorithms manage to read "very diagonally" the data to which they apply.

The historical archetype of such an example is Google. A search on Google must provide results almost instantaneously. But the Google database contains a large portion of the millions of billions of pages on the web. It's out of the question for Google to explore all of its database before returning responses to the user. This would take days! The Google search algorithm must be sublinear.

Google's trick, like the trick of libraries and dictionaries, is to sort and arrange their database beforehand, so that queries can be made quickly. For example, by arranging the words in alphabetical order, dictionaries allow users to quickly find where to look for a word they are looking for. Better still, thanks to the alphabetical order, given a page and a word to search, the user will be able to know if the word to search precedes the page in question, or if it will appear later in the dictionary. More generally, searching for data in a sorted database can be done very quickly. The so-called *dichotomic* algorithm (which is probably roughly the one you would use to search for a word in a dictionary) has a logarithmic computation time in the size of the database.

However, Google's algorithms, librarians, and dictionaries require huge amounts of preprocessing power. Organizing and ordering all websites cannot be done without having visited them all. Can useful information be nonetheless extracted from data sets without exploring everything, and without organizing these data sets beforehand? Surprisingly, for some kinds of useful information and data, the answer is yes. This is particularly the case of the *Fourier transform*.

The Fourier transform was studied at the beginning of the 19th century by, among others, Joseph Fourier. It's a way of changing the description of sound signals, images, videos, or other stock market prices. Let's take the example of sound.

Sound can be described by the variations of air pressure in your eardrums. But in an equivalent way, sound can also be described via the frequencies that it is made of. This is particularly useful when writing music sheets. The Fourier transform is then a kind of bilingual dictionary, which translates the description of the sound by its variations in volume, into a description by its frequencies. However, such translations have a very large number of applications. In particular, it's often more relevant to modify the frequencies of a sound to improve it than to change its volume variations. Also, many sounds that we like are composed of only a handful of frequencies, which makes their frequency description compact and informative.

In fact, the Fourier transform has invaded our daily lives. According to Professor Richard Baraniuk, our computers and phones calculate billions of Fourier transforms a day. As a matter of fact, whenever you listen to music, look at a digital image, or watch a video, Fourier transforms are performed by your machines.

Therefore, any potential acceleration of the Fourier transform calculation algorithms represents billions of euros in terms of computer calculations or electricity required for these calculations, not counting the waiting time of the users. In 1964, James Cooley and John Tukey achieved the feat of significantly accelerating the computing time of the (discrete) Fourier transform. Whereas the naive approach requires a quadratic time in the size of the data, the Cooley and Tukey algorithm, called *fast Fourier transform*, runs in quasi-linear time[13]. In other words, to calculate the Fourier transform of a megabyte of data, the *fast Fourier transform* requires only a few million operations (a few milliseconds for a modern computer), as opposed to millions of millions of operations (almost a minute of calculation) for the naive approach.

However, in the age of *Big Data*, the *fast Fourier transform* is still too slow, especially if it has to handle gigabytes, or even terabytes of data. Can we make the Fourier transform even faster? In 2012, Hassanieh, Indyk, Katabi, and Price[14] discovered that the answer is yes. Or rather, they came up with an astute algorithm that approximately calculates the k main frequencies of a signal, in a time of the order of k times the logarithm of the data size[15]. In particular, this remarkable algorithm, called *sparse Fourier transform*, runs in *sublinear time*. This means that it gets to know the signal to be processed by ignoring it almost entirely.

---

[13]The complexity of the discrete Fourier transform then changed from $O(n^2)$ to $O(n \log n)$.

[14]*Faster than Fast Fourier Transform*. ZettaBytes, EPFL. M Kapralov (2017).

[15]The complexity is actually in $O(k(\log n)^2)$.

14.9 DIFFERENT THINKING MODES

While they are not yet the norm in computing, sublinear algorithms seem to be the ones that our brains prefer. Who has never read documents diagonally, listened to a discussion with one ear, or eaten without savouring the taste of the food? Strangely sometimes, though, something in the text, the discussion, or the food gets our attention. If this is the case, we seem to switch our modes of reflection. Slower and more precise algorithms are used by our brains to decode the meaning of the text, the subtlety of the discussion, or the particular flavor of a local specialty.

What I'm describing here is the heart of the excellent book *Thinking Fast and Slow* by Nobel Prize winner Daniel Kahneman. Kahneman distinguishes two systems of thought: System 1 and System 2. System 1 is akin to a *sparse Fourier transform*. It is fast, efficient, busy, and possibly very wrong. System 2 is like an exact Fourier transform. It is slow, energy-consuming, lazy, and more correct.

According to Kahneman, we tend to identify with our System 2, and we tend to be unconscious of our System 1. However, almost all of the time, System 1 is the one in control. And this leads us to often deceive ourselves. Let's test this: a bat and a ball cost $1.10. The bat costs $1 more than the ball. How much does the ball cost?

There is a good chance that one answer has quickly come to your mind; and that this answer is false. If this is the case, according to Kahneman, it's because your System 1 rushed in and gave the first answer in mind. It's only afterwards, maybe after you read me, that your System 2 begun to question System 1.

One might guess that Kahneman wants us to abandon System 1, and to make use of System 2 as often as possible. This is not entirely the case. After all, even if it's more right, System 2 is also more tiring for the brain. System 2 is costly. And as any actor, pianist, or surfer knows, we cannot afford to think about all our actions. It's necessary that the gestures be executed in a natural and spontaneous way, without major conscious intellectual effort.

In order to make competent gestures without having to think about them for a long time, actors, pianists, and surfers all use the same trick. Their Systems 2 force their Systems 1 to learn the gesture in question. And this is probably the most important remark to remember for whoever wants to learn or teach. Learning is not about bringing information into our brain; it's more about making System 2 educate System 1, so that System 1 discovers a heuristic that quickly solves the problem that

System 2 can already solve, but with a lot of time and energy. In other words, learning is, above all, discovering heuristics that can be computed quickly and (fairly) well.

14.10 BECOME POST-RIGOROUS!

It seems that this is how mathematical learning works. Today, if I am asked questions about addition, exponentials, or the Turing machine, it's not unlikely that I'll be able to answer these questions without resorting to System 2. This is because, over the years, my System 2 has managed to successfully teach my System 1 how to solve these problems without any intellectual effort! I have a lot of heuristics in my brain to solve a very large number of mathematical problems, problems that a young student might find difficult. Some call it mathematical faculties. Others call it a mathematical intuition. I prefer to call it a set of powerful heuristics discovered through the hard work of System 2 and taught to System 1.

But this is not the most important aspect of the mathematical training of System 1. On his blog[16], Fields medalist Terence Tao distinguishes three phases in mathematical learning, which he calls pre-rigorous, rigorous, and post-rigorous. In short, students first play with numbers and basic mathematical concepts without worrying about the validity of their algebraic manipulations. Then, with time and thanks to mathematical education, rigor is learned, which quickly turns into purism. Everything must then be justified in a formal language. Finally, the last post-rigorous phase is that of researchers, who spend most of their time combining heuristic arguments to have the main lines of the proofs of their theorems, as they break away from the formalism learned earlier. Crucially, the second phase seems to be a must to access the last one.

Tao's reasoning can be understood in terms of Systems 1 and 2. The pre-rigorous phase corresponds to a case where neither System 1 nor System 2 has learned rigor. The rigorous phase is initiated when System 2 discovers rigor and its importance. During this phase, System 2 will then educate System 1 to make it aware of the limits of its intuitions. But above all, System 1 will then learn to measure the credence it has in itself, to better determine whether it can trust itself or if it must invoke System 2. It's only after this hard learning that System 1 will be a perfect ally to System 2. System 1 will then disturb System 2 only when necessary. In particular, it would seem that becoming a good mathematician must

---

[16] *There's more to mathematics than rigour and proofs.* T Tao (2009).

rely on a competent System 1 - but that the rigorous phase is a necessary step to reach this state.

This being said, even in the post-rigorous phase, System 2 does not cease to educate System 1 to allow it to progress. This is surely what happened in the brains of all number theorists after Lemke Oliver and Soundararajan's discovery on the last digits of prime numbers. This discovery surprised number theorists' System 1, whose intuitions on prime numbers were based on the heuristic of the prime number theorem. This heuristic predicted that the last digits of the successive prime numbers were roughly independent. This heuristic was wrong in this case. Since then, number theorists have probably made an approximate Bayesian inference to reduce their credence in the applicability of the heuristic of the prime number theorem to consecutive prime numbers.

Daniel Kahneman's two systems of thought are certainly an interesting model for the *pragmatic Bayesian*. Of course, it's wrong. But "all models are wrong". Kahneman's model, however, seems useful - albeit chapter 17 suggests the presence of a third system responsible for the creative process.

In practice, especially to face *Big Data*, most useful algorithms are probably fast heuristics, maybe even sublinear algorithms. Nevertheless, it remains crucial to have slower and more accurate algorithms. Better yet, if we have some slow algorithms that we greatly trust, then we can use them as references to search for fast heuristics that fairly well approximate the slower algorithms.

If I had to bet (and as a good Bayesian, I like to bet!), I would say with great credence that this is what the artificial intelligences of the future will be like.

14.11   BAYESIAN APPROXIMATIONS

The *pragmatic Bayesian* will have to make many approximations on the phenomena that he studies. But the most important approximation he will have to make is that of the equation that allows him to learn: he must find heuristics to approximate Bayes' rule. We can roughly distinguish five approaches to this.

One approach is to consider only a limited number of candidate models. This small number can be large - one can choose to consider millions or billions of candidate models. However, for the exact calculation of Bayes' rule to be feasible, this number of models cannot be exponentially large nor infinite, as it would be for parameterized models. This approach can be typically complemented by the *multiplicative weights update* algorithm discussed above.

A second approach consists in calculating only a high-credence model, if not *the* most credible model. This is called maximum a posteriori (MAP), as we saw in chapter 12. In other words, MAP will look for a predictive theory T such that $\mathbb{P}[T|D]$ is maximized. What's great is that many algorithms to maximize these quantities have been proposed, such as gradient descent, *expectation-maximization* (EM), or *generative adversarial networks* (GANs). However, one shouldn't forget that MAP is quite a crude approximation of Bayes' rule. In particular, by selecting a single model, MAP will *overfit*.

A third approach is to ignore the partition function, that is, the denominator of the Bayes formula that requires the comparison of all imaginable models. In particular, therefore, the sum of the weights of the different models may not be equal to 1. In chapter 17, we'll see how, despite this, with algorithms like MCMC or contrastive divergence, it's still possible to make predictions.

The fourth approach I would like to mention is the strangest. It consists of allowing oneself to violate the laws of probability. This is, for example, what Samuel Rodriques proposed in 2014, by modifying the definition of conditional probabilities, and thus allowing the manipulation of probabilities that do not obey Bayes' rule[17]. In a similar genre, the *Sum of Squares* algorithm, which some researchers like Boaz Barak and David Steurer believe to be a candidate to be, in a sense, an *optimal* algorithm[18], introduces the notion of pseudo-probability. These pseudo-probabilities generalize those that we know well. But they can take negative values! This largesse seems to lead to results which have the advantage of being quick to calculate, and which, when properly interpreted, remain very useful. Still another approach has been proposed[19] by researchers at the Machine Intelligence Research Institute (MIRI). Their approach was to regard the market equilibrium of a polynomially large betting market composed of simulated computationally bounded traders as a probabilistic belief on the truth of logical phrases within a predefined axiomatic system.

Finally, a fifth and last approach consists in considering a limited set of probability laws whose manipulation can be done quickly, even though

---

[17] *Probability Theory without Bayes' Rule*. S Rodriques (2014).

[18] *Sum of Squares: An Optimal Algorithm?* ZettaBytes, EPFL. B Barak (2017).

[19] *Logical Induction*. S Garrabrant, T Benson-Tilsen, A Critch, N Soares & J Taylor (2016).

they do not nicely fit the data. This approach has several subvariants, including *Gaussian mixture models, variational Bayesian methods,* and *expectation-propagation.* Crucially, it's important to have a measure of similarity between probability distributions. It's not an easy task! Quantifying uncertainty properly is not intuitive. In fact, this is the subject of the next chapter.

FURTHER READING

Quantum Computing since Democritus. Cambridge University Press. S Aaronson (2013).

Deep Learning. MIT Press. I Goodfellow, Y Bengio & A Courville (2016).

Computing π(x): the Meissel, Lehmer, Lagarias, Miller, Odlyzko Method. Mathematics of Computation of the AMS. M Deléglise & J Rivat (1996).

Unexpected Biases in the Distribution of Consecutive Primes. PNAS. R Lemke Oliver & K Soundararajan (2016).

Computing Machinery and Intelligence. Mind. A Turing (1950).

Why Philosophers Should Care About Computational Complexity. S Aaronson (2011).

Nearly Optimal Sparse Fourier Transform. STOC. H Hassanieh, P Indyk, D Katabi & E Price (2012).

Logical Induction. MIRI. S Garrabrant, T Benson-Tilsen, A Critch, N Soares & J Taylor (2016).

Deep Learning Works in Practice. But Does it Work in Theory? R Guerraoui & LN Hoang (2018).

Mathematicians Discover Prime Conspiracy. Quanta Magazine. E Klarreich (2016).

There's More to Mathematics than Rigour and Proofs. T Tao (2009).

Primes Are like Weeds (PNT). Numberphile. J Grime (2013).

The Riemann Hypothesis. singingbanana. J Grime (2013).

Pi Is (Still) Wrong. Vi Hart (2011).

The Science of Thinking. Veritasium. D Muller (2017).

Faster than the Fast Fourier Transform. ZettaBytes, EPFL. M Kapralov (2017).

Why Computers Are Bad at Algebra. Infinite Series. K Houston Edwards (2017).

Alan Turing's Lost Radio Broadcast Rerecorded. singingbanana. J Grime (2017).

CHAPTER 15

Wish Me Luck

Absolute certainty is a privilege of uneducated minds and fanatics. It is, for scientific folk, an unattainable ideal.

Cassius J. Keyser (1862-1947)

Mild success can be explainable by skills and labor. Wild success is attributable to variance.

Nassim N. Taleb (1960-)

15.1 FIVETHIRTYEIGHT AND THE 2016 US ELECTION

Breaking news. Against all odds, on November 8, 2016, the candidate Donald Trump is elected President of the United States of America. And if it was a monumental surprise, it's because all the polls had predicted the defeat of Trump. The next day, my colleagues were quick to tease me about the failure of Bayesian models. In particular, Nate Silver and his team at FiveThirtyEight[1] gave only 28.6% chance for the Trump election, against the overwhelming 71.4% for Hilary Clinton. Bayesianism had failed.

Or maybe not. My bet is that many of those who have seen these figures have confused them with the results of an election; namely the percentages of votes received by the different candidates. But that's not what the figures of FiveThirtyEight measure. FiveThirtyEight's numbers are Bayesian credences about the identity of the future president.

Think of Alex and Billie. *A priori*, it's safe to bet that these two children are not both boys. As good Bayesians, you will associate a probability of 25% to the boy-boy case (assuming for simplicity that a

---

[1] *Who Will Win the Presidency?* Election Forecast. Five Thirty Eight (2016).

child has a one in two chance of being a boy). Imagine that you discover that the two children are indeed two boys. Is this really sufficient to reject the Bayesian approach?

I will even go so far as to claim that the prediction of FiveThirtyEight is not a failure, but a success of the Bayesian approach. While many experts were already anticipating Clinton's first presidential projects, a good interpretation of the results of FiveThirtyEight should have been regarded as a call for caution. In particular, a probability of 28.6% is not negligible. If an event with such a probability occurs, there is no reason to call it a general surprise. To have two boys is not to challenge the laws of probability.

Moreover, the *pure Bayesian* never judges the validity of a model in isolation - except maybe to point out, again and again, that "all models are wrong". The *pure Bayesian* will constantly judge the validity of a model relative to other models. However, in the case of the election of Donald Trump, most models univocally predicted his inevitable defeat. Compared to this poor competition, the Bayesian model of FiveThirtyEight has clearly shone. Moreover, the entire history of all predictions of the model should be considered. This includes both the 2016 election, but also the results per state and previous predictions - including when FiveThirtyEight stunned the world with the unprecedented success of its predictions.

Now, you might reply that the *pure Bayesian* sounds like a prophet who never exposes herself to rejection. Sciences, it's sometimes said, do not make approximate predictions that experiments will not be able to reject. That would be a bad analysis of the history of science. "People are looking for certainty. But there is no certainty", warns Richard Feynman. And there is one area in particular where the central role of probabilities has finally become consensual: quantum mechanics.

15.2 IS QUANTUM MECHANICS PROBABILISTIC?

The uncertainty of quantum mechanics is usually illustrated by Schrödinger's cat. In this thought experiment, a cat is trapped in a box, with a device that releases a poison in the event of disintegration of a radioactive atom. Quantum mechanics predicts that, as long as the box is closed, the cat is in a strange quantum superposition. The cat is both living *and* dead - which is mathematically distinct from the cat being living *or* dead.

But let's put these quantum quirks aside. Let's focus on a less strange aspect of quantum mechanics. When you open the box, the cat has a certain probability of being alive; and another probability of being dead. It becomes alive *or* dead. But above all, the quantum state of the cat after observation is unpredictable. It seems fundamentally random.

The physicist Erwin Schrödinger hated this conclusion. "I do not like it, and I'm sorry to have something to do with it," he wrote. He was not the only one. "God does not play dice," Albert Einstein added. To which Niels Bohr supposedly replied: "Einstein, stop telling God what to do."

In 1935, Einstein, Podolski, and Rosen published an astonishing article, in which they proved that quantum mechanics cannot be a local theory. In other words, they proved that quantum mechanics implies that two very distant particles can instantly influence each other. This is what Einstein famously dubbed "spooky action at distance". Einstein rejected this absurdity which seemed to violate the postulate of special relativity that the speed of light is a maximal speed. He concluded that quantum mechanics was incomplete, and in particular that the use of probabilities was a pure artefact of our ignorance.

However, in 1982, Alain Aspect and his collaborators showed experimentally the existence of this "spooky action at distance". That led Stephen Hawking to joke: "Not only does God play dice, he sometimes throws them where we cannot find them."

Today, a large proportion of quantum physicists adheres to the so-called *Copenhagen interpretation,* in reference to the Danish physicist Niels Bohr. According to this interpretation, the quantum state of the cat after opening the box is probabilistic. But not only in the epistemic sense. According to this interpretation, the unpredictability of the quantum state of the cat is not only a failure in our knowledge. It's a mechanism fundamentally encoded in the laws of the universe.

However, we are far from a scientific consensus on this interpretation[2]. There are many alternative explanations. One of them is Hugh Everett's *quantum multiverse* theory. However, Everett's proposal got rejected by Niels Bohr. Disgusted, Everett ended his physics career, generalized the use of Lagrange multipliers in optimization, and became a multimillionaire. Poor guy.

While Everett's multiverse might sound like nonsense, especially for some of Popper's followers, Bayesian Eliezer Yudkowsky defends it on the Less Wrong blog[3], notably by virtue of an algorithmic version of

---

[2] *Quantum Mechanics (an embarrassment).* Sixty Symbols. S Carroll (2013).

[3] *If Many-Worlds Had Come First.* Less Wrong. E Yudkowsky (2008).

Ockham's razor. Although its implications may exceed our limited imagination, Everett's interpretation is based on a single, very simple principle: what if the only law of the universe was Schrödinger's deterministic equation, the one that predicts the evolution of quantum states in the absence of observations?

This Everett proposal had everything to please whoever admires the elegance and simplicity of the equations of physics. Farewell to the appearance of a probabilistic phenomenon whose trigger coincides with the obscure and ambiguous notion of observation (which can very well be the observation of a quantum state by a machine and therefore has nothing to do with consciousness). According to Everett's interpretation, the uncertainty of the observation is in fact the result of a quantum entanglement between the objects that came into interaction at the moment of observation. In particular, we can only observe the quantum state of the universe with which we are entangled. It's sometimes said that one is thus stuck in one of the many universe branches created by the Schrödinger equation at the time of the entanglement [4].

In particular, therefore, the probabilistic phenomenon that occurs at the moment of observation is no longer intrinsic to the laws of the universe. This is an *epistemic uncertainty*, which is due to the fact that we belong to one and only one of the possible branches of the quantum multiverse. Granted, the consequences of Everett's postulate are abysmal. But as good Bayesians, these unobservable consequences do not matter when judging the credence of a theory. What matters is the *thought experiment term*, namely the capacity of the theory to explain the data *observed*, and the *prior*, which Solomonoff measured via the length of the shortest algorithmic description of the theory. Everett's interpretation equals any other interpretation in predictive terms. Admitting, however, a much shorter algorithmic description, it seems to deserve the credences of Bayesians.

It seems to me, however, that Yudkowsky's argument is not fully convincing, especially if we are to believe the Solomonoff formalism of chapter 7. In particular, to be predictive, it seems necessary to combine the Schrödinger equation with a description of the complex physical state of the multiverse. In particular, because of the possible interference between quantum branches, the description of alternative quantum universes seems necessary to allow predictions. However, this description

---

[4] *The Many Worlds of the Quantum Multiverse.* PBS Space Time. M O'Dowd (2016).

seems extremely expensive in memory space. That seems to make it hardly credible *a priori*...

As I am not an expert in such questions and therefore have only limited credence in my own reasoning, I prefer to confess the extent of my ignorance on this subject. Moreover, several other interpretations of quantum mechanics have been proposed, like the deterministic but non-local theory of De Broglie-Bohm[5], *quantum Bayesianism* (QBism), or any of the 10 other variants shown on the English Wikipedia page[6] in 2019.

Be that as it may, whatever your preferred interpretation of quantum mechanics, it's clear that probability theory plays a critical role in quantum mechanics. When it comes to predicting the outcome of the collision between two protons at CERN's Large Hadron Collider, the best description to date is to assign various probabilities to different possible outcomes. And if quantum mechanics is so successful despite the indeterminacy of its predictions, it's because the probabilities that it assigns to the different issues are remarkably in phase with the observed frequencies!

15.3   CHAOS THEORY

The inevitable use of probabilities seems far from being restricted to quantum mechanics and Nate Silver's political predictions. In particular, following the discovery of chaos, mathematicians have finally convinced themselves that, in many cases, it's impossible to do better. This is typically the case of meteorology. While the meteorological equations are fairly well known, despite the growing number of sensors of all kinds on Earth and in orbit, long-term weather predictions remain very unreliable. This inevitable uncertainty of meteorological predictions was predicted by the mathematician Edward Lorenz in the 1960s, when Lorenz laid the foundation for what has been called "chaos theory" since.

This theory makes the following observation: some simple dynamical systems are incredibly sensitive to imperceptible variations of the initial conditions. This is perfectly illustrated by the *double pendulum*, as opposed to the simple pendulum. While Galileo noted with amazement the incredible regularity of the simple pendulum, whose oscillation period is almost independent of the amplitude of oscillation, I was lucky

---

[5] *Do we have to accept Quantum weirdness? Broglie Bohm Pilot Wave Theory Explained.* Looking Glass Universe. M Yoganathan (2017).

[6] *Interpretations of quantum mechanics.* Wikipedia (2019).

to be invited by the YouTuber Dr. Nozman to contemplate the amazing unpredictability of the double pendulum[7].

This device is incredibly simple: attach a pendulum to a pendulum. Put the double pendulum in the vertical instable position. And let go of it. If the double pendulum is sufficiently well oiled, then you can be sure that the trajectory it portrays is unique in the history of the universe. Even if you wanted to repeat it, you could not get that trajectory again, or even a trajectory that would only be partially similar. Indeed, an imperceptible variation in the initial condition will completely upset the behavior of the double pendulum, after only a handful of oscillations.

Since the discovery of Lorenz, mathematicians have discovered that chaos was far from being the exception. It even seems to be the norm. The real world is chaotic, and small imperceptible fluctuations can have dramatic consequences shortly afterwards. This is commonly known as the *butterfly effect*, which answers in the affirmative Philip Merilees' rhetorical question: "Does the flap of a butterfly's wings in Brazil set off a tornado in Texas?" Of course, this does not mean that the butterfly is the sole culprit of the tornado. What this means is that no medium-term weather prediction can be fully reliable unless you measure all the motions of all the butterflies on Earth.

But unpredictability is not only a property of complex systems with potentially very complex states.

15.4 UNPREDICTABLE DETERMINISTIC AUTOMATA

Different mathematicians and computer scientists have turned to *automata* to study the emergence of complexity in simple dynamical systems. An automaton is a virtual universe, whose physical state evolves according to a discrete time. At each time step, a new physical state is calculated from the previous state, according to an often very basic rule.

Among the classic examples of spectacular automata are the *Wolfram automata*, which form an infinite one-dimensional universe, composed of cells next to each other. These cells can be turned on or off. At the initial time, only one cell is on, and all others are off. At each time step, each cell turns on or off, depending on the status of neighboring cells and rules of interaction between neighboring cells. Surprisingly, Wolfram's simulations show that simple rules can lead to unpredictable phenomena.

---

[7] *Cet objet est chaotique! (double pendule)*. Dr. Nozman. LN Hoang and G O'livry (2017).

This is also particularly the case of Wolfram's rule 30, which draws surprising fractal figures.

Better yet, the great mathematician John Conway proposed simple rules for a 2-dimensional automaton called the *game of life*. Again, this is a set of cells, but now arranged as an infinite grid. The cells turn on or off depending on the state of the neighboring cells and according to simple rules. Nevertheless, these simple rules have been proven Turing-complete. In other words, any calculation made by a machine can be simulated by the game of Conway's life. In particular, if we accept Church-Turing's thesis, then our entire universe would be a mere calculation, which could therefore be entirely simulated by an enormous grid of the game of life (whose monstrous size would likely be at least of the order of the googol, that is, 10^{100}).

A last example of a highly studied automaton is *Langton's ant*. An ant is placed on a grid whose cells are on or off. If the cell that the ant is on is off, then the ant turns right and moves one cell forward. Otherwise, it turns left and advances one cell forward. In both cases, when leaving its box, the ant reverses the state of the box. If the box was on, it goes off, and vice versa. Now, switch off all the cells. Put the ant somewhere on the grid, with its head facing upwards. And launch the simulation. Not surprisingly, the first 500 movements of the ant are relatively symmetrical and structured. However, these apparent symmetries seem completely destroyed after a few thousand movements. The movements of the ant then seem random.

But that's not the strangest thing. After 10,000 movements, the ant suddenly begins to follow a regular and quasi-periodic trajectory which makes it deviate in the diagonal left-up direction. This diagonal walk then continues to infinity. It is called the *Langton ant highway*. Surprisingly, it's still an open problem to determine what initial conditions will eventually make the ant get on such a highway.

These automata show that simple rules can easily lead to phenomena that seem unpredictable. For the *pure Bayesian*, however, these rules are in fact completely predictable. Just compute the simulations to determine them! However, for ordinary people, phenomena such as the Langton ant highway are, in practice, very unpredictable. This is because they seem to require a large number of calculations to be predicted. We then speak of *emergence*.

15.5 THERMODYNAMICS

The Langton ant highway seems to *emerge* from the basic laws governing the movement of Langton's ant. In the same way, it seems that, under certain conditions of temperature and pressure, the fundamental laws of molecular interactions lead to the *emergence* of the equations of the dynamics of the fluids. Similarly, in 1872, Ludwig Boltzmann derived the second law of thermodynamics from the atomic hypothesis. In particular, Boltzmann showed that the irreversibility of time was an emergent property. Let's dwell on this point.

One of Boltzmann's key insights was to connect the atomic hypothesis to the notion of entropy. To understand this, we must begin with a trivial observation: mixing warm and cold creates lukewarm. And most importantly, the opposite is not true. Pour yourself a glass of water. It's unlikely that the right side of the glass will freeze while the left side boils. In other words, energy tends to become homogenously spread. It usually does not focus at one point.

This observation may seem trivial. But it took a genius, physicist Rudolf Clausius, to take it seriously. Clausius succeeded in mathematizing this principle by introducing a physical quantity which he called *entropy*. To say that mixing hot and cold creates lukewarm amounts to saying that the entropy of a closed system increases. It's this second assertion that Clausius promoted to the rank of *second law of thermodynamics*.

However, Clausius' entropy remained obscure and misunderstood. Boltzmann's genius was to define entropy from the atomic hypothesis, thereby laying the first bricks of *statistical physics*. Boltzmann was so proud of his definition that he had it inscribed on his grave: $S = k \ln W$. What does this equation mean? First note that our thermodynamical measuring devices do not measure the positions and velocities of the 10^{26} particles that surround us. They cannot. After all, it would take millions of zettabytes of data storage to do this. In contrast, the thermodynamic quantities that we measure, such as pressure, temperature, volume, or mass, are magnitudes that summarize the behavior of a very large number of particles. We speak of *macroscopic* quantities, as opposed to the *microscopic* quantities which correspond directly to the particles.

Boltzmann's genius was to note that entropy exactly quantifies *microscopic uncertainties*, once the macroscopic quantities are known. More precisely, Boltzmann showed that the entropy S studied by Clausius was none other than the logarithm of the number W of microscopic states that are coherent with the macroscopic magnitudes, up to a multiplicative factor k that we now call the *Boltzmann constant*. Entropy was actually a quantification of the microscopic uncertainty that remains, once macroscopic measurements have been made. And Clausius' second law of thermodynamics could be rephrased as the claim that microscopic uncertainties will always increase.

15.6   SHANNON'S ENTROPY

Quantifying uncertainty may seem unnecessary, even absurd. Yet there is a context where managing uncertainty played a major role in the history of humanity: the decoding of Nazi codes. During World War II, Englishman Alan Turing and American Claude Shannon met to discuss their decoding advances. It seems that Shannon and Turing did not talk much about cryptography, though[8]. Yet both understood the importance of quantifying uncertainty. In particular, Turing introduced *banburismus* calculations during the war to infer a credence in the fact that different messages were encrypted with the same configuration of the Enigma machine[9]. Shannon went further.

In 1948, Shannon published one of the most influential articles in the history of mankind. Its title was *A Mathematical Theory of Communication*. This sublime article proposed to model the messages sent by a source by a probability law. In Bayesian terms, this amounts to considering a *prejudice* on the likely messages that this source will emit. A Nazi soldier typically has a good chance to insert *"Heil Hitler"* somewhere in his message, to use German words like "genau" or to only transmit the German translation of "nothing to declare". Nazi messages felt random; but they did not feel arbitrary.

Shannon's first stroke of genius was to define the amount of information in a message as its rarity with respect to some Bayesian belief. For example, the word "Lê" contains a lot of information in Europe. It identifies me almost surely. This is because this word is very rarely used in Europe. It's this scarcity that allows it to communicate a lot of information.

On the contrary, this same word in Hanoi contains little information. It only makes reference to the few thousand or even hundreds of thousands of Vietnamese whose name it is. The fact that this word is very

---

[8] *A Mind at Play: How Claude Shannon Invented the Information Age.* Simon & Schuster. J Soni & R Goodman (2017).

[9] *Maths from the talk "Alan Turing and the Enigma Machine".* J Grime (2013).

common implies that it only conveys a small amount of information. In other words, the information of a message can only be measured with respect to a context. More exactly, it depends on a Bayesian belief on the probability of messages. *Without context, the wrong probabilities get tested.*

Shannon's second stroke of genius was to use the logarithm to quantify the amount of information in a message. Why the logarithm? This is so that the amount of information of two independent messages is the sum of the amounts of information of each message. However, the probability of two independent messages is the product of the probabilities of the two messages. For the product to become a sum, we have to use the mathematical object that translates the products into sums. As we saw in chapter 11, this object is the logarithm.

More precisely, Shannon defines the information amount of a message m whose probability is $p(m)$ by $h(m) = \log_2(1/p(m))$. In other words, the information amount of a message m is the exponent $h(m)$ such that $p(m) = 1/2^{h(m)}$. The messages with very low probabilities will then have a large amount of information $h(m)$. Finally, Shannon deduced from this formula the expected amount of information H from a source of information. This is the average amount of information $h(m)$ that this source emits. In other words, Shannon derived the fundamental equation:

$$H = \mathbb{E}_m[h(m)] = \sum_m p(m) \log_2(1/p(m)).$$

Shannon first wanted to call H the expected information of the source, or the uncertainty function. But he ended up following the advice John von Neumann gave him: "You should call it entropy, for two reasons. In the first place your uncertainty function has been used in statistical mechanics under that name, so it already has a name. In the second place, and more important, no one really knows what entropy really is, so in a debate you will always have the advantage."

But is Shannon's entropy really the same as Boltzmann's entropy? Yes. It's actually a generalization of it. To understand why, note that, given macroscopic measurements, we have a *prior* on the various plausible microscopic states. However, Boltzmann showed that each of the W microscopic states consistent with the macroscopic measurements was equiprobable at thermodynamic equilibrium. Therefore each microscopic state has a $1/W$ probability of occurring. By replacing $p(m)$ with $1/W$ in the Shannon equation, we deduce that, in the case where W microscopic states are equiprobable, the Shannon entropy of a thermodynamic system is then $H = \log_2(W)$. Adjusting the units to make them compatible with the International Systems of Units then forces adding a multiplicative coefficient k.

Shannon had indeed generalized Boltzmann's entropy.

15.7   SHANNON'S OPTIMAL COMPRESSION

Shannon's third stroke of genius was to understand what entropy really measures. As strange as it may seem, entropy measures the optimal compression of messages. In other words, it measures the minimum number of bits needed to store a message on a hard disk, or the minimum time to send this message through a cable with limited throughput. Indeed, Shannon proved that it will be impossible to do better than the fundamental limit calculated by his entropy.

To understand the connection between Shannon's entropy and compression, let's look at the board game "Guess who?" In this game, each player must guess the person that the other player chose out of a set of possible persons. To do this, each player must ask a binary question of the following form: is your person a man? Does your person wear glasses? Does your person have long hair? Each player asks a question, one after the other. The first to guess the character of the other wins.

In his 1948 article, Shannon proved that if there are n possible persons and if the opponent randomly chooses one of the n persons, then, on average, at least $\log_2(n)$ binary questions are needed to determine the other player's person. Better still, let's say that the opponent always chooses a male rather than a female person, and prefers individuals with glasses. In other words, suppose one player has a justified Bayesian belief in what the opponent might have chosen. Then, Shannon's article proves that the average number of needed questions will be at least Shannon's entropy.

Better still, Shannon's entropy corresponds to an idealized case where the sequence of answers provided by the opponent determines the optimal encoding of the opponent's chosen person. Specifically, the encoding proposed by Shannon is to label the character by a sequence of 0s and 1s, where 0 corresponds to no, and 1 to yes. Thus, if the opponent answers yes and then no to the first two binary questions asked, then the optimal encoding starts with 1 and then 0. The opponent's person is then identified with a sequence of 0s and 1s, which Shannon called *binary digits*, or *bits* to make it shorter[10].

More generally, Shannon proved that all communication was reduced to a sequence of 0s and 1s, and that communication had everything to gain from being digitized. Today, this observation may seem trivial. But this was not the case in Shannon's days, as many computers still relied on analogic technology. Thanks to his 1948 article, Shannon started the digital era.

15.8 SHANNON'S REDUNDANCY

Shannon's fourth stroke of genius was to show how to communicate through imperfect channels. In practice, when an electrical message is sent between A and B, this signal can easily be disturbed by various interferences. A 1 may have turned into 0, and vice versa. To address this case, Shannon had the idea of introducing a Bayesian belief about the disturbances that a message may have received. Shannon then proved that, provided that the Bayesian belief is justified, the imperfect channel is then equivalent to a perfect channel whose flow is equal to the flow of the imperfect channel minus a sort of entropy of imperfections of the channel. In particular, any message could still be communicated via an imperfect channel, provided we add sufficient redundancy. Shannon even quantified the necessary redundancy: it must be equal, roughly speaking, to the entropy of the disturbances experienced by the messages that pass through the imperfect channel[11].

All this may seem very obscure. However, redundancy is a phenomenon that is very familiar to us, even if we don't always realize it. When talking with your friends in a noisy bar, you are unlikely to hear everything that is said. Nevertheless, it's usually not necessary to hear *everything* that is said to understand what is being said. Indeed, a large part of the words of the English language play a minor role in the meaning of sentences. Remove unimportant words from a sentence. You see easy guess I try say.

French language is particularly redundant. This is why French translations are often longer than their English versions. Interestingly, this also explains why French people can afford to speak faster than English people. The flow of actual information remains about the same. Indeed,

---

[10]*Entropy as a Fundamental Compression Limit.* ZettaBytes, EPFL. R Urbanke (2017).

[11]*Shannon's Optimal Communication.* ZettaBytes, EPFL. R Urbanke (2017).

while French people say more syllables than English speakers, the French syllables contain more redundancies, and therefore less information, than the English syllables.

Today, all of Shannon's concepts, including bits, entropy, channel capacity, or redundancy, have become central tools of information technology. But their applications far transcend the world of technology. We find them, of course, in statistical physics, to study the evolution of gases. But also in linguistics to understand the evolution of language, and even in (exo)biology to detect intelligent life, whose communications very likely possess a redundancy structure similar to that of communications between adult humans, or between adult dolphins. It's typically this redundancy of our language that allows us to finish each other's _____. And it would probably be the same for the sentences of extraterrestrial intelligent lives, since these intelligent lives will probably also have to communicate through imperfect channels[12].

15.9   THE KULLBACK-LEIBLER DIVERGENCE

But that's not all ! One of the applications of Shannon's concepts was to finally propose a way of judging the validity of probabilistic predictions like that of FiveThirtyEight. Recall that FiveThirtyEight had assigned a probability of 28.6% to the Trump election. If such a probability were to be encoded by Shannon, it would take $\log_2(1/0.286) \approx 1.8$ bit. Well, we can consider that this is the number of points lost by the predictive model of FiveThirtyEight. More generally, we can count the total number of points lost by a predictive probabilistic model by adding the values $\log_2(1/p(m))$, where the $p(m)$'s are the probabilities that the model assigns to different events m that occurred.

Why is it a great idea to measure the performance of probabilistic predictions using a Shannon approach? This is because Shannon proved that if the world were really probabilistic according to a law q, then the predictive model that would minimize the (expected) sum of lost points would be the one that predicts the $p = q$ law. In other words, with this quantification of uncertainty, predicting uncertainty is not detrimental. In particular, when there is a need to be uncertain, as is the case in chaotic systems for example, the best prediction is then a probabilistic prediction. This is the kind of predictions a Bayesian will be making.

---

[12] *The History of SETI (Search for Extraterrestrial Intelligence)* . Art of the Problem (2014).

In particular, since the number of points lost is minimized by the probabilistic prediction q, we can compare the performance of a prediction p with respect to the optimal prediction q. For that, we can calculate the differences of expected scores:

$$D_{KL}(q||p) = \mathbb{E}_{m \leftarrow q} \left[\log_2 \frac{1}{p(m)} - \log_2 \frac{1}{q(m)} \right]$$

$$= \sum_m q(m) \log_2 \frac{q(m)}{p(m)}.$$

Introduced by Solomon Kullback and Richard Leibler in 1951, this quantity is now known as the *KL divergence*[13]. It calculates the prediction error of p with respect to the optimal prediction q, and is always greater than or equal to 0. This is a measure of how prediction p diverges from the optimal prediction.

In many ways, the KL divergence is an excellent quantification of the accuracy of a probabilistic prediction. Or at least it's a much better approach than the intuitive and naive approach of rejecting the prediction of FiveThirtyEight, only because it attributed a less-than-half probability to the event that occurred. Too often, our reaction is to oversimplify probabilistic models and derive a deterministic prediction. Too often, we want to determine who is right and who is wrong, and omit those whose opinion was justifiably balanced. Too often, we ignore the caution that probabilistic models and *Bayesianism* call for.

15.10 PROPER SCORING RULES

Sadly, probabilitistic predictions are often mocked for being unfalsifiable. They seem designed to be indecisive to avoid rejection. Yet, a prediction that minimizes the KL divergence need not be indecisive. If one alternative is far more likely than any other, the optimal prediction is to grant it a very large credence. In other words, the KL divergence makes it perfectly possible to discriminate predictions that do not take a definite position because they do not know the (however very predictable) problem they face, from predictions that do not take a definite position because they know that the problem is fundamentally difficult and full of uncertainties.

---

[13] Although physicists will prefer to calculate it with the natural logarithm.

Unfortunately, in practice, the KL divergence is scarcely used to judge the validity of our predictions. In general, because we often want to (financially or socially) reward those who have made the right deterministic prediction, we are often urged to make deterministic predictions, and to ignore the desirable amount of prudence. The lack of adequate quantification of uncertainty leads us to overfit and to fall into the trap of survivor bias. Experts without clear positions are usually not the most audible ones. They are not the experts that are broadcast in prime time. And they are not the ones that are massively relayed by social networks.

Worse, perhaps because we tend to remember our triumphs more than our failures, we are constantly far too sure of ourselves and of our deterministic predictions. We are often way too overconfident. Changing the way the validity of our predictions is judged seems to be a necessary first step to combat this overconfidence; and applying Bayes' rule would be the ideal way to determine the adequate degree of confidence to have.

To change our poor habits, it may be a good idea to score exams using KL-like scores. Typically, it may be pedagogically instructive to turn classical multiple choice questions into *Bayesian* multiple choice questions. Instead of selecting one single answer, students would be asked to quantify their credences p in the different choices proposed as answers. A student would then lose an amount of points that is equal to $\log_2(1/p)$. Interestingly, because of the properties of this scoring function, the student will then be incentivized to answer his true credences. We say that this Shannon scoring is a *proper scoring rule*[14].

Such *proper scoring rules* may help train the habit to adequately judge the extent of his ignorance by using the language of probability theory. Moreover, they may be regarded as more *fair* than classical multiple choice questions, as they reduce the chances of a lucky guess. In fact, a student that hesitates between answers A and B, but manages to discard answer C, may be rewarded, as instead of expressing probabilities 1/3 for each question, he may be answering 1/2 for answers A and B, and a zero probability for answer C. Thereby, the student would maximize his expected score, where the expectation is computed with

---

[14]More formally, suppose there are multiple answers i, to which the student assigned probabilities p_i's. And suppose i^* is the right answer. The student's Shannon score would be $\log_2(1/p_{i^*})$. Another perhaps more fair and simpler proper scoring rule is the quadratic scoring rule, equal to $2p_{i^*} - ||p||^2 = 2p_{i^*} - \sum p_i^2$. Still another proper scoring rule is the spherical scoring rule, equal to $p_i/||p||$. See more: *Bayesian examination*. LessWrong. LN Hoang (2019).

respect to his Bayesian credences. More importantly, this will motivate him to compute and be fluent in manipulating Bayesian credences[15].

15.11   WASSERSTEIN'S METRIC

However, KL divergence is not the only possible measure of the performance of probabilistic predictions. In fact, there are a large number of them[16]. However, unlike the KL divergence, many of these measures do not lend themselves well to the algorithmic search for a good prediction p.

Nevertheless, there is another interesting way to measure the distance between different probabilities p and q which does lend itself well to algorithmic calculations. This is the *Wasserstein metric*, also known as gunner distance or the optimal transport solution. The advantage of this metric is that it takes into account how much an event m is different from m', which the KL divergence does not. So, if you predict that the color of a spot will be yellow, and if I predict it will be blue and if it's actually only yellowish, then the KL divergence will say that we have both been wrong. However, you were closer to the correct answer than I. The Wasserstein metric gives a precise meaning to the intuition that you were more right than I.

Let's examine. Imagine that you have to predict where coffee stains will appear on the floor of a warehouse in the coming year. Since you are a good Bayesian, your prediction is probabilistic. It says that some locations are more probable than others. Let's make it simple. Say you use a thousand grains of black sand so that the black sand density matches your probabilistic prediction. In other words, you put a lot of grains of sand where you think that coffee stains are very likely to appear.

The year passes and the employees have been particularly clumsy. They spilled a thousand coffee stains. For each spot, place a grain of yellow sand. On the floor of the warehouse there is now a thousand grains of yellow sand, and a thousand grains of black sand.

You now have a thousand ants. Each ant must take a grain of black sand and bring it next to a grain of yellow sand, so that in the end, the piles of black sand and yellow sand are identical. Each ant can walk around very quickly and easily, except when carrying a grain of sand, in which case it moves extremely slowly. Also, assume that ants are

---

[15]The brilliant idea of Bayesian multiple choice questions has been suggested to me by Oliver Bailleux.

[16]*Statistical distance*. Wikipedia (2018).

coordinating to solve the problem of black sand transport as quickly as possible. Well, the average time these ants will put in is precisely the Wasserstein distance between your probabilistic prediction and the empirical data.

Note that, while the Wasserstein metric is generally well suited for computer calculations, it presupposes a measure of similarity between the data. In our case, we could measure the similarity between two grains of sand using the distance between these grains of sand. However, in many cases, identifying a relevant measure of similarity is a major difficulty.

15.12   GENERATIVE ADVERSARIAL NETWORKS (GANS)

Draw several cats on a sheet of paper. Which of the cats you have drawn looks more like a "real" cat? And how could you quantify the similarity between your drawings and authentic cats? How to define a similarity metric between images?

Such questions actually have many applications beyond *Pictionary*. Determining the similarity between complex objects such as text, sound, or images[17] has become one of today's biggest challenges for the *pragmatic Bayesian*. Indeed, the really interesting data are in fact such highly complex objects, such as Laplace's written texts, electrocardiograms, or images of tardigrades under a microscope.

Take the example of cosmology. Nowadays, the data in this area are essentially photographs of the sky in all wavelengths, from radio waves to gamma rays, including microwaves, infrared, visible light, ultraviolet rays, and X-rays. The question that fascinates astrophysicists is then determining the credible parameters θ of cosmological models, given the photographs of the sky ■. This is a typical case of application of Bayes' rule!

$$\mathbb{P}[\theta|■] = \frac{\mathbb{P}[■|\theta]\mathbb{P}[\theta]}{\mathbb{P}[■|\theta]\mathbb{P}[\theta] + \sum_{\omega \neq \theta} \mathbb{P}[■|\omega]\mathbb{P}[\omega]}.$$

This calculation is easy for the *pure Bayesian*. However, it is inaccessible to the *pragmatic Bayesian*. As often, the denominator is much too long to calculate. But that's not all. Because cosmological models are very complex, even the *thought experiment terms* $\mathbb{P}[■|\theta]$ require

---

[17]Or better still, between laws of probability on texts, sounds, and images.

unrealistic computational times. In fact, especially because they often are Bayesian networks[18], these cosmological models are mainly designed to allow for simulations. In other words, all that can be done in a reasonable time is to generate probable galaxy images, assuming that the parameters of the cosmological models are θ. We speak of *generative models*[19].

Therefore, the *pragmatic Bayesian* will have to rely on methods that bypass the direct calculation of the *thought experience terms*. We talk about *inferences without likelihood*, or *likelihood-free methods*. Several methods to do so have been proposed, like *Approximate Bayesian Computation* (ABC) and *parametric Bayesian Indirect Likelihood* (pBIL). But since 2014 and the work of Ian Goodfellow and his collaborators[20], a method without likelihood in particular seems to have won much of the pragmatic credences of many researchers. This method is known as *Generative Adversarial Networks* (GANs).

Intuitively, the idea of GANs is to replace the human who would guess words at *Pictionary* by an algorithm called *adversary*, or *teacher*. This *adversary* will be in charge of measuring the similarity between images generated by a model and authentic images. To train the *adversary*, toss a coin. If the coin lands heads, we choose an authentic image from a large bank of authentic images. Otherwise, the generative model is asked to generate an image. Then, no matter how the image was obtained, we ask the *adversary* to yield a Bayesian belief p in the authenticity of the image. Intuitively, if the generative model is good, then the *adversary* should be confused. The *adversary* would then assign a probability of $p = 1/2$ regardless of the image.

Critically, to make sure the *adversary* has incentives to respond with a Bayesian belief, Goodfellow and his collaborators propose to assign a cost to the *adversary* equal to $\log_2(1/p)$ if the image is authentic, and $\log_2(1/(1 - p))$. In other words, the scoring of the adversary's performance is done by the KL divergence[21].

---

[18]We'll talk about this again in chapter 17.

[19]In other words, a generative model is a law of probability on a complex set like the set of images, and that one can only study via model simulations. Or, to put it another way, it's a model designed to be sampled. This will be discussed further in chapter 17.

[20]*Generative Adversarial Nets*. NIPS. I Goodfellow, J Pouget-Abadie, M Mirza, B Xu, D Warde-Farley, S Ozair, A Courville & Y Bengio (2014).

[21]Variants of Goodfellow's GANs use other measures of distance. For instance, using the Wasserstein metric yields Wasserstein GANs, also known as WGANs.

In addition, the generative model may then seek to determine parameters for which the data it generates will be most similar to the authentic data. And crucially, thanks to the existence and competences of the *adversary*, there will be a precise meaning to this. It will be about generating data so that the *adversary's* choice of p will always be as close as possible to $p = 1/2$, whether the *adversary* is given authentic or model-generated images. In fact, the *adversary's* malus points can be used to measure the performance of the generative model[22].

At the time of writing, GANs are particularly on the rise. Their amazing performance continues to improve, by taking advantage of *deep learning*. Indeed, in most frameworks, the generative model and the *adversary* of the 2019 GANs are deep (convolutional) neural networks[23]. Thanks to their approximation of Bayes' rule and the usual tools for neural network learning, GANs have acquired the ability to generate photographs whose indistinguishability with authentic photographs is undetectable[24]. In many ways, the spectacular performance of these modern artificial intelligences was allowed by the advent of new ways to measure the credibility of probabilistic predictions.

In fact, to conclude this chapter, I'd like to emphasize once again the importance of taking uncertainties into account in judging the quality of predictions. To underestimate the role of chance in the analysis of phenomena is unfortunately too common. As advised by the editors of the 2019 American Statistician special edition[25], it's essential to learn to constantly reason with uncertainty, rather than trying to impose deterministic predictions on contexts that are fundamentally unpredictable. In our complex universe, no knowledge can be certain, whether because of a physical nature of reality, our lack of empirical data, the presence of chaotic phenomena, or our bounded computational power.

---

[22]The generative model and the *adversary* then play a zero-sum game!

[23]The model is typically a deep neural network $\mathcal{G}$ that is fed with variables z drawn from some "simple" distribution. The generative model then outputs model-generated data $\mathcal{G}(z)$. The *adversary* is another deep neural network $\mathcal{D}$. The advantage of this structure is that the backpropagation algorithm can be applied to train both $\mathcal{D}$ and $\mathcal{G}$! This backpropagation algorithm identifies why $\mathcal{D}$ chose $p \neq 1/2$, and feeds back this information to the generative model $\mathcal{G}$ to tell it how to improve. Thereby, $\mathcal{D}$ acts more like a *teacher* than an *adversary*.

[24]*How an AI Cat-and-Mouse Game 'Generates Believable Fake Photos'*. New York Times. C Metz & K Collins (2018).

[25]The editorial's first advice was to "accept uncertainty". See: *Moving to a World Beyond "p < 0.05"*. The American Statistician. R Wasserstein, A Schirm & N Lazar (2019).

Almost all prediction questions have no simple, unambiguous answer. They require caution. And this prudence can be justified only from the moment when we finally accept that many events are best explained by luck. To get there, the judgment of our models and our predictions must absolutely make use of the tools to quantify uncertainty. *The quantification of uncertainty is too important to be left to chance.*

FURTHER READING

The Signal and the Noise: Why So Many Predictions Fail–but Some Don't. Penguin Books. N Silver (2015).

The Black Swan: The Impact of the Highly Improbable. Random House. NN Taleb (2007).

A Mind at Play: How Claude Shannon Invented the Information Age. Simon & Schuster. J Soni & R Goodman (2017).

A Mathematical Theory of Communication. The Bell System Technical Journal. C Shannon (1948).

On Information and Sufficiency. Annals of Mathematical Statistics. S Kullback & R Leibler (1951).

Maths from the Talk "Alan Turing and the Enigma Machine". J Grime (2013).

Generative Adversarial Nets. NIPS. I Goodfellow, J Pouget-Abadie, M Mirza, B Xu, D Warde-Farley, S Ozair, A Courville & Y Bengio (2014).

Moving to a World Beyond "p < 0.05". The American Statistician. R Wasserstein, A Schirm & N Lazar (2019).

Who Will Win the Presidency? Election Forecast. FiveThirtyEight (2016).

If Many-Worlds Had Come First. Less Wrong. E Yudkowsky (2008).

Shannon's Information Theory. Science4All. LN Hoang (2013).

Entropy and the Second Law of Thermodynamics. Science4All. LN Hoang (2013).

Statistical Distance. Wikipedia (2018).

How an A.I. 'Cat-and-Mouse Game' Generates Believable Fake Photos. New York Times. C Metz & K Collins (2018).

Bayesian Examination. LessWrong. LN Hoang (2019).

Inventing Game of Life. Numberphile. J Conway (2014).

Quantum Mechanics (an Embarrassment). Sixty Symbols. S Carroll (2013).

The Many Worlds of the Quantum Multiverse. PBS Space Time. M O'Dowd (2016).

Do We Have to Accept Quantum Weirdness? De Broglie Bohm Pilot Wave Theory Explained. Looking Glass Universe. M Yoganathan (2017).

The Higgs Mechanism Explained. PBS Space Time. M O'Dowd (2015).

What Is Information Entropy? (Shannon's Formula). Art of the Problem (2013).

Entropy as a Fundamental Compression Limit. ZettaBytes, EPFL. R Urbanke (2017).

Shannon's Optimal Communication. ZettaBytes, EPFL. R Urbanke (2017).

CHAPTER 16

Down Memory Lane

Our memory is a more perfect world than the universe: it gives back life to those who no longer exist.

Guy de Maupassant (1850-1893)

Don't trust your memory. If someone asks you if you can remember something, say no.

Julia Shaw (1987-)

16.1 THE VALUE OF DATA

The last few years have seen the emergence of a new *buzzword*: *Big Data*. However, for many experts in the field, this *buzzword* does not reveal any phase change in our economies. *Big Data* has been here for long. And it's been growing at an exponential rate for a while. We often talk about *information science*, and the word *information* is nothing other than a synonym for the word *data*. However, while it may not fundamentally change the nature of information technology, the notion of *Big Data* seems to adequately emphasize the growing role of data in our industries, economies, and societies.

I was particularly struck by it when moving to Switzerland. The search for a home, the collecting of rental documents, and the contract between landlord and tenant are dreadful administrative paths. To search the market for apartment rentals in Lausanne, I had to register on many mailing lists, Facebook groups, and rental agency websites. This research would correspond to hours and days of work and cognitive efforts, which I could have easily delegated to a powerful recommendation algorithm.

Worse still, I then had to collect numerous documents from many different entities such as my employer, the prosecution office, and the bank. I also had to take care of the transfer of these documents to the real estate agency. Somehow, it still is necessary today for these documents to come through me, even though these documents could have been simply queried from the databases of the interested entities - up to an agreement that I could have provided electronically. The real estate agency (or even the owner) could have directly asked my employer, the prosecution office, and the bank to access the information they wanted to have.

I am baffled by the inefficiency of the current administrative procedures. I am also baffled by the cost it causes. For example, the printing of an administrative document certifying that I had initiated a work permit procedure cost me a whopping 20 Swiss francs (about $20 USD), even though the cost of access to the database, which would generate a more reliable certificate, does not exceed one cent.

Finally, the drafting and signing of the contract still requires paper support today. I had to fill out a pile of forms, one after the other, and sign a lot of documents that I only skimmed through. What a loss of time! Why is it still necessary to write and rewrite surnames, first names, and dates of birth on so many documents? The irony of it all is that, meanwhile, I was working on ZettaBytes videos with the EPFL on electronic signature[1] and *Blockchain*[2], which are modern solutions for the development, signature and management of data and contracts.

The day the data management systems will be fully digitized - this day is coming! - all these procedures will be triggered in a few clicks. Many industries have already arrived at this day. The purchase of music, the reading of books, and the viewing of videos have already gone digital. The companies behind the digitization of these services have flourished. In fact, they have become giant multinational firms. They are Apple, Amazon, and Netflix. These companies, like others in Silicon Valley, including Google, Facebook, and Twitter, have exploited the value of data before anyone else. They are now investing millions, if not billions, in the collection, management and analysis of their data. It's the emergence of their business models that triggered the *Big Data* hype.

16.2 THE DELUGE OF DATA

We are entering an intriguing phase of the information world, as the amount of data produced is starting to exceed our capacity to analyze

---

[1] *Bitcoin*. ZettaBytes, EPFL. R Guerraoui & J Hamza (2017).
[2] *The Blockchain*. ZettaBytes, EPFL. R Guerraoui & J Hamza (2017).

it, and even to store it. This is exemplified by the Large Hadron Collider (LHC) at CERN. This immense underground structure, which is tens of kilometers wide, produces billions of proton collisions every second. The amount of information produced is so huge that most of this information is immediately discarded. Initial filters make it possible to select only the data that seem to be of interest to physics. But despite this stringent sorting, the stored data still represents petabytes. It fills entire rooms of calculating machines.

What CERN is forced to do could be the future of all companies facing *Big Data*. Indeed, the amount of *data* grows even faster than the storage space! For the moment, it's still possible to make several backups of our data. But that will not be the case for long. Soon, it will be necessary to throw away a large fraction of our data, perhaps even the overwhelming majority of the data collected. At a time when, moreover, we are seeing the proliferation of sensors of all kinds and the emergence of the *Internet of Things* (IoT), the question of the choice of data backup is about to become an unprecedented question in the history of information.

In addition to posing serious storage problems, *Big Data* also poses big computation time problems. Imagine looking for something among millions of billions of pieces of data. Even at the rate of one billion processed data per second - the speed of the microprocessors in your computers - it would take several days to get there! To overcome these problems of storage and responsiveness of computers, many data scientists already prefer to imagine how to solve problems without the storage of raw data.

16.3 THE TOILET PROBLEM

Imagine yourself at a festival. You have to go to the bathroom and 300 toilets are lined up in a very large driveway. However, these toilets are all horribly dirty. You aim to use the best of all toilets. However, there is a line behind you, so that once you have closed the door of a toilet and advanced to test the next toilet, the one you have just visited will no longer be accessible; someone else will have accessed it. In other words, you have to decide once and for all if you are going to use the toilet the moment you visit it. How can you maximize your chances of using the cleanest toilet?

This problem has become a classic of mathematics[3]. Introduced by Martin Gardner in 1960, it is known under various names, including the *secretary problem*, the googol game, the marriage problem, the fussy suitor problem, or the sultan's dowry problem. All these formulations are equivalent and are based on the following dilemma: data arrives sequentially, and you must make a decision before all of it arrives, because the opportunities of the past disappear over time.

If this problem is so famous, it's also because it has a pretty counterintuitive response. Indeed, if you follow the optimal strategy, then the probability that you choose the best toilet is about[4] 37%. This optimal strategy is to first visit about 37% of the toilets and to reject them all. Then, you are to accept the first toilet that is better than all toilets visited so far. Amazingly, this simple strategy guarantees a 37% probability of finding the best toilet, even when considering thousands, millions, or googolplexes of toilets[5].

In particular, what's remarkable is that this optimal strategy requires almost no memory. The only data that must be memorized is the cleanness of the best toilet visited so far. All other toilets can be safely forgotten.

That being said, I strongly discourage the use of this strategy in practice. Indeed, it maximizes the probability of finding the best toilet; but this says nothing of the cases where the strategy fails. In fact, this strategy will often make you reject all the toilets! You would then have to accept the very last toilet. This potentially catastrophic case even has a 37% chance of happening!

16.4   EFFICIENT *BIG DATA* PROCESSING

The problem of toilets has inspired a very large number of variants. Such variants model a wide range of problems, especially since the advent of the web. Each variant then gives rise to different resolution algorithms. Nevertheless, intriguingly, the solutions developed by such resolution

---

[3] *Mathematical Way to Choose a Toilet*. Numberphile. R Symonds (2014).

[4] In fact, in the limit of infinitely many toilets, it is equal to $1/e$, where $e \approx 2.718$ is Euler's constant.

[5] For a small number of toilets, the optimal exploration period is lower, and the probability of finding the best toilet larger. For 2 or 3 toilets, the probability of the best strategy is $1/2$. Between 4 and 10 toilets, this probability decreases from 46% to 40%. For 26 toilets, it drops to 38%. For 150 toilets, it drops to 37%, but still stays above 36.78%!

algorithms seem to rely on a few general principles. These principles seem key to decision making given a deluge of data and a limited memory.

One of the variants places more emphasis on the dilemma between accepting a present opportunity or hoping for better. In practice, this setting is particularly relevant and interesting when there are many opportunities with diverse and varied characteristics; and when it's possible to accept several of these opportunities. The study of these problems is often called *online optimization*. Applications include many resource allocation problems, from selling concert tickets to advertising on the Internet. Many of the proposed solutions rely on the quantification of the value of resources from past data, either through the study of the constraints of the problem[6] or via variants of the quasi-Bayesian algorithm by *multiplicative weights update* (see chapter 11).

Other variants of the problem focus more on the uncertainty that one has about the probabilistic law of the data. In other words, these variants are not so much about the uncertainty of prejudices; they are about the uncertainty *on* our prejudices. In particular, this gives rise to the *exploitation versus exploration* dilemma, and is typically captured by the *online learning* framework. Exploration involves doing more potentially expensive tests to collect more data and improve our Bayesian credences, while exploitation is about making optimal decisions, given our present credences. Thompson's Bayesian sampling algorithm makes it possible to propose a solution to this dilemma. Applications include decisions to either pursue or terminate medical tests prematurely, or test new products on the web (also known as A/B tests).

Finally, a third class of variants puts more emphasis on limitations in computation time - but allows itself to re-examine past data. Let's say you've come back from your vacations in Nigeria. You now want to pick a top 10 of the 2,000 vacation photos you've taken. A difficulty is added to the problem: we must now be careful to avoid duplicates. A variant of the toilet algorithm that addresses this problem and its variants is the *greedy algorithm*. This algorithm will first retain only the best picture it sees. Then it will retain the best photo that's compatible with the selected photo, then the best that's compatible with the two photos selected, and so on. Unfortunately, this approach is generally suboptimal, because it does not anticipate the synergy of a retained photo with future photos. The greedy algorithm is said to be *myopic*. Nevertheless, it has been shown that this myopia, especially when combined with some

---

[6]This corresponds to variables called *dual* and Lagrange multipliers.

appropriate randomization[7], guarantees that the greedy approach will be a good heuristic, in some rigorous sense[8].

In recent years, algorithms based on the estimates of resource values, on the adequate management of uncertainty on knowledge, or on the greedy approach have gained a lot of interest, especially among web giants. Their research groups have been actively engaging with the theoretical study of the properties of such algorithms. The advent of the *Big Data* will probably only accelerate this trend.

16.5   THE KALMAN FILTER

However, the variants around the toilet problem generally assume that the future will be similar to the past[9]. This explains why it's often not necessary to have a lot of memory to solve them. To address dynamical contexts, learning must anticipate change. One classical model to achieve this feat is the *Kalman filter*, named so after Rudolf Kalman, and its generalizations like *hidden Markov models*.

Imagine that your car wants to know its position and speed. Like any good Bayesian, it begins by describing the extent of its ignorance. It can model this ignorance by using mean estimates plus errors. By virtue in particular of the central limit theorem, Kalman supposed that these errors were distributed according to a Gaussian law. And this hypothesis was very useful, especially because the sum of Gaussian variables is a Gaussian variable, and because the product of Gaussian densities is a Gaussian density. But don't worry. I'll spare you the details of the calculations here.

What's unfortunate to observe is that at each time step, new uncertainties are added. The car may have accelerated. And any measure of this acceleration will be accompanied by error. This mistake will be added to the uncertainty that we already had. But then, the more time passes, the more errors are added, and the less we can be sure of the position and speed of the car.

To reduce these uncertainties, we can take advantage of measurements provided by other devices like GPS. However, these measures

---

[7]This randomization may consist, for example, of randomly drawing a photo among the 10 best photos that are compatible with those already selected.

[8]*Weakly Submodular Maximization Beyond Cardinality Constraints: Does Randomization Help Greedy?* L. Chen, M. Feldman and A. Karbasi (2017).

[9]Data are typically assumed to be independent and identically distributed (iid), according to frequentist tradition! Or sometimes, they are assumed to be randomly shuffled.

are uncertain. Nevertheless, by combining the positions and velocities derived from the different measurements of the different devices[10], the Kalman filter allows us to deduce the Gaussian law which describes the position and the speed of the car *a posteriori*. In other words, the Kalman filter allows to derive the posterior belief, given the data provided by the measuring devices. Thanks to this additional data, the posterior will be more precise.

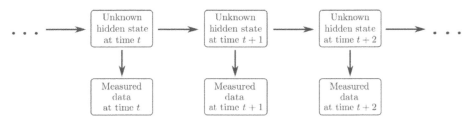

Figure 16.1 The Kalman filter is a special case of *hidden Markov models*. These models describe a causal evolution of a hidden internal state. At any given moment, this internal state *causes* observable data.

This Kalman filter is today used in a very large number of areas. It's of course found in many problems of navigation or trajectory control, but also in signal processing, econometrics, battery charge estimation, computer interfaces, particle detectors, computer vision, tomography, seismology, medical monitoring, and weather prediction. And crucially, it's just Bayes' rule, applied under several useful hypotheses, such as Gaussian uncertainty, linearity between the variables, and the causal structure between the variables.

This causal structure is called a *hidden Markov model* and is represented graphically in Figure 16.1. It has variables called *hidden*, which take unknown values at every moment. In the case of the car, these hidden variables are the position and speed of the car. At each moment, moreover, we suppose that observable variables are deduced from the hidden variables. In the case of the car, these observable variables are the data provided by the measuring devices. Intuitively, the position and the speed of the car *cause* the measured data. Bayes' rule then allows to determine the probable hidden variables at a given instant, given the probable hidden variables of the previous instant and the data measured by the apparatus at the present time.

---

[10] And provided that these measurements are functions of the position and the speed of the car.

Like many other graphical models with hidden variables, as will be discussed in more detail in the next chapter, *hidden Markov models* have a very large range of applications; these include, of course, the applications of the Kalman filter that they generalize. Most importantly, these models can react in real time to massive data, without ever requiring more memory. Indeed, only the description of the probability law of the hidden variable must be kept in memory. However, especially in the context of the Gaussian laws of the Kalman filter, or in a case where the number of possible values of the hidden variables is small, the description of the probability law of the hidden variable remains succinct. This makes *hidden Markov models* promising to face the deluge of *Big Data*.

16.6   OUR BRAINS FACED WITH *BIG DATA*

All these algorithms may seem very far from your everyday life. You may think that you are able to remember the information you are exposed to. This would be a serious mistake.

In particular, what we do not always realize is the monstrous amount of information that floods our cerebral cortex at any instant. Our eyes, our ears, our noses, our touch, our thermoception, and our numerous other senses send us nearly a gigabyte of information every second[11]. If we combine all the data collected over the course of decades, we can see that the total amount of data to which our senses have exposed us is of the order of exabytes, that is, millions of terabytes! This is astronomically huge.

Like modern information technologies, our brain cannot store all the data it is exposed to. In fact, it should not want to store everything; most of this data is irrelevant. The brain must forget almost all of the information it is given.

This is typically the case of visual data processing. According to neurologist Marcus Raichle[12], at every second and with minimal energy consumption, our visual cortex has the impressive ability to transform the gigabyte of visual raw data into a few kilobytes of relevant information[13].

More generally, the brain will try to only remember "big ideas", rather than the details of the raw data it is given. Typically, I bet you

---

[11] *How much bandwidth does each human sense consume relatively speaking?* Quora. R Rapplean (2012).

[12] *Two views of brain function.* Trends in Cognitive Sciences. M Raichle (2010).

[13] Raichle believes that, every second, about 10^{10} bits are collected by our retinas, but only 10^4 bits reach layer IV of the V1 region of the visual cortex.

are unable to recite any of the sentences from the first 15 chapters of this book (except maybe "all models are wrong"!). Yet you remember, I hope, that these chapters discuss Bayes' rule, its logical foundations, its application to the induction problem, its history, Solomonoff's demon, its use in game theory, and its connections with Darwinian evolution. You have, I hope, retained abstract representations of various fundamental concepts around what you have read, even though you may remember almost none of the sentences you have read.

Far from being a weakness, this ability to retain only a compressed representation of what our senses expose us to is a strength of our brains. We are usually able to remember things that matter to us, and to completely forget things we do not care about.

16.7   REMOVING TRAUMATIC SOUVENIRS

Well, that's not quite true. Unfortunately, we sometimes forget things we would like to remember, or remember things we would like to forget. Veterans returning from war and exposed to atrocious images often suffer psychological traumas that continue to haunt them. On another scale, we all have our own phobias. And we would often like to just forget them.

What if forgetting our phobias was possible? Surprisingly, psychologist Merel Kindt has shown that this could be done! In a 2016 NOVA PBS documentary entitled *Memory Hackers*, Kindt is shown to heal the arachnophobia of one of her many patients. To do this, she forces the patient to look at a tarantula. This activates fear in the patient's brain. The patient is then given propranolol. Propranolol is a molecule that fixes itself between neurons and thereby interrupts the communication between these neurons. According to Kindt, this disturbs the fear memory, which makes fear disappear. And this works! The next day, the subject begins to caress the tarantula as if it were a harmless hamster.

Research in this area is still in its infancy. But the work of Kindt and other researchers could offer new solutions for dealing with drug addiction and even post-traumatic stress disorder. They also point to a counterintuitive fact that is now well established: memories are stored in synaptic connections.

Contrary to what neurologists at the beginning of the twentieth century may have thought, long-term memory does not correspond to information bits stored inside dedicated neurons. When we think of a memory, torrents of neural activations propagate through our cerebral cortex. It's in the furrow of these torrents that our memory is engraved. This memory is stored in the way neurons are connected, not in the physical state of the neurons themselves. In particular, far from being located at a specific place in the brain, the information associated with a memory is diffused through the connectivity of the neural network that constitutes our brain.

Unfortunately, accessing such information can be incredibly hard. In fact, the relationship between the brain's active memory and its long-term memory is somewhat similar to the relationship between your web browser and the web. To access a web page of interest to you, you need to find the URL link for your web page. Once this link is known, the browser will easily explore the Internet to retrieve the information needed to display the page you wanted to access. However, if you have lost this link, finding it can be a horrible difficulty - just as remembering a memory, which you know you have but cannot retrieve, can be a frustrating experience.

A corollary of this observation is the fact that remembering a memory reactivates the torrent of neural activations associated with it, which allows better engraving of the memory. Remembering helps remembering. However, from one recollection to another, like a stream, the torrent of neural activations varies slightly. Even worse, propranolol can greatly affect the trajectory of this torrent, for example by dissociating it from the part of the brain associated with fear. It seems that this forced deviation of the torrent of neural activations is what causes a modification of the memory, and the disappearance of a phobia.

16.8   FALSE MEMORY

If a memory can be changed for the better, it can also be so for the worse. This is what psychology researchers like Julia Shaw discovered over and over again. Shaw in particular performed a formidable experiment that will undoubtedly greatly question the functioning of our judicial system. This experiment shows how easy it is to make people believe that they have committed a crime they did not commit!

To begin, Shaw tells her subjects that the purpose of the experiment is to study childhood memories. Shaw then tells them an invented story, and states that the story was reported by the parents of the subjects. The story is that of a crime committed by the subjects. This story is made up. But it's not absurd. In particular, it includes elements of the subjects' past, such as the place or names of the people involved in the story. The subjects begin by retorting that they do not have such memories. However, Shaw then asks them to relax and subtly appeals to their imagination. Shaw adds that the other subjects were able to remember their childhood memories, just to encourage the subjects in this effort. In doing so, Shaw activates the neurons of subjects associated with childhood memories; the mechanism useful for the consolidation (and deformation) of memories has been triggered! Shaw then asks her subjects to continue to think about it, but not to talk to anyone about it.

A week later, Shaw interviews her subjects again. The subjects then begin to tell a story that seems plausible to them, with hesitant words. Two weeks later, the hesitant words disappear and the subjects are more sure of themselves. Surprisingly, in 70% of cases, Shaw's subjects got persuaded to have committed a crime! Shaw's experiment was so overwhelming for the subjects that she had to stop it prematurely.

The experiments of Kindt and Shaw show that one cannot trust one's memories. Our memories are imprecise, vague, but mostly readjusted, reconstructed, and readapted every time we ponder them. And the worst thing is that we are largely unaware of these reconstructions! Yet, it's on the memories of judges and jurors of the facts reported by the memories of witnesses, victims, and defendants that our judicial system rests. The memories of judges, jurors, witnesses, victims, and defendants are then readjusted, reconstructed, and readapted by the persuasive and emotionally engaging speeches of the various lawyers. What credence can we really attach to such memories? Undoubtedly much less than the credence that is given to them in practice. In fact, based on the study of hundreds of cases, psychologist Elizabeth Loftus showed that, in three quarters of the cases, those exonerated by DNA tests after being found guilty were convicted because of false visual memory.

Visual testimony cannot be trusted, even when the witnesses seem confident. In fact, it's incredibly easy to get witnesses to be convinced that they have seen something very wrong. In 1999, in a well-known experiment, Daniel Simons and Christopher Chabris proposed to count the number of passes made by a set of basketball players. Their video is available on YouTube[14] and I urge you to go and check it out before continuing reading.

When asked about the number of passes, the subjects often gave the correct answer. Then Simons and Chabris asked the subjects if they

---

[14] *Selective attention test.* D Simons (2010).

had seen the gorilla cross the basketball court. The subjects answered no. Worse, many said they were convinced that no gorilla had crossed the field. *I would have seen it*, they said. But the replay is univocal. A gorilla did cross the playground. It even took the time to take a few dancing steps! Subjects with a demanding cognitive task were affected by *inattention blindness*. Even worse, they are unaware of their inattention. In this experiment as in many others[15], they are blatantly overconfident in their perception abilities.

Other cognitive biases make our case worse. As previously discussed, psychologist Jonathan Haidt says that we are always trying to justify our first intuitions with the help of our reason. This means that we are fine with adjusting our memories. Worse still, Derek Muller showed that exposing students to videos that only rigorously explain a counterintuitive scientific phenomenon tends to increase their confidence in their erroneous non-scientific beliefs. This explains why the scientific community so often rejects any testimony and personal experience, including when these testimonies are told with great conviction.

These major defects of our memories show the limits of our brains, and invite us to greatly diminish our credences in our memories, and the convictions that are based on these memories. We live in a world full of uncertainties. These uncertainties are about the future, of course, but also about the past - and even the present. This led Descartes to methodically doubt all that he could doubt, before concluding that only one thing was undoubtable: the fact that he was thinking. *Cogito ergo sum* (I think therefore I am), he said.

However, taken at its extreme, this radical approach also leads to doubt science and the well-established consensus of the scientific community such as the effectiveness of vaccines or the anthropogenic origin of global warming. For the *pure Bayesian*, the fallacy of extreme skeptics is their quest for undoubtable truths. "All models are wrong." Knowledge, therefore, does not consist in characterizing real facts or undoubtedly true theories. It's more about calculating degrees of credence in different facts, theories, and memories. To know is to determine the appropriate levels of uncertainty. To this end, the appropriate language does not seem to be the language of formal logic, of true and false; the appropriate language is the language of probabilities - and the essential reasoning tool is Bayes' rule.

---

[15] *Test Your Awareness: Whodunnit?* dothetest (2008).

16.9   BAYES TO THE RESCUE OF MEMORY

We have seen that one of the advantages of the human brain was its ability to compress a very large amount of raw data into a handful of ideas. Inspired by this, researchers in artificial intelligence invented the architecture of the *auto-encoder.*

The role of an auto-encoder is to condense large amounts of information, such as a high-definition image or a whole movie, to retain only the core content. In other words, these neural networks aim to do exactly what you have been asked to do at school: write summaries. And to test the quality of a summary, we also ask the auto-encoder to decompress the summary, and imagine what high definition image or other raw data is likely to be associated with this summary.

This problem typically falls under the umbrella of Bayes' rule. Indeed, to reconstitute the raw data from a summary corresponds to determining which raw data are likely given the summary. This is a typical case of Bayes' rule! It's a problem of determining the probable causes given observations. We have

$$\mathbb{P}[\text{data}|\text{summary}] = \frac{\mathbb{P}[\text{summary}|\text{data}]\mathbb{P}[\text{data}]}{\mathbb{P}[\text{summary}]}.$$

We will see in the next chapter how, by relying on this formula and introducing unusual summaries, researchers in artificial intelligence have equipped their machines with creative powers.

What's more, beside the decoding of encoded memory, Bayes' rule can also be used to determine an adequate coding of memories. After all, by virtue of Ockham's razor, a credible theory given the data is almost always much shorter than the data itself! In particular, by keeping only a few *a posteriori* very likely models in memory, the memory can effectively summarize the massive data and thus optimize the management of the memory.

What I say here may seem abstract and reserved for artificial intelligences. Yet it's the crux of a theorist's frustration when confronted with an exhaustive collection of unrelated incompressible facts. This can be exemplified by Claude Shannon, who once said of chemistry that "[it had] always seemed a little boring" to him. "Too many isolated facts and too few general principles for my taste."

A Bayesian approach to knowledge will, on the contrary, seek models capable of summarizing a large number of isolated facts by a handful of key principles with the hope that keeping in mind the few general principles will suffice to infer, via Bayes' rule, much of the isolated facts. This greatly justifies the usefulness of Bayesianism for the study of the sciences of the past, like history, Darwinian evolution, or cosmology. According to Bayesianism, the purpose of these sciences is not so much to distinguish true from false. It would be more about constructing simple, structured, and easy-to-remember models, which are nevertheless capable of sufficiently explaining the multitude of observations made - though one can also make the case that these sciences also try to predict not-yet-observed vestiges.

16.10   SHORTER AND LONGER-TERM MEMORIES

Kindt and Shaw's studies are about long-term memory. It involves encoding information into the topological structure of the brain's neural network. The problem is that accessing this information can be difficult. It's necessary to test different propagations of possible signals through the network. This is what happens to us when we try to remember the lyrics of a well-known song. The first time we think about it, we often stutter after a few words, as if the flow of electrical signals in our brain were stopped, or deviated in the wrong direction. However, after a few attempts of repeating this flow, we sometimes finally find the right way, and find the lyrics of the song.

The same problem arises when searching for information in large databases. In fact, what made Google's fortune is, among other things, the organization of web data to accelerate the search for information. But this is not just a matter of organizing information. The support of the information also poses a tradeoff problem between the storage capacity and the speed of access to the stored information. Some media are remarkably fast, such as RAM, or, better yet, microprocessor registers, but of limited capacity. Others are slow, but of enormous capacity, such as hard disks, tape storage used at CERN, or DNA storage[16].

Another data storage limit is actually the finiteness of the speed of light derived from Einstein's relativity! If the information is stored remotely, as is increasingly the case with *cloud computing*[17], then there is a necessary delay to access the information, called *latency*. Remember that the speed of light is about 10^5 kilometers per second. Suppose we access information located in Stockholm from Lausanne, a few

---

[16] *All of Humanity's Data in a Backpack.* ZettaBytes, EPFL. C Dessimoz (2017).
[17] Cloud Computing. ZettaBytes, EPFL. B Falsafi (2017).

thousand kilometers away. Then the information will necessarily take at least $10^3/10^5 \approx 10^{-2}$ seconds, that is, a few tens of milliseconds.

This may seem fast. However, this already prevents several round-trips without testing the user's patience. In particular, if the user's query must first be sent to a New York server, which will request it from a Tokyo server that will then relay the information to Stockholm, and if the information from Stockholm must then travel through Tokyo and New York before reaching the user in Lausanne, then the application of the user will necessarily lack responsiveness.

For most users, such a delay is a small discomfort. But in the world of finance, especially in high-frequency trading, such delays can make all the difference and represent millions of dollars. As a result, many companies have discovered that they can make millions by communicating between New York and Chicago via microwaves rather than fiber optics. After all, via the optical fiber, the speed of the signal is only the speed of light *in the fiber*, which is slightly lower than its speed in air or vacuum. Some precious milliseconds can be gained by communicating via microwaves.

Slow access to data from high-capacity media and unavoidable communication delays has led computer engineers to use caches. A cache is a fast memory close to the computing unit. In your computer, it is the RAM, or, better yet, registers L1, L2, and L3, whose memory contains only a few bytes, but whose access time is in microseconds, if not nanoseconds.

The cache principle has other applications. For example, when you browse the web with Mozilla Firefox, Google Chrome, Safari, or other web browsers, your browser will cache the information that you often download in the computer's memory. As a result, this information will be directly accessible. You will not have to suffer the delays of communications that Internet requires.

In the same way, some researchers think that our brains treat in a different way short-term memory and long-term memory. While long-term memory is engraved in the connectivity of the neural network, short-term memory could be rather controlled by neurotransmitters - even if it seems to be a hypothesis that is not sufficiently experimentally substantiated and which may not deserve all our credences.

16.11 RECURRENT NEURAL NETWORKS

In any case, artificial neural network research has opted for a third alternative: neural propagation loops. The architecture that includes such loops is called the *recurrent neural network* (RNN). It allows making part of the information of the immediate past available to process the present data. This has become the state of the art for processing data whose structure is fundamentally sequential, such as the text of a book or the sound of a speech.

The trick of recurrent neural networks belongs to a larger family of internal state algorithms. This family includes the *Kalman filters* and the *hidden Markov models* that we discussed previously. The objective of the internal state is to synthesize information from the immediate past, in order to better understand the last data analyzed. However, this raises the question of the rate at which the internal state evolves, in comparison with the reading of the data.

This is something that I noticed in my own way of learning mathematics, as well as in my students' way of learning. When reading a document for the first time (like the book you are reading!), it's useful not to dwell too much on the details in order not to lose the key points of the text. This is because too-slow reading, or, similarly, too many changes of the internal state of our short-term memory, desynchronizes what the text wants us to take away from where our minds go. Thus it may be useful to have a first quick read of a text, then to re-read it in depth, before doing a quick re-reading. Each reading will bring something new, because each reading will be associated with a different dynamic of the internal state of our short-term memory.

So far, we talked about linear reading of data. But the way we read and listen is not quite linear. Indeed, when a sentence is very convoluted, we tend to read it several times before resuming the linear reading. Likewise, we sometimes understand others only once their sentence is over. This is particularly the case in Sanskrit, Hindi, and Japanese where the verb is at the end of the sentence. In the same vein, in mathematical papers, the explanation of the calculations is often provided after the calculation. Or similarly, the punchline of a joke, of an advertisement, or of a movie is often given at the end. And it's this punchline that gives sense to the joke, advertising, or film.

To take advantage of both past and future data to understand the present, another neural architecture has been proposed. These are so-called *bidirectional* recurrent neural networks. To be able to include the future, these networks also require a response time. It's a safe bet that similar structures are at work in the human brain. This would explain our ability to understand jokes with a delay that sometimes causes the mockery of others.

Finally, a recent advance in artificial intelligence is the introduction of a neural device that forces us to forget. This is the LSTM architecture for *long-short-term memory*. In addition to the usual neural loops of recurrent neural networks, LSTMs have an additional loop that, if activated, will force any signal from neural loops to disappear. That would explain why, after a break in a discussion, it can sometimes be very difficult to retrieve the thing we were discussing just before the break.

16.12   ATTENTION MECHANISMS

Over the last two years, still another way to store, manage, and access memory in neural networks is being investigated with remarkable success, especially in natural language processing. Instead of managing efficiently a small amount of data in memory as LSTMs and variants do, this alternative approach proposes to work with a large memory, but to exploit it efficiently by optimizing the extraction of relevant information in the memory. This information extraction is known as an *attention mechanism*.

Typically, one task at which attention mechanisms have been very good at is *image captioning*. Interestingly, we can better understand what attention-based neural networks caption by reverse-engineering their attention mechanisms. For instance, a neural network might be captioning "a man is lifting dumbbells", even though the image featured a man drinking tea. By reverse-engineering the neural network's attention mechanism, it might be possible to figure out that the neural network was actually paying attention to the man's muscular biceps. This would help debug the neural network. It probably thought that "lifting dumbbells" meant "having large biceps", because in all images with "a man lifting dumbbells" that it was exposed to, the man had large biceps[18].

In 2017, a group of Google scholars found out that you could beat all previous recurrent neural networks by simply combining several attention mechanisms within a neural network[19]. They called their neural network the *transformer*. Two years later, this was generalized into a structure called a *universal transformer*, which essentially allowed the neural network to choose by itself the amount of attention it should pay to different sorts of data[20].

[18] *AI Language Models & Transformers*. Computerphile. R Miles (2019).
[19] *Attention is all you need*. NeurIPS. A Vaswani et al. (2017).
[20] Universal Transformers. ICLR. M Dehghan et al. (2019).

In 2019, OpenAI also stunned the world by introducing an attention-based neural network called gpt-2, which was able to write impressively coherent page-long stories[21]. While they did not release their models, it seems that attention allowed to greatly scale their language models. In particular, the large memory of attention-based models allowed them to much better handle *longer-term dependencies*, while the parallelizability of attention mechanisms allowed them to distribute their computations over a large number of processing units[22].

Nowadays, universal transformers and their variants seem to be becoming the state of the art of artificial intelligence for sequential data processing. But I gladly acknowledge my lack of understanding of this success and my lack of expertise in predicting which memory management systems will make up the future.

16.13 WHAT SHOULD BE TAUGHT AND LEARNED?

Like human memory, the memory of artificial neural networks, whether it is encoded in the connectivity of the neural network or in the signals that propagate in loops of this network, has no guarantee of reliability. The computers around us dominate us largely for at least two algorithmic tasks: computational speed and reliable storage of information. It's a safe bet that tomorrow's artificial intelligences will be able to take advantage of these literally superhuman capabilities, probably by combining some aspects of neural networks with other much more reliable and optimized algorithms for narrow tasks.

New technologies may have already changed our cerebral cortex. Our addiction to our phones, Google, and Wikipedia seems to affect the management of our memory. It's not necessarily a bad thing. The preceding generations did not hesitate to glorify those who could recite verses of Shakespeare, to give the date of the abolition of slavery, or to state Maxwell's equations by heart. Memorizing facts used to be necessary to have access to them. However, many teachers complain about the excess of knowledge required to succeed in school, as opposed to a real understanding of the underlying concepts. Some argue that in the modern world, know-how is much more important than knowledge. Thus, according to this logic, it would be much better to know how to find or retrieve information than to simply know it.

---

[21] *Unicorn AI*. Computerphile. R Miles (2019).
[22] *Better Language Models and Their Implications*. OpenAI. A Radford et al. (2019).

Personally, it seems to me that there may be far too much knowledge and know-how to learn at school. Worse, knowledge is often taught as an absolute truth, and know-how as a solution *sine qua non*. But for the *pure Bayesian* and the *pragmatic Bayesian*, "all the models are wrong". Yet, this may not be the greatest issue.

Most importantly, it seems to me that the excess of knowledge and know-how is detrimental to the understanding of the most useful and credible notions and models to understand the world around us. This teaching too often ignores the reasons for the credibility of our models and the limits of their applicability. I'd be inclined to teach much less, and to restrict ourselves to what is important, counterintuitive, and instructive. I typically believe that cognitive biases, theoretical computer science, and moral philosophy should be taught, as opposed to, say, trigonometry or quantum mechanics.

Moreover, Bayes' rule seems to invite us to learn more by example, as opposed to through the memorization of theories - we'll see in the following chapters that our brains seem relatively Bayesian and that they are very quick to generalize from examples. In particular, it seems that the relevance of theories only comes into play once enough data are known and "readily available" for the *thought experience terms* of Bayes' rule to be readily estimable. Thus, it seems desirable to learn these theories only after having studied several examples where the *usefulness* of these theories will become obvious. This invites us, for example, to first introduce mathematics in the form of games, puzzles, or logical paradoxes that attract students' attention, before explaining to them that these are cases of more general theories - this is what I attempted to do throughout this book.

But, in my humble opinion, the most important thing to teach is epistemology. "Today having power means knowing what to ignore", Yuval Noah Harari writes. This requires questioning much more what *is* knowing and what is worth knowing. But importantly, epistemology should be taught along with statistics, without which it is inapplicable. Of course, given how much of an Bayesian extremist I am, I think in particular that Bayes' rule and its numerous counterintuitive consequences should form one of the pillars of education. It's time, I believe, to stop dogmatically accumulating supposedly true knowledge, and start teaching what knowledge is, how to acquire it, and how to distinguish credible theories from those that do not deserve our credit. Unfortunately, nowadays even the great scientists have a very incomplete understanding of epistemology; many ignore the very existence of Bayesianism.

FURTHER READING

The Memory Illusion: Remembering, Forgetting, and the Science of False Memory. Cornerstone Digital. J Shaw (2016).

Deep Learning. MIT Press. I Goodfellow, Y Bengio & A Courville (2016).

Two Views of Brain Function. Trends in Cognitive Sciences. M Raichle (2010).

Weakly Submodular Maximization Beyond Cardinality Constraints: Does Randomization Help Greedy? L Chen, M Feldman & A Karbasi (2017).

Attention Is All You Need. NeurIPS. A Vaswani et al. (2017).

Universal Transformers. ICLR. M Dehghan et al. (2019).

Memory Hackers. NOVA PBS (2016).

Mathematical Way to Choose a Toilet. Numberphile. R Symonds (2014).

When To Quit (According to Math). Up and Atom. J Tan-Holmes (2017).

Selective Attention Test. D Simons (2010).

Bitcoin. ZettaBytes, EPFL. R Guerraoui & J Hamza (2017).

The Blockchain. ZettaBytes, EPFL. R Guerraoui & J Hamza (2017).

Arm Up for the Big Data Deluge. ZettaBytes, EPFL. A Ailamaki (2017).

All of Humanity's Data in a Backpack. ZettaBytes, EPFL. C Dessimoz (2017).

Why Blockchain Is a Revolution. ZettaBytes, EPFL. EG Sirer (2018).

AI Language Models & Transformers. Computerphile. R Miles (2019).

Unicorn AI. Computerphile. R Miles (2019).

How Much Bandwidth Does Each Human Sense Consume Relatively Speaking? Quora. R Rapplean (2012).

The Secretary/Toilet Problem and Online Optimization. Science4All. LN Hoang (2015).

Better Language Models and Their Implications. OpenAI. A Radford et al. (2019).

CHAPTER 17

Let's Sleep on It

The art of doing mathematics consists in finding that special case which contains all the germs of generality.

David Hilbert (1862-1943)

Nothing in life is as important as you think it is when you are thinking about it.

Daniel Kahneman (1934-)

17.1 WHERE DO IDEAS COME FROM?

"Thought is only a gleam in the midst of a long night. But it is this gleam which is everything," wrote mathematician Henri Poincaré. He illustrated this gleam by the story of one of his great discoveries: "At the moment when I put my foot on the step, the idea came to me, without anything in my former thoughts seeming to have paved the way for it, that the transformations I had used to define the Fuchsian functions were identical with those of non-Euclidean geometry."

Poincaré adds a telling of another of his discoveries: "One day, as I was walking on the cliff, the idea came to me, again with the same characteristics of conciseness, suddenness, and immediate certainty, that arithmetical transformations of indefinite ternary quadratic forms are identical with those of non-Euclidian geometry."

These stories of Poincaré echo the experience of many mathematicians. In his book *Birth of a Theorem*, Cédric Villani writes that after working hard until 3 am to complete a gaping hole in a 150-page proof and after going to bed with despair, he woke up with the solution of the problem! The brains of these mathematicians seem capable of running in spite of them.

Personally, I have also experienced this kind of epiphany many times - even though my best ideas do not come close to Villani's or Poincaré's! I'll even go so far as to postulate with great credence that the mathematician's ability to make his or her unconscious continuously work is the primary reason for his or her ease with mathematical objects. For two years, Bayes' rule has never seemed far away. And it's often without warning that it whispered to me its secrets.

I can only suggest that you follow Poincaré's footsteps. If you really want to progress in mathematics, I'd advise you to become passionate about this discipline, to the point where your unconscious never leaves maths aside, even when you go to sleep. In the language of Kahneman's psychology, which we used in chapter 14, it's as if, in doing so, a System 3 was born. This System 3 somehow educates System 1, without System 2 being aware of it.

This, however, raises an intriguing question for the psychologist. What is this System 3? What happens in the brain of a mathematician so that it continues to progress without being aware of it? Do dreams help? And more generally, why do we dream? These are difficult questions. I do not claim that I'll provide reliable answers. Nevertheless, I'd like to present the hypothesis put forward by Francis Crick, who won the 1962 Nobel Prize in Medicine, and Graeme Mitchison. Why? Because it's based on an elegant Bayesian argument....

But before getting there, I'll dwell on the creative process that machines are now capable of.

17.2 CREATIVE ART BY ARTIFICIAL INTELLIGENCES

Recently machines have been able to compose music and draw paintings. As briefly suggested in the previous chapter, the key to this creative process seems once again to be closely related to Bayes' rule.

Indeed, in many models of *deep learning*, it's possible to excite some so-called *deep neurons*, to create a combination of abstract concepts in the neural network. Some of these neural networks will then be able to imagine which raw data could have led to the excitation of excited neurons. In other words, while neural networks usually seek to derive from the data the abstract concepts that summarize them, we can also ask them to guess the probable data, given the abstract concepts.

By then choosing usually unrelated abstract concepts, we can then make the neural network create raw data that are both unusual and relatively credible. This is the artistic process of machines. It's a safe bet that this process shares at least some similarities with the creative process of our brains.

In 2015, Google published images generated by its artificial intelligence *DeepDream*[1]. These images seem psychedelic. Clouds turn into fish, trees into temples, and tree leaves into birds. Better still, you can now ask another artificial intelligence, called *DeepArt*, to reinterpret your photographs with the style of some famous painter, be it Van Gogh, Picasso, or Kandinsky. My Twitter profile picture is, for instance, the result of a few seconds of free work by *DeepArt*.

One of the important steps in the creative processes of these artificial intelligences is the ability of the neural network to find credible data, given abstract summaries of these data. The network must be able to *sample*, according to the probability law $\mathbb{P}[\text{data}|\text{summary}]$. This corresponds to the probability of some raw data conditioned by excited abstract concepts. In this sense, it seems that *creativity is the sampling of contextual beliefs.*

But more importantly, sampling allows the creation of a large set of representative examples. It allows reasoning with examples, rather than reasoning with very complex probability distributions. This can be very useful. After all, the natural learning of us humans, including in mathematics, often seems to rely more on representative examples than on formal theory. Our cerebral apparatus seems much more optimized to infer rough rules from examples than to make its neural network coincide with a formal theory. This seems to be why sampling is essential for humans. Curiously, this seems crucial for machines too. But this may not be that surprising, as it's also what Bayes' rule says. We learn by inferring from data, not by memorizing formal theories.

To better understand, we need to discuss adequate models to represent such relationships between raw data and abstract concepts. Several architectures of *machine learning* have been proposed to do so. These architectures fall into two main categories (which can be combined in more complex models): these are *Bayesian networks*[2] and *Markov fields*. Let's discuss Bayesian networks first.

---

[1] *Deep Dream - A code example for visualizing Neural Networks.* Google Research Blog. A Mordvinster, C Olah & M Tyka (2015).

[2] Also called *Bayes nets* or *forward models*.

17.3   LATENT DIRICHLET ALLOCATION (LDA)

In addition to *Kalman filters* and *hidden Markov models*, one of the major successes of Bayesian networks is the latent Dirichlet allocation (LDA), invented in the early 2000s. LDA aims to classify texts by category. Typically, a computer could use LDA to store your emails in folders called "personal", "work", "holidays", and "spam". Better still, LDA is even able to detect combinations of categories, and say that some document is half-work half-vacation, or that it is two thirds-personal and one third-work.

To do this, LDA exploits the fundamental notion of Bayesian networks: *causality*. What's nice about the causality of Bayesian networks is that it fits our intuition. This makes Bayesian networks often relatively simple to interpret. It's by virtue of this correspondance between Bayesian networks and our intuitions that, as discussed in chapter 3, Neil and Berger advocate for their use in the judicial field.

Back to LDA. LDA postulates that every word in a document is obtained by the following causal process. First, a certain combination of categories is drawn randomly for the document[3]. Then, for each word of the document, a category of the combination of categories is drawn at random. Finally, the word is drawn randomly depending on the category drawn. This causal description of a writing process is depicted in Figure 17.1.

Figure 17.1   LDA is a typical example of a Bayesian network. It has a causal structure that aims to deduce the observed data from abstract concepts. The diagram above is a simplified representation of LDA.

LDA is simple. It's also very wrong. A document written by LDA would be a messy bag of *buzzwords*. This is clearly not how I wrote this book! But "all models are wrong, some are useful". While it's *wrong*, LDA is *useful*. This technique is used throughout the web and in bioinformatics to classify data into categories. And if this technique is so effective, it's because it can constantly improve with Bayesian inferences. Thus, every time a new document is presented, LDA is able to analyze this document to self-improve.

Better, LDA does not need to be told which category the document is supposed to belong to - even if this would help. We say that LDA is

---

[3]This draw is done precisely according to *Dirichlet's law*.

an *unsupervised* learning algorithm: give it plenty of documents without specifying their category, and it will improve anyways!

Even better, LDA also does not need to be given a list of the categories to be considered, nor even the number of categories. In particular, there is a variant of LDA, called hierarchical LDA, where the number of categories can be automatically increased if it seems that there are now documents of a new type. Hierarchical LDA is said to be a non-parametric method because its complexity can grow indefinitely.

17.4 THE CHINESE RESTAURANT

To achieve this feat, hierarchical LDA, like LDA, has a profoundly Bayesian touch. This means that, in particular, it's essential to have a *prior* on how the number of categories will have to increase according to the quantity of documents. The commonly used *prior* for LDA is called the *Chinese restaurant process.*

Imagine ourselves in a Chinese restaurant. At every moment, a new client arrives. The new customer will then assign a number to each of the $n-1$ customers already present, and the number n to an unoccupied table. The new customer then draws a random number between 1 and n, and goes to the table of the customer he drew the number of (or if he obtained n, he will sit alone). Hierarchical LDA assumes that each new document is, *a priori*, a kind of new customer, and that each category is a restaurant table[4].

This process is undoubtedly relevant for us in practice. If we try to organize our documents by category, it's reasonable to consider inventing a new category for the n-th document with probability $1/n$ only. In doing so, we guarantee that the number of categories will never become too large. Indeed, I let the mathematicians among you check that the number of categories will be logarithmic in the number of data - in particular, hierarchical LDA is perfectly adapted to *Big Data!*

What's amazing with hierarchical LDA is that a computer that applies it invents its own new concepts! Moreover, in practice, it often happens that the categories invented by hierarchical LDA, although relevant, cannot easily be interpreted by humans. The computer will have invented a concept totally absent from our vocabulary.

---

[4]The fact that this restaurant is Chinese is a mystery. But amusingly, there is a variation called the *Indian buffet process*, which allows for combinations of categories.

According to Sean Carroll's *poetic naturalism* described in chapter 13, we are forced to acknowledge that this concept *exists* (for LDA) because it's indeed useful to the computer. More pragmatically, what LDA shows above all is that the usefulness of abstract concepts does not require the physical existence of these concepts. To best explain data sets in practice, inventing these abstract concepts is an inescapable step. We'll come back to this fundamental remark in the next chapter.

For now, note how easy it is to sample data from a Bayesian network and deep concepts. Indeed, by definition of the Bayesian network, the way in which fictitious raw data is generated corresponds to a very precise causal process (even if it includes randomization). In the case of LDA, you can even easily ask LDA to generate a half-personal, half-work document. Admittedly, LDA is not sophisticated enough to produce a meaningful text, but the words that will be generated will have a good chance to indeed mix work and personal life.

17.5 MONTE CARLO SIMULATIONS

If sampling does not seem very successful in the case of LDA, its application in different settings can be dramatic. Take a box of needles and a large sheet of paper. Draw horizontal lines spaced at a distance equal to 4 times the length of a needle. Throw a very large number of needles. Count the proportion of needles that intersect a horizontal line. This proportion will be approximately the inverse of the fundamental constant τ of geometry, equal to the ratio of the circumference of a circle by its radius.

This is the *Buffon needle experiment*[5]. It probes the nature of a mathematical constant with the help of experiments - which is very different from scientific experiments that seek to discover properties of our universe! Strangely enough, randomness is essential to the experiment.

Other similar ways of estimating τ have been proposed. Draw a disk of radius $1/2$, and frame this circle within a square of side 1. The area of the disk will be then of $\tau/8$, while that of the square will be of 1. Then throw points (uniformly) randomly on the square. The proportion of points in the disk is then expected to be about the ratio of the disk area to the area of the square, that is, $\tau/8$. This gives us another way of experimentally estimating τ: just (randomly) throw a large number of points on the square, and multiply by 8 the proportion of points that fell in the disk. Such sampling methods are illustrated in Figure 17.2.

---

[5] *Pi and Buffon's Matches*. Numberphile. T Padilla (2012).

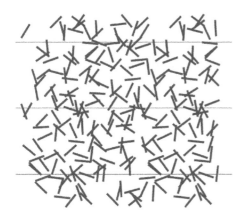 

Figure 17.2 These two images illustrate two methods of approximating τ by the Monte Carlo method. On the left is Buffon's needle. The fraction of needles that cut a horizontal line will be about $1/\tau$. On the right, the fraction of points in the square on the inside of the disk will be about $\tau/8$.

This experiment was conducted by Dianna Cowern and Derek Muller, better known as Physics Girl and Veritasium[6]. Cowern and Muller threw a large number of darts at a square target. However, after a first day of experiments, they found out that their throws were not uniformly distributed. They touched the center of the target more than the corners, which led to an overestimation of τ. The next day, they modified the experiment by drawing on the back of the target several disks circumscribed by squares. They finally got the excellent estimate $\tau \approx 6.28$.

Buffon, Cowern, and Muller's experiments are part of a larger family of experiments, the accuracy of which depends on the quality of the randomness. Such experiments are called *Monte Carlo methods*. These methods were formalized in the 1940s by Stanislaw Ulam and John von Neumann, who sought to calculate the probability of winning at a certain game of cards. While Ulam and von Neumann spent considerable time performing difficult combinatorial calculations, Ulam wondered if it would be simpler to repeat the card game a large number of times, and to estimate the theoretical probability of winning by using the empirical frequency. Von Neumann immediately understood the genius of Ulam's approach and took care of programming it on the ENIAC computer he had just created. The works of Ulam and von Neumann had immediate applications to the Manhattan project, which would lead to the invention of the nuclear weapon.

---

[6] *Calculating pi with darts.* Physics Girl. D Muller & D Cowern (2015).

Since then, applications of Monte Carlo simulations have flourished and invaded many fields. I personally used them to make predictions for the 2006 World Cup. But they are also used in quantum physics, aerodynamics, thermodynamics, statistical physics, astrophysics, instrumentation analysis, electronic engineering, geostatistics, energy, ecology, robotics, telecommunication, study of risks, signal processing, climatology, phylogeny (the study of the tree of life), molecular biology, computer graphics (especially to calculate the trajectories of light rays), or in finance (especially in the management of portfolios). These simulations are particularly useful for studying the sensitivity of systems to variations in initial conditions.

The most telling case of study is perhaps meteorology. Because of the infamous *butterfly effect*, the accuracy of meteorological measurements is bound to always be insufficient to make deterministic predictions. A tiny measurement error, or the lack of measurement of a small phenomenon, can lead to a very erroneous prediction. Meteorologists have taken this into account, and rather than attempting deterministic predictions, their predictions are now probabilistic. To generate them, they will typically simulate several possible future weather forecasts, as they slightly vary the initial conditions according to the inaccuracy of the measurements. In other words, they perform Monte Carlo simulations for a set of credible initial conditions. Their bet is then that the empirical frequency of the simulations will correspond to a relevant Bayesian prediction.

17.6   STOCHASTIC GRADIENT DESCENT (SGD)

In the era of Big Data, another application of Monte Carlo simulations may be the creation of a small representative set of data. This simple approach is also at the heart of one of the most important algorithms of today's machine learning, namely *stochastic gradient descent* (or SGD).

Instead of trying to make a theory fit a whole set of data, SGD will randomly draw a few data, and take a step towards explaining these few data. In terms of neural networks, this corresponds to slightly adjusting the synaptic connections so that the calculation of the neural network better corresponds to the last data drawn at random. SGD will then repeat this approach a large number of times, until its explanations of the randomly sampled data are sufficiently accurate.

One might think that the sequential treament of data in random order makes SGD subefficient. This is not really the case. From a theoretical point of view, one can prove that the performances of SGD are not significantly worse than an exact gradient descent. On the application side, SGD's computing time savings have made SGD the preferred solution for today's *deep learning* at Google and Facebook[7]. SGD is today's state of the art in machine learning. And many researchers seem willing to bet that it will always be so.

But there is much better. In 2017, Mandt, Hoffman, and Blei managed to reinterpret SGD as an approximated Bayesian inference[8]. In particular, each random draw of SGD then makes it possible to vary the parameters of the model. By adjusting the parameters of this fluctuation, the three researchers even showed that these fluctuations made it possible to adequately explore a set of credible models, rather than restricting themselves to MAP (the most credible model). Strangely, instead of a weakness, the random fluctuations of SGD could be an asset!

Like *dropout* that we discussed in chapter 12, the stochastic feature of SGD, and in particular its non-convergence to a MAP, could be a key solution to better approximate Bayes' rule. In particular, in doing so, we obtain a form of averaging. Such averaging is extremely useful to avoid *overfitting*. In fact, this discovery has greatly changed my credences in the use of a form of SGD by our brains - even if, again, I gladly acknowledge the extent of my ignorance!

From a technical point of view, though, SGD sampling is very basic. Just draw a given item from a known list. But in more sophisticated cases, sampling can become a whole field of research in itself.

17.7 PSEUDO-RANDOM NUMBERS

How would you randomly draw a point in a square? As we have seen, Cowern and Muller struggled to evenly distribute their darts on their square targets.

Very quickly, John von Neumann realized the difficulty of generating randomness. Before him, some statisticians had chosen numbers found in complex tables of numbers, such as logarithmic tables. In 1939, the *RAND Corporation* published a book containing 100,000 digits obtained

---

[7]Another important explanation of SGD's success lies in the fact that neural networks are well suited to compute gradients with respect to single data only, through backpropagation. So to treat n data, they need to rerun backpropagation n times.

[8]*Stochastic Gradient Descent as Approximate Bayesian Inference*. S Mandt, M Hoffman & D Blei (2017).

via an electronic device measuring roulette numbers. But this was not enough for von Neumann's Monte Carlo simulations.

To better automate his Monte Carlo simulations, von Neumann looked for a way to generate random numbers by a machine. Except that, as von Neumann himself remarked, "any one who considers arithmetical methods of producing random digits is, of course, in a state of sin". But von Neumann also understood that *pure* randomness was not necessary for his simulations. It was sufficient that these numbers had "sufficiently random" properties. Thus were born *pseudo-random numbers*[9].

Using these pseudo-random numbers, von Neumann was able to write deterministic algorithms that pseudo-randomly generate an independent series of numbers uniformly distributed between 0 and 1. These numbers form the basis of any sampling of the probability distributions. For example, draw two pseudo-random (and pseudo-independent) numbers between 0 and 1. You will then obtain the coordinates of a pseudo-random point uniformly distributed in the square. We did it!

Let's now move on to a more difficult problem. How can we pseudo-randomly uniformly draw points in a disk? And what about more general distributions?

17.8 IMPORTANCE SAMPLING

To draw a pseudo-random point in a disk, there is in fact a surprisingly simple method. Draw a pseudo-random point uniformly distributed in the square. If this point is outside the disk, discard it and try again. If not, accept this point. It can then be proved that the accepted points will be uniformly distributed in the disk!

The example of the disk is, in fact, a special case of a more general approach called *importance sampling*. Importance sampling allows a weighted sampling of a target distribution, using a reference distribution that we already can sample. This is what we did to sample the disk, using a sampling of the square.

More generally, to sample the target distribution, one samples according to the reference distribution. Then, crucially, the drawn point is given an importance proportional to the probability of this point according to the target distribution[10]. In the case of the disk, this

---

[9]*How to Generate Pseudorandom Numbers.* Infinite Series (2017).

[10]If the reference distribution is not uniform, then the importance should be the ratio of the probability assigned by the target distribution, divided by the probability assigned by the reference distribution.

importance was either 0 (if the point was outside the disk) or 1 (if the point was inside). Yes, because granting zero importance to a point is equivalent to rejecting it.

Importance sampling is particularly useful when we do not know the precise probability law of a variable, but we are able to calculate the relative probabilities of two values of the variable. This is typically the case of the disk, where we do not know the probability (or rather, the density of probability) of a point inside the disk, but we know that it will be just as likely as any other point inside the disk. This is also the case for many hidden variable models.

17.9   IMPORTANCE SAMPLING FOR LDA

Let's go back to LDA. Imagine reading a set words of words in a text. What other words x are we likely to find in this text? To find out, LDA will first try to determine which category this text belongs to. Then LDA will deduce which words are likely in a text of this category. To do this, LDA must first determine the likely categories Cat, given the words. Since LDA considers that categories *cause* words (remember, this is a causal model!), we will have to infer causes given the consequences. We need Bayes' rule! In fact, this is what LDA will have to calculate:

$$\mathbb{P}[\mathsf{Cat}|\mathsf{words}] = \frac{\mathbb{P}[\mathsf{words}|\mathsf{Cat}]\mathbb{P}[\mathsf{Cat}]}{\mathbb{P}[\mathsf{words}]}.$$

The great difficulty of this equation is the denominator. This denominator, also called *marginal* or *partition function*, requires combining all conceivable category combinations that could produce the set words of words of the text. The only problem is that there are infinitely many such category combinations! Remember, a text can be one third-personal, two thirds-work. But these fractions can actually be any combination of positive numbers whose sum is equal to[11] 1. Without additional hypotheses, it's hopeless to calculate exactly the probability that a text belongs to such and such combination of categories, given the words of the text!

However, as in the case of the disk, we can compute the relative probabilities of two different categories (or combinations of categories) Cat$C$ and Cat$D$, without knowing the denominator. Indeed, we have

---

[11]The exact calculation requires the computation of an integral over all possible combinations of categories!

$$\frac{\mathbb{P}[\mathrm{Cat}C|\mathrm{words}]}{\mathbb{P}[\mathrm{Cat}D|\mathrm{words}]} = \frac{\mathbb{P}[\mathrm{words}|\mathrm{Cat}C]\mathbb{P}[\mathrm{Cat}C]}{\mathbb{P}[\mathrm{words}|\mathrm{Cat}D]\mathbb{P}[\mathrm{Cat}D]}.$$

Importance sampling can then be applied to construct a weighted representative sample of the categories to which the text is likely to belong, and to then deduce other probable words.

If completing a text using LDA does not seem of great interest, some variants of this problem represent billions of dollars. At the time of *Big Data*, one of the most lucrative problems in computer science is recommendation. Billions have been invested to best answer the following question. Given your Facebook, iTunes, or Amazon history, which posts, music, or products are you likely to like?

This problem became glorious when, on October 2, 2006, the Netflix streaming company launched the Netflix Prize[12]. Designed for *data scientists*, this competition consisted of predicting the ratings that users gave to certain films, given a large number of ratings that were publicly released. Specifically, Netflix had a database of 100 million ratings given by 480 thousand users to 18 thousand films. A certain fraction of these notes was made public. The problem was to guess the rest.

While LDA does not tackle this problem in particular, I imagine you will not be indifferent to the similarity with LDA. However, improving the predictions by just a few percent could increase user retention by several percent, which would increase revenue by several percent, potentially representing millions or even billions of dollars!

LDA was not a key tool to solve the Netflix prize. But among the best tools that can be used, there are other kinds of Bayesian networks. Today, the most successful model seems to be the *Generative Adversarial Networks* (GANs). The principle of the GANs is the construction of deep Bayesian networks with little or no intermediate stochastic perturbation[13]. The main source of probability is then the uncertainty on the very deep hidden variables. But like the elementary pseudo-random numbers of von Neumann, this uncertainty generally corresponds to a very simple probability law to sample[14].

---

[12] *The Netflix Prize.* ZettaBytes, EPFL. AM Kermarrec (2017).

[13] The use of *dropout* corresponds precisely to adding intermediate stochastic perturbations.

[14] However, Bayesian inference is then untractable. The trick then is to build another neural network whose task is to help calculate an approximate Bayesian inference, as discussed in chapter 15.

But to return to the Netflix prize, it turns out that the other architecture of probabilistic model with hidden variables did play a key role. This other model comes from physics.

17.10 THE ISING MODEL*

In the 1920s, Wilhelm Lenz and his student Ernst Ising sought to understand phase transitions. In many ways, this is a research problem that is still largely open today. Rather than attacking it in its generality, Lenz and Ising restricted themselves to the case of the ferromagnetic phase transition: whereas iron is magnetic at low temperature, at high temperature, these magnetic properties disappear.

To understand the origin of this phenomenon, Lenz and Ising sought an explanation with a microscopic origin. The magnetic moment of iron is the sum of the magnetic moments of its atoms, called *spins*, whose value can be up or down. Lenz and Ising assumed that the spins of neighbor atoms tended to line up. They described this by assuming that the energy of a pair of aligned spins is −1, while that of opposite spins is +1. The total energy E of the iron is then the sum of the energies of these local interactions between neighbor spins.

The question that Lenz and Ising asked themselves was whether, at a given temperature T, the *spins* would line up or not. In other words, is the set of configurations where the *spins* are aligned, at temperature T, more likely than the set of configurations where *spins* do not align?

Fortunately for Lenz and Ising, a similar problem had already been solved half a century earlier by Ludwig Boltzmann. Boltzmann discovered that at thermodynamic equilibrium and at temperature T, the probability of a configuration i with energy E_i is proportional to $\exp(-kE_i/T)$, where k is known as the *Boltzmann constant*. Specifically, Boltzmann's law says that the probability of configuration i is

$$\mathbb{P}[i|T] = \frac{\exp(-kE_i/T)}{\exp(-kE_i/T) + \sum_{j \neq i} \exp(-kE_j/T)}.$$

The denominator of this equation is the famous *partition function*, whose calculation is impossible because the number of configurations is typically exponential in the number of atoms. In particular, according to Boltzmann's law, configuration i is exponentially rare in its energy E_i. Crucially, this is especially the case at low temperature T.

This turned out to be precisely what explains the role of temperature in the ferromagnetic phase transition. At low temperatures, high energy configurations, those where spins do not align, are exponentially rare very quickly; likely configurations are therefore those where the spins line up. This makes iron magnetic. Conversely, at high temperature, and assuming that kE_i is much smaller than T, then the quantities $\exp(-kE_i/T)$ are all close to 1. However, the configurations where *spins* are not aligned are exponentially more numerous than those where *spins* align. Indeed, if you randomly (evenly and independently) draw the values of the spins, there is almost no chance of them being aligned. As a result, at high temperature, the set of configurations where *spins* do not align is exponentially more likely than the set of configurations where they align, hence the disappearance of iron magnetism.

The Ising model is fascinating for several reasons. First, it's one of the simplest models explaining phase transitions. Second, the understanding of Ising's model goes through the Boltzmann distribution, the one that links the energy E of a configuration to its probability. Finally, and most importantly, the Ising model is a wonderful example of *random Markov fields*, which are at the heart of many modern models of *machine learning*.

17.11 THE BOLTZMANN MACHINE

Markov fields are described by variables linked together by undirected edges. They are thus very similar to Bayesian networks. However, as opposed to Bayesian networks, the edges of the Markov fields do not represent cause-and-effect relationships. They rather represent some sort of correlations between variables, as opposed to unrelated variables that are intuitively almost independent. This is roughly what we call the *Markov property of the random fields*[15].

There is one particular case of Markov field which lends itself particularly well to *machine learning*. In fact, it was one of the key ingredients for solving the Netflix challenge. This particular case is called the *restricted Boltzmann machine*. Similarly to a Bayesian network with hidden variables, the principle of the Boltzmann machine is to find correlations between observable variables and hidden variables. In other words, the restricted Boltzmann machine is a Markov field whose edges

---

[15]The Hammersley-Clifford theorem proves that the Markov property is equivalent to saying that the joint probability factorizes into a product of clique functions, that is, is written $\mathbb{P}[X = x] = \prod_{Clique\ c} f_c(x_c)$, where x_c is the vector of x_i for $i \in c$.

link each observable variable to hidden variables[16]. What's more, like the Ising model, the Boltzmann machine associates each edge with a correlation energy[17]. These correlations are graphically depicted in Figure 17.3.

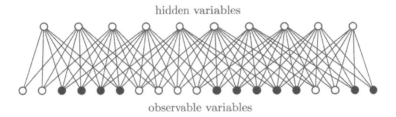

hidden variables

observable variables

Figure 17.3 A Boltzmann machine links observable variables via hidden variables. When certain observable variables are observed (here in black), we can then use the hidden links via hidden variables to guess the likely values of the unobserved observable variables.

The application of the restricted Boltzmann machine to the Netflix challenge requires two phases. At first, the data are used to learn the parameters of the machine. These parameters are the equations of the correlation energies of the edges[18]. Once the parameters are determined, the Boltzmann machine then uses the Boltzmann distribution to make probabilistic predictions that address the Netflix challenge. Given the observed observable data, the Boltzmann machine infers credible hidden variables, from which it infers likely values of unobserved observable variables.

As usual, the *pure Bayesian* could perform this computation without difficulty, and derive the appropriate credences on the possible values of the unobserved observable variables. But this reasoning is out of reach of the *pragmatic Bayesian*, because it requires the exact calculation of the monstrous partition function. This is why the *pragmatic Bayesian* will have to be content with sampling. It seems that he too is forced to reason with (representative) examples.

---

[16]In other words, it's a bipartite graph, with the observable variables on one side and the hidden variables on the other.

[17]It's then possible to generalize this configuration by considering a layer of observed data, connected to a first layer of hidden variables, itself also connected to a second layer of hidden variables, and so on. One would thereby construct a *deep Boltzmann machine*.

[18]It's generally assumed that the energy of an edge is bilinear in the observable and hidden variables bound by the edge. What's left is to determine the coefficient associated with this bilinear form.

17.12 MCMC AND GOOGLE PAGERANK

Unfortunately, importance sampling is often unable to provide a good sampling of Boltzmann machines in a reasonable time. In general, if the reference distribution of importance sampling is too distinct from the target distribution (which is often the case in large dimensions), then importance sampling may require too many iterations to be representative of the target distribution. In fact, determining a good way to perform a fast and representative sampling is a difficult field of research.

Oddly, to solve this problem, it's often useful to replace the repetitions of independent Monte Carlo experiments by so-called *Markov Chain Monte Carlo* (MCMC) methods. These methods consist in carrying out a random walk in the set of possibilities. The steps of this random walk will be absolutely not representative of the target distribution. However, provided that each transition of this walk is appropriately chosen, it remains possible to guarantee that, at infinity, the frequency of visit of any point[19] in the set of possibilities converges to the probability of this point according to the target distribution. Put simply, at infinity, the sampling yields perfectly representative samples of the target distribution.

At first glance, you might think that MCMC is a very bad idea. This is what I first thought. Yet, surprisingly, the algorithm that made Google's fortune is an MCMC algorithm! This algorithm, called *PageRank*, is the heart of the first versions of the search engine. The trick of PageRank was to calculate the importance of each web page, depending on the importance that other web pages give it, and the importance of these other web pages. For instance, a Wikipedia page can be considered important because many other web pages, including important web pages, have links to this Wikipedia page.

However, because the importance of each web page depends on the importance of other web pages, calculating the importance of a web page amounts to solving a horribly complicated equation: this equation has as many unknowns as there are web pages! The trick found by Larry Page and Sergey Brin, the founders of Google, was to solve the problem via MCMC. Their PageRank algorithm will imagine an fictitious surfer of the web who randomly clicks on one of the links on the page he is visiting. Our surfer will jump from page to page. Intuitively, we can

---

[19]In the case of continuous distributions, we should rather speak of the frequency of visit of any measurable set.

expect that the pages he will visit the most are the ones that matter, since many paths lead to these pages.

We can in fact prove that if the web is strongly connected, then the empirical frequency at which the surfer arrives at a given page will converge[20] exactly to the importance that we have sought to compute previously. In particular, after a sufficiently long simulation, this empirical frequency will be a good approximation of the importance of the page[21].

This remarkable principle is what allowed Page and Brin to create one of the most powerful companies on the planet!

17.13 METROPOLIS-HASTING SAMPLING

Although PageRank was excellent for organizing the web, it does not seem to be suitable for sampling the distribution of a restricted Boltzmann machine. There is another approach to do so, called *Metropolis-Hasting sampling*. Like PageRank, Metropolis-Hasting will take us on a walk. However, this walk will now take place in the space of imaginable values of observable and hidden variables.

At every moment of this random walk, we consider a random step to be made, which we can refuse if it leads us to a state whose probability is too small. Specifically, let's call i the current position, and assume that the random step takes us to position j. To find out whether to accept or reject this random step, Metropolis-Hasting tells us to calculate the *acceptance rate* A. We define A as follows:

$$A = \frac{\mathbb{P}[j]}{\mathbb{P}[i]} \frac{\mathbb{P}[\mathsf{Step}(i \to j)|i]}{\mathbb{P}[\mathsf{Step}(j \to i)|j]}.$$

Intuitively, the acceptance rate of a step from i to j is large, if j is a state more likely than i and if this step is probably reversible, in the sense that the probability that the opposite step is proposed by the random walk is not much lower than the probability of the random step. If the acceptance rate A is greater than 1, then the sampling of Metropolis-Hasting tells us to take that step. If not, we need to toss a random coin whose probability of falling on "yes" is the acceptance rate A.

---

[20] With probability 1 (almost surely).
[21] *Can a Chess Piece Explain Markov Chains?* Infinite Series. K Houston-Edwards (2017).

Crucially, A can be calculated without using the partition function. In the case of the Boltzmann distribution, we have $\mathbb{P}[j]/\mathbb{P}[i] = \exp(-kE_j/T)/\exp(-kE_i/T) = \exp(k(E_i - E_j)/T)$. Nevertheless, despite the non-use of the partition function, under certain reasonable assumptions[22], after enough time and as for PageRank, this random walk will provide a representative sample of the target distribution.

Better yet, it's then possible to condition the sample to observed observable variables, thereby forcing the random walk to never change their values. We will then obtain a representative sample of the unobserved variables, given the observed variables.

Metropolis-Hasting sampling comes in many variants useful for the *pragmatic Bayesian*, for all distributions whose partition functions are too long to estimate correctly. On the one hand, there are so-called *adaptive* variants, which make it possible to optimize the properties of the random steps during sampling. On the other hand, there are Metropolis-Hasting approximations, which make it applicable even when the ratios $\mathbb{P}[j]/\mathbb{P}[i]$ of the probabilities of the two positions i and j are not easily computable. This is typically the case for generative models of complex data, such as the simulations of the universe discussed at the end of chapter 15. In particular, we may replace these probabilities with performance measures $\mathsf{perf}(i)$ and $\mathsf{perf}(j)$ of positions i and j, in the hope that these quantities will be sufficiently correlated with the probabilities of these positions.

17.14 GIBBS SAMPLING

However, it's still another MCMC sampling method that is most often used in the case of the Boltzmann machine. This other method, called *Gibbs sampling*, is based on a remarkable property of the Boltzmann machine. Remember, in this machine, the observable variables are only connected to hidden variables, and vice versa. In particular, two observable variables are never connected, just as two hidden variables are never connected either[23].

The idea of Gibbs sampling is to alternate the sampling of observable variables and that of hidden variables. Crucially, given the values of the observable variables, the total energy of the restricted Boltzmann

---

[22]Especially the fact that the walk has a non-zero probability to bring us everywhere.

[23]This is due to the bipartiteness of the Markov field graph, and there is a nice and easy generalization of this remark to deep Boltzmann machines.

machine is then a linear function of the hidden variables [24]. In particular, the contribution of a hidden variable to the total energy is a term that does not depend on the values of other hidden variables. So, given the observed variables, each hidden variable can be sampled independently[25]. And independent variables are very easy to sample.

Given some observed observable variables, Gibbs sampling then consists in first assigning arbitrary values to the unobserved observable variables. This will be the arbitrary starting point of the random walk. Then, Gibbs sampling will sample all hidden variables conditionally to observable variables. Then, it will sample all unobserved observable variables, conditionally to hidden variables. We will then have made a random step in the space of the observable variables. Gibbs sampling then consists of repeating such random steps. After sufficiently many iterations, this calculation will generate a representative sample of the credible values of the unobserved observable variables, given the observed observable variables.

In all cases of MCMC that we discussed, be it PageRank, Metropolis-Hasting, or Gibbs, it's crucial for the sampling to last long enough. Otherwise, the sampled data will be strongly dependent on the starting points of the random walks, which have no reason to be representative of the distribution that we want to sample.

Worse, it's often the case that most imaginable data have in fact a cumulative negligible probability. In other words, only a handful of data within a huge set have a non-negligible credence. This is typically the case of credible theories, which often form a handful of islets in a very large ocean filled with non-sense theories[26]. Therefore, as long as MCMC has not come across one of the few credible data or theories, then it will be impossible for MCMC to realize that the data and theories explored so far are not credible. Remember. MCMC only knows the credences of visited data relative to the other visited data. As a result, if no really credible data has been visited, then all the visited data will seem credible!

In short, the MCMC intermediate samples are not representative of the distribution that MCMC is trying to sample. Even worse, it's

---

[24] This requires to assume that the total energy $E(v, h)$ is bilinear in the observable variables v and the hidden variables h.

[25] In other words, conditionally to the observable variables, the hidden variables are independent. This is actually a direct consequence of the fact that the set of hidden variables form a stable set of the Markov field graph.

[26] This is particularly the case in high dimensions, as evidenced by phenomena like the concentration of measure. This is often referred to as the *curse of dimensionality*.

impossible to anticipate or determine the representativity of a sample obtained by MCMC. Certainly, at infinity, MCMC will become valid. But MCMC will take a long time to become valid enough. And it will never offer any provable guarantee.

Nevertheless, because it allows a sampling that no other method seems able to provide with the same efficiency, MCMC has become a key tool for the *pragmatic Bayesian*.

17.15 MCMC AND COGNITIVE BIASES

Given the complexity of MCMC, I could have spared you its detailed description. In fact, if I spent so much time explaining MCMC, it's not really so that you will use it in practice. My goal was to convince you that Darwinian evolution probably made our brains MCMC calculators; and that, as a result, you really should pay attention to the properties of MCMC computations.

Psychologist Daniel Kahneman says that, one day during the period of repeated terrorist attacks on buses in Israel, he rushed to get away from any bus he saw when he was driving. He became ashamed of this behavior. As a good statistician, he knew that the probability of dying from a road accident remained far greater than the probability of witnessing the attack of a bus. Terrorism causes a negligible number of deaths compared to the death toll on the roads. But even a well-informed and well-educated brain like Kahneman's failed to avoid the overestimation of the danger of terrorism.

This cognitive bias is what Kahneman calls the *availability bias*. We tend to give too much importance to the first ideas that come to mind. If I ask you to visualize Linda, the 31-year-old anti-nuclear activist, it's likely that you will soon have an image in mind, and that you will overfit this image. This bias might be revealing the use of MCMC by our brains. Unless one thinks long enough, MCMC is heavily biased by its starting point - and media coverage often makes terrorism the starting point of our thoughts.

If this bias may seem relatively obvious, it's partly because it's not too hard to be aware of what you think when you are a victim of availability bias. However, many of the signals that cross our brains are unconscious. But this does not prevent MCMC from making them the starting points of its random walk into the world of ideas. Therefore, it's without our knowledge that what we think is incredibly affected by the context in which we find ourselves. This is known as the *priming effect*.

A rather unbelievable example of this effect is the experiment by Gary Wells and Richard Petty in 1980. Wells and Petty invited 72 students to test headphones in various contexts of use. Twenty-four students had to listen to an editorial about raising tuition fees from $587 to $750, with this headset and without moving their heads. Twenty-four other students were asked to listen while shaking their heads up and down (as if to say "yes"). Finally, the last 24 students listened while shaking their heads left and right (as if to say "no"). At the end, in the last question of a questionnaire about the quality of the headphones, the researchers asked the students their opinions about tuition fees. The results are amazing. On average, those who shook their heads left and right answered $467, those who did not move their heads said $582, and those who nodded declared $646. Unbelievable! Our judgment is determined by our movements, and we are completely unaware of it!

Another amazing example of our reliance on the initial point of our MCMC reasoning is the astonishing *anchor effect*. Kahneman and Tversky performed a troubling experiment to reveal it. Kahneman and Tversky first drew a random number before the eyes of a subject, which was either 10 or 65. Imagine that the number was 10. Kahneman and Tversky then asked the subject if more or less than 10% of the countries in the world were in Africa. If the number drawn was 65, they asked the same question with 65% instead of 10%. Then they asked the subject to estimate the percentage of countries that are African. Surprisingly, on this second question, the subjects answered on average 25% when the randomly drawn number was 10, and 45% when it was 65! Other variants show that this staggering effect persists, even if the numbers proposed by the researchers are absurdly small or large!

A third example of MCMC-caused bias, among many others, is *loss aversion*. This aversion was theorized by Kahneman and Tversky's theory of perspectives, which explains a large number of cognitive biases at once. The premise of this theory is that our preferences are always strongly influenced by a reference. The gains vis-à-vis this reference are good, but the losses vis-à-vis this reference are catastrophic. This is why silver medalists of the Olympic Games look sad, as opposed to smiling bronze medalists[27].

---

[27] Another (still Bayesian!) explanation for this anchoring effect could be the mechanism of anticipation and correction of our cerebral cortex proposed by Karl Friston. According to this theory and consistently to Shannon's optimization of communication, our brain would constantly make predictions, and only react when observations contradict these predictions, thus initiating a learning process only if necessary.

MCMC is not a problem in itself. The problem is that MCMC is relevant only after a very large number of random steps. Worse, the first ideas suggested by MCMC can all be very unlikely, especially as opposed to a handful of credible ideas in the vast space of ideas. As long as none of the credible ideas has been sampled by MCMC, the conclusions of MCMC will be very wrong.

Knowing that our brain probably applies MCMC, we must absolutely recognize the extent of our ignorance, and the need to take the time to think - much more than is the case in practice. In particular, the benefits of long meditations and sleep may partly lie in an extended MCMC computation.

17.16 CONSTRASTIVE DIVERGENCE

What we have seen so far, however, is only part of Crick and Mitchison's explanation. For these two researchers, the utility of dreams in particular may rather reside in another algorithm useful for learning called *contrastive divergence*. This algorithm aims to calculate the maximum a posteriori of a hidden-variable model, especially when we only know the probabilities of the data relative to others, as was the case for the Bayesian inference of LDA or for the Boltzmann machine.

Let me spare you the calculation steps. The important equality is

$$\partial_\theta \log \mathbb{P}[\theta|D] = \partial_\theta \log \mathbb{P}[\theta] + \partial_\theta \log \tilde{p}(D|\theta) - \mathbb{E}_{x|\theta}\left[\partial_\theta \log \tilde{p}(x|\theta)\right].$$

In other words, to know how to adjust parameters θ to obtain a more credible theory, we need to understand three terms. The first is the effect of the variations of θ on the *prior*. The second is the effect of the variations of θ on the non-normalized probability of the model (which is typically $\exp(-kE_i/T)$ in the case of the Boltzmann distribution). The third, finally, is the effect of the variations of θ on the non-normalized probabilities of the alternatives x to D.

For many models, especially Boltzmann machines, the first two terms are very easy to calculate. However, the last term will typically require (long) representative sampling. The crucial remark, however, is that this last term does not depend on the data! We do not have to observe the world to calculate it. Dreams suffice.

According Crick and Mitchison, this may be precisely the reason why we dream. Dreams would correspond to a maximum a posteriori computation, by sampling the effect of a change in brain parameters on the non-normalized probability of alternatives to observed data.

This would be one more reason to take the time to reflect.

FURTHER READING

Deep Learning. MIT Press. I Goodfellow, Y Bengio & A Courville (2016).

Thinking Fast and Slow. Farrar, Straus and Giroux. D Kahneman (2013).

The Function of Dream Sleep. Nature. F Crick & G Mitchison (1983).

Stochastic Gradient Descent as Approximate Bayesian Inference. S Mandt, M Hoffman & D Blei (2017).

How a Kalman Filter Works, in Pictures. Bzarg. T Babb (2015).

Deep Dream - A Code Example for Visualizing Neural Networks. Google Research Blog. A Mordvinster, C Olah & M Tyka (2015).

Pi and Buffon's Matches. Numberphile. T Padilla (2012).

Calculating Pi with Darts. Physics Girl. D Muller & D Cowern (2015).

Inside Google: Page Rank. Wandida, EPFL. R Guerraoui (2013).

What's a Random Number? ZettaBytes, EPFL. P Blanchard (2016).

How to Generate Pseudorandom Numbers. Infinite Series (2017).

CHAPTER 18

The Unreasonable Effectiveness of Abstraction

Philosophy is written in that great book which ever lies before our eyes - I mean the universe - but we cannot understand it if we do not first learn the language and grasp the symbols, in which it is written.

Galileo Galilei (1564-1642)

18.1   DEEP LEARNING WORKS!

On March 10, 2016, AlphaGo made the headlines. To everyone's surprise, this artificial intelligence from Google had just beaten Lee Sedol, then regarded by many as one of the best players in the world at the game of go.

Go is a two-player game roughly similar to chess. But go is harder. It's more complex, more combinatorial, and more unpredictable. The best go players sometimes rely on intuitions that they have trouble explain in - some even claimed that this human faculty was out of reach of the machines. And for a long time, the best algorithms were barely able to compete with weak human go players.

However, in recent years, a certain model of *machine learning* has been becoming increasingly successful. Object detection, face recognition, optical character reconstruction, natural language processing, automated translation, and recommendation systems had all long been inaccessible to the machines. But all were suddenly solved by *deep learning*. Suddenly, *deep learning* was all Silicon Valley investors could talk about. And all the giants of the web were proudly bragging about the new features that they were now able to offer.

The success of *deep learning* is in sharp contrast with the development of many conventional engineering methods. Usually, we build a theory that guarantees the proper functioning of a technology. Then, we get frustrated that the technology fails *despite* theory. That's when theorists are usually asked with an accusing tone: where does this difference between theory and practice come from? *In theory, it's the same thing*, says the joke. *But in practice....*

However, the case of *deep learning* seems to be the exact opposite. There is a technology whose success was predicted by no theory. Theorists are puzzled. The performance of *deep learning* is amazing, but nobody seems to know why. *Deep learning* works very well *in practice*; but does it work *in theory*? For once, this question has become of interest to practitioners. This is because they too feel that any progress in the theoretical understanding of *deep learning* could lead to significant, perhaps even dramatic, advances in artificial intelligence.

But what is *deep learning*? Nowadays, *deep learning* research is so abundant and dynamic that it's difficult to define the boundaries of this field. Roughly, however, we can consider that *deep learning* is the study of models stacking a large number of layers of hidden variables, like LDA and (deep) Boltzmann machines. However, unlike the examples in the previous chapter, a *deep learning* architecture does not necessarily seek to describe a probability distribution. In fact, most are neural networks designed to approach deterministic functions[1]. All these structures, however, have one thing in common: they manipulate *deep* hidden variables, that is, variables intuitively far removed from the observable variables. More rigorously, any signal that propagates from observable variables to deep variables passes through a large number of intermediate variables.

This invites us to consider that the level of abstraction of a hidden variable is measured by its depth. Indeed, intuitively, abstraction is opposed to what is tangible. It's natural to consider that the observable

---

[1] In particular, the GANs discussed in chapters 15 and 17 consist of fixing a distribution Z of deep variables, and generating credible data $X = f(Z)$, by adjusting the function f.

variables are precisely what is tangible in a model; and that deep variables are abstract. Hence the title of the chapter - which will be related later to mathematical abstraction!

The theoretical problem posed by the success of *deep learning* is then the explanation of the unreasonable effectiveness of abstraction. We shall roughly distinguish three arguments to justify this effectiveness. The first is the necessary preliminary synthesis of raw data. The second is the particular expressiveness of deep models (particularly deep neural networks). Finally, the third, and most promising explanation in my opinion, lies in the algorithmic properties of the data that we want to study, measured in terms of the *Solomonoff sophistication* and the *Bennett logical depth*.

But let's start with the need to synthesize raw data.

18.2 FEATURE LEARNING

Before the success of *deep learning*, *machine learning* algorithms required the intervention of competent *data scientists* to "clean" the raw data and extract useful features of the raw data. One of the main motivations of *deep learning* research was to bypass the expertise of *data scientists* by automating the feature extraction of data. This is called *feature learning* or *representation learning*. This approach allowed *deep learning* to simultaneously analyze different media, be it images, sounds, videos, texts, or other real-time measures of sensors.

To understand this, let's take a quick look at artificial neural networks. These networks are a collection of (usually virtual) neurons interconnected via so-called synaptic connections, much like the computers on the Internet, or like the neurons of our brains. Some of these neurons are directly connected to data sensors, like cameras, microphones, or other measuring devices. These neurons correspond to the observable variables.

A network of neurons will then link these observable neurons to hidden neurons. This process is very similar to the confounding and conciseness variables discussed in chapter 13, which helped us choose between the hospital and the clinic, or explain the correlation between the colors of a sheep's hair. In other words, a data processing by a hidden neuron can be interpreted as the computation of an abstract and relevant concept to explain the correlations between observed variables. Typically, such a hidden neuron may determine the presence or absence of guidelines in an image captured by observable neurons. The stack of neuron layers can then be used to explain the correlations of the layers of intermediate neurons. Typically, the deep layers of the neural network infer from guidelines the presence or absence of sheep in the image.

This process is inspired by our cerebral cortex. Indeed, neuroscience has discovered that our brains analyze what our eyes see through a growing abstraction. At the level of the eye, the sensors, called cones and rods, detect the brightness and the color of the ray of light that excite them. A computer scientist would talk about the brightness and color of each pixel in an image. It's this kind of raw data that will be the observable variables of a deep neural network.

Then, the second computing layer typically measures correlations between neighboring pixels. In humans, each neuron in this second layer binds the signals of a few cones and neighboring rods of the eyes. For example, it could get excited when these cones and neighboring rods are both excited, or when both are extinct[2].

The third layer will then combine these correlations to determine the outlines of the images that the eyes see. The fourth layer will then combine these outlines to determine elementary objects in the image, such as the ears, eyes, or legs of a sheep. The following layers will then combine these elementary objects to determine deeper structures, such as the presence of a sheep.

Nowadays, the state of the art in artificial intelligence corresponds to so-called *convolution neural networks* (CNN) (because they are connected to a mathematical operation called the *convolution product*). These networks are roughly inspired by the architecture of the visual cortex. Guess what. While we have primarily sought to make the neural network globally efficient, the neural network, once trained with photographs, calculates increasing levels of abstraction of images, like the human neural network. It seems that the performance of an image analysis strongly relies on this stack of levels of abstraction that only sufficiently deep architectures allow.

18.3   WORD VECTOR REPRESENTATION

A similar phenomenon has been discovered empirically in *natural language processing*. One of the difficulties is the very large number of natural language words. Rather than having a neuron dedicated to understanding each word, it's often best to interpret each word in relation to

---

[2]This is an example with no biological ground.

other words. In mathematical language, strangely enough, this can be done nicely by embedding the space of words into a large vector space, typically of dimension 50 to 300. In less obscure language, each word will correspond to a certain combination of activations of different neurons.

Remarkably, in 2013, a group of researchers at Google around Tomas Mikolov managed to teach a neural network such as a neural representation of the words of the English language. They called this representation word2vec. The word2vec algorithm transforms any English word into a vector in a large space. This allows to compactify and to structure the representation of all words. Indeed, from an informational point of view, the vector representation of words is more compact. It takes fewer bits to be described.

But there is better! The vector representation of the words allows to reveal many relations between the words. In particular, one of the great discoveries of Mikolov's research group was that the vector addition between words matches human intuition. For example, they discovered that if one carries out the *king − man + woman* vector addition, one obtains approximately the vector representation of the word *queen*[3]!

What's even more fascinating is that this phenomenon has also been observed in *Generative Adversarial Networks* (GANs). Take pictures of individuals with sunglasses and individuals without sunglasses. Using a GAN, calculate the average vector of individuals with glasses, and the average vector of individuals without glasses. Then take the difference between these two vectors. You then get a *glasses* vector that, intuitively, corresponds to having sunglasses. Now take a picture of an individual without glasses. Calculate the *individual* vector representation of this image, and add to this vector representation the *glasses* vector. You then get a new vector *individual + glasses*. Now use the GAN to generate an image that matches this vector. Amazingly, you will get an image of the original individual where he or she is wearing glasses[4]! Unbelievable!

This strange compatibility between the abstract addition of neural networks' abstract concepts and the intuitive combination of concepts according to our brains is, to my knowledge, still very poorly understood. This is a wonderful mystery of the current research in artificial intelligence! Weirdly enough, it seems to suggest that there is a natural

---

[3] *Machine Reading with Word Vectors.* ZettaBytes, EPFL. M Jaggi (2017).

[4] *Unsupervised representation learning with deep convolutional generative adversarial networks. A Radford, L Metz & S Chintala (2016).*

algebra to manipulate abstract concepts that both humans and machines stumbled upon.

Now, let's imagine how a machine could read a physical book made of paper. It would likely need to be somewhat similar to a deep neural network whose first layers form a convolutional neural network that recognizes the characters of an image. Then, above these layers dedicated to the vision, one would find a layer of neurons gathering the characters into words. Then another layer would turn these words into vectors, typically like word2vec. Finally, other layers would interpret the vectors that represent the words to link them to other concepts from other image analysis or something else.

In particular, a powerful deep neural network must be able to activate the same neurons if it sees a photograph of a cat or if it reads the word "cat". To do this, extracting the key features of raw data seems crucial. More generally, it seems that depth is essential for synthesizing raw data, either to reduce the massive size of the raw data and to allow calculations in reasonable time, or to reveal relevant explanations of the correlations in the raw data.

18.4 EXPONENTIAL EXPRESSIVITY*

On June 16, 2016, a collaboration between researchers from Stanford, Cornell, and Google Brain uploaded two amazing articles on ArXiV. These explanations of the success of deep learning seduced me greatly. In this book, I'll take the time to describe the more elegant, in my opinion, of these two articles. This article is entitled *Exponential expressivity in deep neural networks through transient chaos*. It combines many sophisticated, deep, and poorly connected mathematical concepts, such as chaos theory, mean field theory, and geometric curvature.

Let's go slowly. The article begins by considering that a network of neurons is nothing but a complicated mathematical function. This function takes a collection of measured data as input and transforms it into a combination of deep concepts. The collection of measured data mathematically forms what is called a *vector*, which can be seen as a point in a very large space. The same goes for the combination of deep concepts, which is also a point in a very large space.

As a result, the neural network can be seen as a geometrical transformation of the first space to the second. The question that the article raises is the following: how does a "typical" neural network distort space? Does a "random" neural network move the points in all imaginable directions? Or will it retain the geometric properties of the curves it transforms?

The intriguing answer of the article is the exponential complexification of geometric structures with the depth of the neural network. More precisely, the article focused on a certain measure called *global curvature* of geometrical figures[5]. The circle, intuitively, is the closed figure with least *global curvature*, since it curves only to loop back on itself. Moreover, the overall curvature of a circle is always equal to $\tau \approx 6.28$, regardless of its size. Conversely, a very convoluted curve that oscillates in all the dimensions of space would have a very large global curvature.

What the article shows is that, for random neural networks sufficiently "agitated", the curvature of a geometrical figure increases polynomially with the width of the network, but exponentially with its depth. In other words, the depth allows the neural networks to much more quickly complexify their geometrical transformations than the width.

In particular, what this suggests is that depth is crucial for the detection and analysis of fractal structures, that is, whose behavior is not always smooth and regular. Yet, fractal structures seem ubiquitous in the world around us, including in numerous fields like biology, cosmology, or finance!

In the same way, even if it may seem abstract and far-fetched, there is a good chance that all the images containing a cat will be a fractal structure in the immense set of all the images. Depth would then be key for cat detection, just as for many more sophisticated tasks.

18.5   THE EMERGENCE OF COMPLEXITY

This amazing discovery resonated with other readings to which I was exposed at about the same time, including that of an article[6] co-authored by Sean Carroll, Scott Aaronson, and Lauren Ouellette. The paper discusses the intriguing temporary complexification of the universe. The physical principle that this article sought to reveal is a phenomenon that is observed both at the level of the cosmos, of life, and of coffee with milk. Originally, these structures were simple. The universe was a quasi-homogeneous plasma, the cup was a superposition of a layer of coffee and a layer of milk, and life did not exist. These structures also had low entropy, a notion that was introduced in chapter 15.

---

[5]The *global curvature* integrates the norm of variations of the unit vector.

[6]*Quantifying the Rise and Fall of Complexity in Automated Systems: The Coffee Automaton.* S Aaronson, S Carroll & L Ouellette (2014).

Figure 18.1   Fractals in coffee with milk. By Pexels on Pixabay.

But if the entropy is weak at first, it can only increase with time[7]. This principle is the second law of thermodynamics. However, the increase in entropy can be interpreted as a homogenization. In the very long term, the cup will be a perfect blend of coffee and milk, life will be gone, and the universe will likely disappear into a perfectly homogeneous interstellar void (called the *Big Chill*). However, Sean Carroll had the intuitive idea that, between the initial and final moments, all these structures, the cosmos, life, and coffee with milk, necessarily go through phases of great complexity, be it galaxies structured along cosmic filaments, the extreme complexity of plants, animals, and our human brains, or the strange phase where milk forms fractal figures in coffee, as displayed in Figure 18.1.

Sean Carroll then received help from computer scientist Scott Aaronson to formalize this intuitive notion. The first stage of this formalization was a digitalization of physical phenomena. Since everything can be reduced to a series of 0s and 1s (for example by encoding the image of the coffee in PNG or JPG format), Carroll and Aaronson, then joined by Ouellette, sought a formalization of the algorithmic complexity of finite binary sequences x that was consistent with Carroll's intuition.

The three researchers then presented four definitions already introduced in the scientific literature: *sophistication, apparent complexity, logical depth,* and *complexity of the cone of light* (which I won't discuss here). Intriguingly, all of these definitions are in fact subtly connected to each other.

---

[7]The argument is more subtle than that, but I won't dwell on this subject.

18.6   THE KOLMOGOROV SOPHISTICATION*

Let's begin with *sophistication*. Sophistication is a concept proposed by Kolmogorov, which builds upon the concept of *Solomonoff complexity*.

In fact, one might think that the *Solomonoff complexity* would be a good candidate for measuring the complexity of a finite binary sequence. Unfortunately, this complexity is in fact maximal when this sequence is so random that it has no regularity. Yet, such states without regularity seem to correspond more to a maximal entropy which, in the case of the cosmos, life, and coffee, corresponds to a state that is simple to describe: it's then a homogeneous nothingness.

Like Boltzmann nearly a century before him, Kolmogorov had the brilliant idea of separating as much as possible the "macroscopic" easily describable structures from the "microscopic" structures that cannot be better described than by a perfectly random noise. It's this intuition that the *Kolmogorov sophistication* formalizes.

Let's call S the macroscopic structure. Roughly, to say that the microscopic structures of the binary sequence x are sufficiently random amounts to saying that the *Solomonoff complexity* of the binary sequence x knowing S is (almost) maximal among the set of binary sequences which have the macroscopic structure S. In other words, x is considered to be a "typical instance" of S, if it can be accurately described as S plus a quasi-random noise. What's more, importantly, for this macroscopic description S to be "valid", the description of x as $S + noise$ must be almost as compact as the most concise description of[8] x. Any macroscopic structure having this property is then in accordance with Kolmogorov's prerequisites.

Kolmogorov then defines the *Kolmogorov sophistication* of the binary sequence x as the lowest *Solomonoff complexity* among the macroscopic structures S that satisfy Kolmogorov's prerequisites for "validity".

---

[8]Let's identify S with the set of binary sequences compatible with the macroscopic structure S. To say that *noise* is "random" is tantamount to saying that its description length is essentially the length of the naive identification of elements of x among S, that is, $K(noise) \approx \log_2 |S|$. For the macroscopic structure S to be "valid", the description of x as $S + noise$ must be nearly as efficient as the shortest description of x, which formally corresponds to $K(S) + \log_2 |S| \le K(x) + c$, for some "small" number c. The description length $K(S)$ of the minimal "valid" macroscopic structure S is called the *Kolmogorov c-sophistication*. Note that this is slightly different from a variant called *naive c-sophistication*, where $K(x) - K(S)$ is replaced by the conditional *Solomonoff complexity* $K(x|S)$.

Intuitively, this is the best way to describe x as a simple macroscopic structure plus (almost) uniformly random noise.

Unfortunately, like the *Solomonoff complexity*, the *Kolmogorov sophistication* of a binary sequence is usually incomputable. To estimate the *Kolmogorov sophistication* of their simulations, Aaronson, Carroll, and Ouellette then turned to a heuristic estimate of the *Kolmogorov sophistication*, which they called *apparent complexity*. The principle of this heuristic is to restrict itself to macroscopic descriptions S which correspond to a kind of smoothing. This smoothing is also inspired by Boltzmann's approach which, instead of considering each particle individually, prefers to describe coarse statistical properties of all the particles located at approximately the same place at a given instant. It's this *apparent complexity* that the three researchers used to analyze simulations of coffee with milk[9].

18.7 SOPHISTICATION IS A SOLOMONOFF MAP!*

While the *Kolmogorov sophistication* has a real mathematical aesthetic and indeed seems to be related to the intuition of our actual description of physical objects, one can nevertheless wonder whether this is really the right way of defining sophistication. In particular, is it not somewhat arbitrary to demand that the sequence x be of near-maximal randomness, given its macroscopic description S? Wouldn't this be like the case of Boltzmann's entropy, which ended up being generalized by a more general and fundamental concept by Shannon? Shouldn't we too invoke a probabilistic formalism to derive a deeper notion of sophistication?

The *pure Bayesian* answers yes! One of my most exciting Bayesian reflections was my discovery that Kolmogorov's elegant notion of sophistication could be interpreted as the *Solomonoff complexity* of the maximum to posteriori (MAP) of a subset of Solomonoff demon's theories!

Let's call T the algorithmic description of the macroscopic structure S. T is none other than a Solomonoff predictive theory that tries to explain the sequence x. In particular, the credence in T given x is calculated via Bayes' rule:

---

[9]In a way, this amounts to replacing the theoretical *Kolmogorov sophistication* that unfortunately involves algorithms with unreasonable computation time by a sort of *pragmatic sophistication* that uses only a handful of fast algorithms.

$$\mathbb{P}[T|x] = \frac{\mathbb{P}[x|T]\mathbb{P}[T]}{\mathbb{P}[x]}.$$

Now, we have seen that the *prior* $\mathbb{P}[T]$ on a Solomonoff theory was necessarily exponentially weak in the *Solomonoff complexity*[10] $K(T)$. Let's write it $\mathbb{P}[T] = \exp(-\alpha K(T))$. To determine Solomonoff demon's MAP, we only need to maximize the logarithm of the numerator, which is equivalent to minimizing the negative logarithm of the numerator. We have:

$$MAP(x) = \arg\min_{T}\{\alpha K(T) - \ln\mathbb{P}[x|T]\}.$$

Guess what! If we now restrict ourselves to the set of theories T for which $\mathbb{P}[\cdot|T]$ is a uniform law over a set of binary sequences, then the theory of duality in optimization[11] shows that there is precisely a value of α for which $MAP(x)$ is no more than an optimal macroscopic description in the sense of the *Kolmogorov sophistication*[12].

In particular, just as Shannon's generalization of entropy extended our conception of information and entropy to non-uniform distributions, Solomonoff's demon invites us to generalize the macroscopic description of the data via the *Kolmogorov sophistication*, to descriptions for which the microscopic uncertainties are non-uniform! In particular, what the *Kolmogorov sophistication* really seeks to measure seems to actually be the *Solomonoff complexity* of credible *Solomonoff theories*, given data x.

Of course, Solomonoff's demon tells us to go further than that. After all, the MAP is only a rough approximation of Bayes' rule. Thus, instead of being restricted to a single macroscopic description of MAP, Solomonoff's demon invites us to consider all macroscopic descriptions, and to weigh them according to their adequate credence levels. In other words, she invites us to introduce the *Solomonoff sophistication* as the

---

[10]In fact, here, I consider rather that $K(T)$ is the length of the algorithmic description of T, not that of its optimal compression.

[11]The Lagrangian associated with the calculation of the *Kolmogorov sophistication* is $\mathcal{L}_c(S,\mu) = (1 + \mu)K(S) + \mu\log_2|S| - \mu c - \mu K(x)$. By strong duality in linear programming, one deduces the existence of $\mu^* \geq 0$ such that the *Kolmogorov sophistication* is equivalent to finding a distribution on S which maximizes $\frac{(1+\mu^*)\ln 2}{\mu^*}K(S) + \ln|S|$. Now, this is exactly Solomonoff's equation of MAP with $\alpha = (1 + \mu^*)(\ln 2)/\mu^*$ and theories T which are uniform distributions on S.

[12]Interestingly, the subjectivity of c in the *Kolmogorov sophistication* corresponds to the subjectivity of Solomonoff's prejudice (and in particular the discount factor α of the *Solomonoff complexity* of predictive theories).

expectation of the *Solomonoff complexities* of credible theories *a posteriori*. In other words, I propose to define:

$$\text{Solomonoff sophistication}(x) = \mathbb{E}_T[K(T)|x].$$

To my knowledge, this quantity has never been studied before.

In particular, the intuitive remark of Carroll, Aaronson, and Ouellette seems to relate to the surprisingly great *Solomonoff sophistication* of the current physical state of our universe. While Leibniz was surprised at the existence of something rather than nothing, what amazes me even more is the huge *Solomonoff sophistication* that we observe. Or to refer to Descartes, besides the fact that I think, there is something else that cannot be doubted: the existence of a great *Solomonoff sophistication*. To me, this is the most fascinating and puzzling mystery of the universe.

And its explanation could partly lie in the equally mysterious second law of thermodynamics....

18.8   THE BENNETT LOGICAL DEPTH

Another definition also caught my attention: the *Bennett logical depth*. Roughly, this *Bennett logical depth* measures the computational time required to compute an observed structure.

So, for Bennett, the initial and final milk coffees are both "shallow" because they can be calculated very quickly by an algorithm. In the first case, the algorithm says: white at the top, black at the bottom. In the second case, the only algorithm[13] able to calculate the positions of the particles of coffee and milk must be a table listing all the information of all these particles. But then, since these positions are in the memory of the algorithm, it suffices to read its memory to give these positions, which does not require much time[14].

On the other hand, when coffee with milk is being mixed, complex structures emerge. These structures seem to be describable by a relatively succinct algorithm. However, this algorithm will have to make

---

[13]Or rather, it's the algorithm of minimal *Solomonoff complexity*.

[14]In fact, if the world is deterministic, it suffices to let the simulation of the world run from the initial state, which could be done with few bits if this initial state is simple to describe. However, it can be argued that, knowing the final state only, this initial state is impossible to determine in a reasonable time. As a result, the *pragmatic logical depth* remains weak, because the shortest algorithmic quickly identifiable that generates the final state is the one that knows this final state. In some sense, both initial and final states have no easy-to-find computation subtlety.

long calculations to determine the positions of the coffee particles. The intermediate state of the mixture between coffee and milk would therefore have a great *Bennett logical depth*. Aaronson, Carroll, and Ouellette suggest that this great *Bennett logical depth* of an intermediate state is not specific to their cup of coffee; the whole universe would be in an intermediate state of enormous *Bennett logical depth*.

In 2018, Rachid Guerraoui and I used this argument to try to justify the success of *deep learning*[15]. We began by noting that almost all *machine learning* algorithms were very fast algorithms. Moreover, roughly, shallow neural networks form the set of fast algorithms (once parallelized). If we believe that the universe around us and the data we collect have a great (non-parallelizable[16]) *Bennett logical depth*, we are led to conclude that the inevitable weakness of these shallow algorithms is their speed!

Add to this argument the *Solomonoff sophistication* of raw data. Our prediction problems now seem to necessarily require both many parameters and sufficiently long computing times. This is precisely what deep neural networks offer! In addition to this, the neural networks have a structure that allows them to learn in real time, for instance through stochastic gradient descent (SGD) discussed in chapter 17. This explains why deep neural networks have reached unequaled performances. Both human-developed algorithms and alternatives of *deep learning* failed because their codes were not complex enough, because their learning was too slow, or because their calculations were too fast.

In particular, if *Bennett logical depth* is necessary to model the world and solve the problems of artificial intelligence, then the intermediate stages of computation seem indispensable. These intermediate calculation steps, especially when they are recurrent, then correspond to hidden variables. And the longer the calculations are, the more these intermediate steps look like deep variables - and are therefore abstract.

The *Bennett logical depth* of our universe thus seems to be the key to explain the unreasonable effectiveness of abstraction.

---

[15] *Deep Learning Works in Practice. But Does it Work in Theory?* LN Hoang and R. Guerraoui (2018).

[16] The parallelizability of algorithms is also related to the open *P versus NC* problem in theoretical computing, which asks if any polynomial algorithm can be parallelized into a polylogarithmic algorithm with a polynomial number of computing machines. Like the majority of computer scientists, we conjectured that the answer is no.

18.9   THE DEPTH OF MATHEMATICS

If *deep learning* is becoming the pinnacle of abstraction in *machine learning*, it's, however, a small hill compared to the mountains that mathematics built. Of all human endeavors, mathematics is by far our most abstract and deepest creations. Thousands of books go deeper and deeper always further in this abstraction, so much so that even the best of us seriously struggle to delve into the abstract creation of others.

A handful of equations may require years, even decades of meditation, for a fraction of its secrets to be revealed. Some of the greatest mathematicians have spent a large portion of their careers focusing on a single equation. Villani said: "The Boltzmann equation, the most beautiful equation in the world [...]! I fell in when I was little, that is during my thesis." Dirac supposedly said of the equation that now bears his name that it was much smarter than he, especially when he was still young, and did not measure its physical implications[17]. As for me, I hope that, in this book, I'll have managed to share my fascination with Bayes' rule and its incredible unintended consequences that have kept me intrigued and excited for the last two years - and are likely to continue to fascinate me for many more years!

In fact, mathematics is far too deep to be followed step by step by our limited cerebral cortex. To apprehend mathematical objects, we must constantly find crude interpretations. We need to imagine an arrow to reason about a vector, we have to think of distorted sheets to think about non-Euclidean geometry and we have to focus on the known properties of prime numbers to prove theorems about them.

And often, when one is confronted with piles of calculation steps in the middle of mathematical reasoning, at first, it may be desirable to abandon all efforts of reflection and to follow mechanically the rules of calculation to get through it. *Shut up and compute*, said David Mermin to summarize the Copenhagen interpretation of quantum mechanics. One might think that this is the wrong approach to science. Are we not trying to understand the world around us? If this is the goal, we might want to reject the excessive use of mathematical abstraction.

But the role of Bayes' rule is not to adapt credible theories to the cognitive abilities of the human brain. Its goal is prediction. And if the universe has a great *Bennett logical depth*, then the best predictions will have a good chance of requiring a very large number of reasoning steps;

---

[17] *Anti-Matter and Quantum Relativity.* PBS Space Time. M O'Dowd (2017).

and because these reasoning steps correspond to deep calculations, they will necessarily be out of reach of our intuition.

In particular, the depth of mathematics is unmatched by intuitive reasoning. After all, our intuition only seems capable of quick calculations. Intuitive reasoning therefore has no great *Bennett logical depth*. This, I think, is one of the main explanations for the unreasonable effectiveness of mathematics. In other words, this effectiveness does not come from some mathematical nature of the universe (I personally have trouble understanding this notion). It rather seems to come from the *Bennett logical depth* of the present physical state of our universe, and in particular from the presence of some deep phenomena with relatively weak *Solomonoff sophistications* and from our cognitive limits.

18.10   THE CONCISION OF MATHEMATICS

Another explanation for the unreasonable effectiveness of mathematics is its incredible concision. After all, the overwhelming majority of this book can be summed up by Bayes' rule, which stands in a handful of characters. In other words, this book admits a remarkably more concise description than the book itself. It is filled with redundancy. Its *Solomonoff sophistication* is relatively weak. Moreover, I'd be willing to bet that anyone who seeks to optimize the pedagogical aspect of his teaching of Bayes' rule would have written a book quite similar to the one you are reading. I hope that what I'm saying in this book is, according to this person, a bunch of banalities that are hugely information-theoretically compressible banalities - but that are hopefully pedagogically instructive for humans.

One of the greatest advances in mathematics, which can be attributed in particular to Al-Khwarizmi, was the synthesis of a concise mathematical language. But that's not all. In addition to being concise, Al-Khwarizmi's language is extremely mechanical to read. There are not thirty-six thousand possible interpretations, and we must not dwell on the meaning of each of the symbols of this language. In fact, to determine the validity of a formal proof, it suffices to read it stupidly (but with great attention). In computer terms, this reading requires an algorithm of low *Solomonoff complexity*, even though its computation time will be potentially long.

One of the most striking examples of mathematical conciseness is the case of the equations of electromagnetism. When the physicist James Maxwell introduced them in 1861, these equations were not concise. However, the increasing abstraction of mathematics has reduced the long description of these equations to a handful of symbols: $\mathcal{L} = -\frac{1}{4\mu_0} F^{\alpha\beta} F_{\alpha\beta} - A_\alpha J^\alpha$, with $F_{\mu\nu} = \partial_\mu A_\nu - \partial_\nu A_\mu$. Of course, making a prediction from these equations will require a whole algorithmic arsenal; but this arsenal, if restricted to the purely computational aspects, is not so long to describe.

This contrasts sharply with non-formal theories. These rely heavily on an interpretation of natural language and the "common sense" that we humans possess. But the algorithmic description of natural language and common sense is likely to require billions of lines of code to approach human performance. As we have seen in chapter 14, this is what, according to Turing, explains the need for *machine learning* to gain mastery of natural language and common sense. As a result, non-formal theories don't work so well, not so much because they are imprecise, but rather because they require algorithms of huge *Solomonoff complexity* (our brains) to become predictive. However, if one believes Solomonoff's prejudice, theories of large *Solomonoff complexity* are exponentially improbable *a priori*.

Of course, natural language and its interpretation by our human brains are not arbitrary. Natural selection has favored languages and cognitive processes that can predict the environment and social relations of primitive tribes. However, this selection did not favor languages and cognitive processes that could describe particle physics, the economics of globalized markets, or the impact of new technologies. For these problems, it's not surprising that mathematical approaches, even simplistic ones, are favored by the prejudices of pure Bayesianism.

In particular, mathematical elegance seems to lead mathematicians to thoroughly explore and understand the set of concise algorithms, those whose *a priori* credences are large, according to the Solomonoff paradigm. It's not surprising then that the best predictive theories based on mathematical language are more credible *a posteriori* too.

18.11 THE MODULARITY OF MATHEMATICS

I would like to conclude this chapter with a third and final explanation for the unreasonable effectiveness of mathematics, namely the modularity of mathematics. Elegant mathematical theorems are often at the intersection of many subdisciplines. They form bridges between different perspectives. They are sort of Swiss knives that, if handled cleverly, can solve a very wide range of problems. Thus, typically, the notions of derivatives, vector spaces, and graphs are omnipresent in geometry, optimization, and probability, but also in physics, computer science, biology, chemistry, and economics. These notions are like bits, list structures, and sorting algorithms. Mathematical objects form the basis of predictive theories, in the same way that elementary algorithms form the basis of any sophisticated source code.

The reason why programmers decompose their algorithms is so that their most basic codes can be constantly used and reused in different places in a more general code. Similarly, addition and multiplication are often used again and again in a physical model, in the same way that the concept of derivatives is often applied to many distinct physical quantities. It's therefore much simpler and more elegant to define the derivative only once in a very abstract and general way, rather than redefine it each time it is reused. The mathematical language makes it possible to study a large number of distinct models without constantly having to reinvent the wheel.

Here is an example that has become unavoidable in recent decades. A very large number of practical problems, in mathematics, machine learning, in material physics, or in economics, can be written in the form of the minimization of an objective subject to various constraints. This formalism is that of *optimization*. It unifies many areas. The tools that dissect and solve this formalism, like gradient descent, local search, or genetic algorithms, are then Swiss knives often useful to address any of the many problems that can be modeled by this formalism.

The case of theoretical physics is even more impressive. In particular, quantum field physics, far from being a single rigid theory, is based more on a principle of *quantization of a Lagrangian formula*. Indeed, since the use of the *least action principle* by Richard Feynman, physicists have become accustomed to define their quantum theories by a single formula, called the *Lagrangian*, typically of the form $\mathcal{L} = i\hbar\bar{\psi}\gamma^\mu D_\mu\psi - \frac{1}{2}\mathrm{Tr}(F^{\mu\nu}F_{\mu\nu})$. Whatever the exact value of the Lagrangian is, physicists will then be able to use systematic methods to transform this Lagrangian into a partial derivative equation of motion (called the *Euler-Lagrange* equation). Then, these equations can be *quantized*. The predictions of this quantization are then deduced from the equations. In other words, to transform the Lagrangian formula into a set of predictions is only a simple (but long) calculation.

Better still, *gauge theory* makes it possible to derive much of the exact formula of the Lagrangian, from only the symmetries of the Lagrangian. Astonishingly, much of the nature of physical objects and their interactions have been reduced to the abstract study of groups of symmetries.

Indeed, Noether's theorem deduces the conservation of energy from the symmetry of the Lagrangian by translation in time, and the conservation of the momentum from its symmetry by translation in space[18]. Better still, a whole new quantum field theory can be constructed, simply by postulating, for example, that the Lagrangian is invariant under the action of a group, say, $SU(5)$. This is absolutely remarkable! Theoretical physics has managed to detach itself from its elementary objects, such as photons and electrons, only to be interested in incredibly abstract concepts, such as the symmetry group of the Lagrangian.

In fact, it's by restricting itself to this strange formalism that two of the greatest theoretical discoveries of modern physics have been able to anticipate experimental results by far. In 1964, Murray Gell-Man and George Zweig independently postulated that the Lagrangian was invariant under the action of the group $SU(3)$. They discovered that this symmetry implied the breakability of protons and neutrons into even more elementary particles, called *quarks*. After decades of theoretical work, experimental discoveries, and controversies, the Gell-Man and Zweig model finally got accepted. It's now part of the standard model of particle physics. However, by that time, Gell-Man had already received a Nobel Prize for some other work. But the Nobel committee didn't want to reward Zweig without rewarding Gell-Man a second time, and they didn't want to reward Gell-Man a second time. Zweig never received the Nobel Prize.

Even more surprisingly, in 1964, three groups of physicists (François Englert and Robert Brout, Peter Higgs, Gerald Guralnik, Carl Hagen and Tom Kibble) independently discovered that the relativistic formalism of the Lagrangian formulation was incompatible with the existence of massive particles. To save the Lagrangian formalism, the six physicists introduced a new quantum field, now called the Higgs field, whose Lagrangian formulation respects so-called *gauge* symmetries, but whose physical state breaks these symmetries. Remarkably, the interaction of classical particles with a Higgs field whose broken symmetry is broken is indistinguishable from the case where these particles have masses! Better still, the quantification of the Higgs field and its excitation led these researchers to predict a new particle, called the *Higgs boson*. As you probably know, this Higgs boson was discovered experimentally by

---

[18] *The most beautiful idea in physics - Noether's Theorem*. Looking Glass Universe. M Yoganathan (2015).

the CERN LHC in 2012. The following year, Higgs and Englert were awarded the Nobel Prize.

Abstraction won again. It's definitely a matter of luck. But, in light of our discussion about the *Solomonoff sophistication* and the *Bennett logical depth*, it seems that this luck may not have been that improbable....

FURTHER READING

The Unreasonable Effectiveness of Mathematics in the Natural Sciences. Communications on Pure and Applied Mathematics. E Wigner (1960).

Logical Depth and Physical Complexity. The Universal Turing Machine: A Half-Century Survey. C Bennett (1995).

Efficient Estimation of Word Representations in Vector Space. T Mikolov, K Chen, G Corrado & J Dean (2013).

Unsupervised Representation Learning with Deep Convolutional Generative Adversarial Networks. A Radford, L Metz & S Chintala (2016).

Deep Learning. Nature. Y LeCun, Y Bengio & G Hinton (2015).

Exponential Expressivity in Deep Neural Networks through Transient Chaos. NIPS. B Poole, S Lahiri, M Raghu, J Sohl-Dickstein & S Ganguli (2016).

Quantifying the Rise and Fall of Complexity in Closed Systems: The Coffee Automaton. S Aaronson, S Carroll & L Ouellette (2014).

Deep Learning Works in Practice. But Does it Work in Theory? LN Hoang & R. Guerraoui (2018).

Artificial Neural Networks in Machine Learning. Wandida, EPFL. EM El Mhamdi (2016).

The Rise of Machine Learning. ZettaBytes, EPFL. M Jaggi (2017).

Machine Reading with Word Vectors. ZettaBytes, EPFL. M Jaggi (2017).

4 Big Challenges in Machine Learning. ZettaBytes, EPFL. M Jaggi (2017).

Google Cars versus Tesla. ZettaBytes, EPFL. B Faltings (2017).

Achieving Both Reliability and Learning in AI. ZettaBytes, EPFL. B Faltings (2017).

The Most Beautiful Idea in Physics - Noether's Theorem. Looking Glass Univers. M Yoganathan (2015).

Anti-Matter and Quantum Relativity. PBS Space Time. M O'Dowd (2017).

CHAPTER 19

The Bayesian Brain

Bayesian inference successfully accounts for perception mechanisms: given ambiguous inputs, our brain reconstructs the most likely interpretation.

Stanislas Dehaene (1965-)

The Bayesian learner can draw out much more information about a concept's extension from a given set of observed examples [...] and can use this information in a rational way to infer the probability that any new object is also an instance of the concept.

Joshua Tenenbaum (1972-)

19.1 THE BRAIN IS FORMIDABLE

In September 2017, I thought I had finished a first full version of this book. I sent it to Julien Fageot, a mathematician friend. Julien strongly advised me to watch the course of neuroscientist Stanislas Dehaene at the prestigious *Collège de France* entitled *The statistician brain: the Bayesian revolution in cognitive sciences*. "I think the Bayesian brain deserves its own chapter," Julien added. That annoyed me. The book already seemed so long.

Still, I began to listen to Stanislas Dehaene. What a delicious treat! In two days, I devoured his two years of classes on the subject - I would have gone faster if I did not have a job! At each class, I was like a kid discovering a new candy: I was ecstatic! Even more surprisingly, despite all my credences in Bayesianism, or even my pro-Bayesian hooliganism, I kept saying to myself: "Come on! Bayes' rule can't possibly be *that* central to human cognition!" As Bayesian extremist as I thought I was, it seemed that I was still not Bayesian enough!

Yet, come to think about it, I should have expected it. If Bayesianism is truly an optimal form of learning, then natural selection should have selected it when choosing the smart species most likely to survive and reproduce. It's even a prediction of Bayesianism combined with Darwinian evolution: if Bayes' rule is really the solution to all our epistemological troubles, then Mother Nature has necessarily found pragmatic ways of approximating it by natural processes. Well, in the last fifteen years, this prediction has been confirmed, again and again, by the cognitive sciences! *Our brain is a formidable calculator of all sorts of approximations of Bayes' rule.*

This statement may seem troubling to you. After all, I spent a lot of this book criticizing our inability to apply Bayes' rule to simple cases like the Monty Hall problem. I have consistently emphasized the relentless overconfidence that accompanies our inability to understand Bayesianism. Daniel Kahneman and Amos Tversky seem to have spent their careers demonstrating it. Yes, we fail to understand Bayes' rule. Yes, we are unable to apply it to mathematical problems. Yes, we are bad Bayesian thinkers.

But Mother Nature did not select our ability to think about abstract problems; she has selected our ability to adapt to our environment. As a result, the Bayesian inferences that our brains make are the *unconscious* processing of the data our senses expose us to, especially when the processing of these data may have been crucial to survival in the animal kingdom or to our understanding of our social environment.

"Our human brain is based on ancient skills in evolution. We inherit intuitive skills and representations in areas that were and still are very important to the survival of our species. So all children are born with a concept of space, with a concept of numbers, with, probably in the case of the human species, a circuit specialized in language," says Stanislas Dehaene. "Obviously Education aims to go beyond that knowledge. Education invites us to conceive new disciplines, such as reading, writing, formal symbolic arithmetic, which have not been anticipated by evolution. But we will 'recycle' old brain systems for these new cultural uses."

It is this recycling that can be faulty and violate the laws of probability. However, pre-existing, often very unconscious, processes and their pragmatic applications seem to coincide with Bayesian calculations in a stunning way.

19.2   MOUNTAIN OR VALLEY?

Open a topographic map, that is to say a map on which the elevations are represented. For example, you can open Google Map on your phones, go to options, and activate the "terrain" mode. Zoom into a mountain range or valley, like the Chamonix Valley in France. Now look at the map upside down. If you are on your phone, turn it without turning the image on your phone. Something weird should intrigue you....

You may have the impression that the mountains have turned into valleys, and that the valleys have turned into mountains! In particular, what is on the map at the top of a shaded area looks like a mountain, while the valleys seem to be at the bottom of the shaded areas.

But where does this perception come from? What makes us distinguish valleys and mountains in a map?

It turns out that, like many other unconscious perceptions, our conclusions are deduced from a Bayesian calculation. In particular, the prejudice essential to our interpretation of maps is the origin of the lighting. Indeed, the shadows of the maps are those of a lighting coming from the top of the map - although this lighting is physically impossible in regions of the northern hemisphere like Europe or the United States, where the Sun always illuminates from the south[1]!

While it's incorrect in the case of maps, this prejudice is in fact perfectly justified in our daily life. The lighting often comes from above, be it the lighting of the sun or that of our electric lamps. As a result, when we observe the faces of others, we usually see the nose emerge above a shaded area, in contrast to eyes that are located below a shaded area. In fact, reverse lighting can seem scary, which is why it's so often used in horror movies.

More generally, our visual cortex possesses the remarkable and unconscious faculty of guessing the origin of illumination in an image, and then of better interpreting the content of the image. This process is very similar to the Bayesian networks and Boltzmann machines discussed in a previous chapter: our brain seems to quickly use hidden variables to understand the observed variables.

19.3   OPTICAL ILLUSIONS

The prejudices we have about lighting, which lead us to misinterpret upside-down maps, are extremely effective in our analysis of many images

---

[1] *The "Mountain Or Valley?" Illusion.* Minute Physics. H Reich (2017).

of the natural world. However, our prejudices can also mislead us when we are dealing with images built according to rules that are unusual in the natural world.

Take two identical pens. Lay one horizontally. Put the other just above, vertically, to form the symbol ⊥ that mathematicians use to designate perpendicularity. Look at your pens. You should have the impression that the vertical pen is longer than the horizontal pen - even though, by construction, these two pens are actually of the same length! You can also test yourself by watching Figure 19.1.

Here again, this optical illusion is perfectly explained by involving a justified prejudice of our Bayesian brain. Accustomed to seeing images in perspective, it often sees vertical lines that actually correspond to horizontal lines seen in perspective - typically the rails of a train line. But the perspective will then modify how we see these lines, so that these lines will appear shorter than they really are. Our Bayesian brains seem to unconsciously integrate this justified prejudice on perspective. This amounts to inferring that the vertical lines are longer than they appear. This is probably why we instinctively think, erroneously, that the vertical line of our perpendicularity symbol ⊥ is longer than the horizontal line.

Figure 19.1 Some classical illusions. On the left, the two segments have the same length. In the middle, the two guys are of the same size. On the right, squares A and B have the same brightness (image by Edward H. Adelson).

Such perspective effects are used in other optical illusions, where a character is typically copied and pasted in the foreground and in the background. If the character in the background seems larger, it's because our Bayesian brains have subconsciously applied (an approximation of) Bayes' rule, to deduce that the character in the background is very probably much larger than he appears to be. The optical illusion is again explained by a Bayesian inference.

In the same way, another classical illusion is a checkerboard, part of which is shaded. The white squares in shade seem brighter than the black illuminated squares. In fact, this is not the case. But our Bayesian brains unconsciously make approximate Bayesian inferences that account for the effect of lighting, so as to obtain the more *useful* conclusion that illuminated black squares are darker than shaded white squares.

19.4 THE PERCEPTION OF MOTION

Take a very flattened diamond and tilt it so that its main axis is slightly diagonally up-right and down-left. Then, move the diamond from left to right. When the diamond is well contrasted with the background (typically black on white), we can see the diamond moving from left to right. However, intriguingly, when we decrease the contrast (light gray on white), something strange happens. Many subjects then begin to see the diamond go in the lower-right direction. I myself reproduced the experience on Twitter, and according to my survey[2], 39% of the 376 respondents saw the diamond going down!

How is it possible that our brilliant Bayesian brains come to this erroneous conclusion? In a remarkable article[3], Weiss, Simoncelli, and Adelson show that this erroneous conclusion is precisely that of a Bayesian brain whose calculation integrates the uncertainty caused by the weakness of the contrast.

But before coming to the Bayesian explanations of the three authors, let's first discuss an astonishing faculty of our brains that we tend to underestimate: when a well-contrasted diamond moves, we are able to determine its movement. It's a remarkable feat! After all, from a sensory point of view, all we see are pixels turning on and off. How does one translate the dynamics of the lighting of the pixels of a video into a dynamics of the object represented in the video?

As we have seen, our cerebral cortex is mostly able to detect the guidelines of images. When a diamond moves, the brain basically sees the edges of the diamond move. Except that each edge of the diamond is inclined. So, when the diamond moves from left to right, each edge of the diamond actually seems to move in another direction. In fact, any moving straight line seems to move in a direction perpendicular to the line. For infinite lines, such a motion is in fact indistinguishable from any other translational motion of the line.

---

[2] `https://twitter.com/science__4__all/status/911943074049396736`.

[3] *Motion illusions as optimal percepts*. Nature Neuroscience. Y Weiss, E Simoncelli & E Adelson (2002).

Nevertheless, as a good Bayesian, our brain knows that a motion of the line in a direction orthogonal to the line is only one of the possible explanations of how we see the line moving. While it's *a priori* the most probable motion, other motions have a non-zero probability. According to Weiss, Simoncelli, and Adelson, the brain will then combine the probable motions of all the edges of the diamond, and suppose that the diamond as a whole has only one motion. In other words, the brain will determine the most credible motion *a posteriori*, given the likely motions of the different edges. This allows the brain to infer the motion of the diamond from the motions of the edges - even if the motions of these edges are uncertain!

This explains why the brain makes the right inferences when the contrast is high enough. What happens when the contrast is low? The brain then sees the motion of the edges poorly. It has uncertainty as to the speed with which these lines move (which adds to the uncertainty of the direction of motion of these lines). Surprisingly, this additional uncertainty leads to a different Bayesian calculation. When this uncertainty is large enough, Bayesian inference then leads to preferring the hypothesis of a diagonal down-right motion - this motion is then the *maximum a posteriori*.

Unbelievable! The erroneous prediction of the brain is explained by a Bayesian calculation of the brain, when it's subjected to an additional uncertainty to which the reduction of the contrast leads! This prediction is bad. But it's bad for good reasons: it was the optimal way to handle the uncertainties to which the Bayesian brain was subjected.

And if the highly artificial experimental framework of this experiment was chosen to fool our Bayesian brain, the Bayesian prediction is probably much more relevant in empirical cases!

19.5 BAYESIAN SAMPLING

One of the most impressive phenomena in cognitive science is the ability of our brains to perform representative sampling, probably through *Markov Chain Monte Carlo* (MCMC), a technique we discussed in chapter 17. As we have seen, sampling is often an effective approach to describe a probabilistic phenomenon, especially when probability distributions at play are difficult to describe in mathematical language. Yet, almost all laws of probability are, in practice, too difficult to describe and manipulate directly.

Our Bayesian brains seem to have understood it. Rather than reasoning with several incompatible theories at the same time like the *pure*

Bayesian, our brains prefer to reason sequentially, first with a very credible theory, then, if time permits, with another credible theory.

This is typically how one will try to interpret ambiguous images. You have probably already seen this ambiguous image which, from one angle, seems to be the image of a duck, whereas, from another angle, it seems to represent a rabbit. I've shown it in Figure 19.2. The strangest thing is that it seems impossible to see both credible interpretations of the image at the same time. Our Bayesian brains seem able to see only one interpretation at once.

Figure 19.2 Duck or Rabbit?

In 2011, Moreno-Bote, Knill, and Pouget[4] studied this phenomenon. They showed subjects two black grids on a white background moving relative to each other. The subjects are then asked which of the two grids is above the other. The subjects then alternate between the two possible interpretations. The Bayesian brain hypothesis postulates that the proportion of time during which the subject adopts an interpretation is his Bayesian credence in this interpretation.

To test this hypothesis, the genius of the three researchers was then to study the effect of two different variables on these proportions of time devoted to the two interpretations, and to verify that the combination of the two effects was the product of each effect. The two effects studied were the tightening of the lines of one of the grids and its relative velocity. Surprisingly, the multiplication of the two effects predicted by the Bayesian brain hypothesis coincided tremendously well with the

[4]*Bayesian sampling in visual perception.* PNAS. R Moreno-Bote, D Knill & A Pouget (2011).

combination of the two effects[5]! The sampling of our Bayesian brains seems to perfectly obey the laws of probabilities! Better still, Wong and Wang[6] determined credible neural mechanisms to justify the physical ability of our neurons to implement such Bayesian samplings!

A strange corollary of this conclusion is that we can improve our predictions by making multiple predictions for different interpretations. Remember. The *pure Bayesian* improves her predictions by averaging the predictions of credible models. To test this corollary of the Bayesian brain hypothesis that samples via MCMC, Vul and Pashler[7] asked 428 subjects: "What percentage of global airports are in the United States?" They asked them to give two answers to this question. The second answers were often worse than the first ones. But surprisingly, the average of the two answers was nevertheless significantly better than the best of both answers!

Better yet, for half of the subjects, Vul and Pashler waited 3 weeks before asking them to provide a second answer, to allow the subjects to really change the interpretation of the problem before they respond. Guess what! The improvement of the average of the answers was then even better than in the case where the second answers were given just after the first ones. "It's not bad to ask yourself the same question a second time," concludes Dehaene.

19.6 THE SCANDAL OF INDUCTION

In 2011, Josh Tenenbaum and three collaborators[8] introduced a new word in the English language: the concept of "tufa". To explain what it is, Tenenbaum offers three images of examples of tufas. Of course, the purists among us want to say that this is a very bad way to define a new concept.

And yet. The four researchers then pointed out that with only three examples of what a tufa is, and no example of what a tufa isn't, we are nevertheless all pretty much in agreement about what a tufa is. Indeed, Tenenbaum then offers 39 images, and it seems obvious to us that 6 of

---

[5]This is based on certain hypotheses of independence of the effects.

[6]*A recurrent network mechanism of time integration in perceptual decisions.* Journal of Neuroscience. KF Wong & XJ Wang (2006).

[7]*Measuring the crowd within: Probabilistic representations among individuals.* Psychological Science. E Vul & H Pashler (2008).

[8]*How to grow a mind: Statistics, structure, and abstraction.* Science. J Tenenbaum, C Kemp, T Griffiths & N Goodman (2011).

these images are tufa images, and that the other images are not images of tufas! Unbelievable!

This amazing phenomenon is sometimes called the *scandal of induction*. It is far from the standards of the *p-value* and Fisher's methods. However, according to Tenenbaum and his collaborators, it is perfectly justified by Bayesian principles. In particular, it's sufficient to start from the prejudice that the three examples of tufa are representative of tufas, and to suppose that what does not resemble these examples sufficiently is not a tufa. Better still, this learning by example seems to be the way we really learn the meaning of words. We have never had a formal definition of what a cat is. We just saw similar shapes, and our parents told us they were called "cats".

A simple but convincing explanation of the scandal of induction is to assume *a priori* that all things are structured in the form of a tree, like the phylogenetic tree of life. In particular, therefore, a definition will be admissible if it's compatible with the structure of the tree, that is to say, if it corresponds to a node of the tree. This is the approach of phylogeny, which defines a family of species, such as mammals, as the descendants of a certain common ancestor.

The set of possible definitions is then all the nodes of the tree. We know that a tufa is necessarily one of the nodes of the tree. It remains to be determined which one. Tenenbaum and his co-authors then propose to study the simple case where, a priori, everything is equiprobable. The maximum *a posteriori* will also be the likelihood maximizer, namely the definition of tufa for which the probability of the three examples of Tenenbaum is maximum. A simple calculation shows that it's the lowest node of the tree which contains all the examples provided.

Tenenbaum and his co-authors postulate that it's thanks to such a calculation that with only three examples, a consensus was formed around the meaning of the word "tufa". And, more generally, they postulate that this is how babies learn the meaning of words.

19.7 LEARNING TO LEARN

This explanation, however, seems quite incomplete. In particular, one may wonder how the brain could determine the tree that classifies objects. Better yet, where does the choice of tree structure come from? The answer of Tenenbaum and his co-authors is fascinating: the learning of the relevance of the tree structure, and that of the structure of the tree of objects, both seem to be obtainable by Bayesian calculations on several levels. In other words, the brain seems to perform *hierarchical* Bayesian calculations.

To understand this, let's imagine that an object A is round and heavy, an object B is round and very heavy, and an object C is very round and heavy. Imagine that you are told that object A is perfect for cooking tufas. Can you generalize this to objects B and C? And if you only have objects B and C, which of these should you use to cook your tufas?

To answer this question, it's necessary to know if, to cook tufas, the roundness of objects is a more important characteristic than their weight, and to what extent the variations of roundness and weight affect the cooking of tufas. The hierarchical Bayesian calculation will then use other similar examples, for example by taking into account the effect of roundness and the weight of objects on the cooking of other foods.

This approach is also the one used to solve Stein's paradox. Remember, to judge the level of a pilot from statistics concerning him, it was then useful to invoke the levels of other pilots. In the same way, this principle is crucial to solve the problem of the Scottish sheep. To generalize the blackness of a sheep, it's useful to know how the colors of other species vary geographically.

What's fascinating is that this hierarchical Bayesian approach can be seen as a way of learning to learn. After learning the effects of roundness and weight on the kitchen in general, we are now more able to determine the usefulness of objects B and C for tufa cooking, even though we have only been given information about object A! Hierarchical learning allows us to ignore irrelevant variables. It enables us to focus on what matters.

Of course, I have given you here only an extremely basic example. But in a more general way, the hierarchical Bayesian approach makes it possible to quickly determine the right way of structuring our models of the world, like the choice of the structure of a graph for the classification of the objects or that of the causal principles to study physical phenomena. Once the good structures of the models are discovered, learning can be (computationally) accelerated, since it can now be done (only) within relevant restricted models.

In fact, we have already seen a more specific example of such learning, namely latent Dirichlet allocation (LDA). This Bayesian structure allows us to learn as we go the relevant categories to classify documents, which will make the classification of future documents much simpler. In fact, it's a similar model that Tenenbaum and his co-authors studied[9] in 2011, to conclude with what they called the *blessing of abstraction*.

19.8 THE BLESSING OF ABSTRACTION

Tenenbaum and his co-authors illustrate this blessing of abstraction as follows. They first consider a hierarchical structure. In other words, they consider several very general theories, each subdivided into particular causal models, which are then differentiated by particular cases. Then, they train a Bayesian artificial intelligence that applies Bayes' rule to all levels of this hierarchy.

This approach has a nice interpretation in terms of Solomonoff induction. Remember. Solomonoff's approach was to learn theories that can explain past data and make predictions for future data. It's a safe bet that the best theories are the ones that use Bayes' rule to make predictions from the past. The Bayesian calculation of Solomonoff's demon, which takes place at the inter-theory level, will then have good chances to put the credence on the Bayesian theories that will again apply Bayes' rule to a lower level. Better still, these Bayesian theories will then study various subtheories, to finally prefer Bayesian subtheories. And so on.

In any case, the simulations of Tenenbaum and his co-authors show that hierarchical Bayesian learning is initially slow at all levels. But after a few hundreds or thousands of samples studied, the hierarchical structure ends up putting its Bayesian credences on good general theories. Intriguingly, learning the good general theories is much faster than learning the lower levels. A crucial corollary of this empirical observation is that starting the learning directly with good general theories does not save much time!

We'll come back to this, but some psychologists stunned by the prowess of young children ended up postulating that many models were genetically encoded in their brains. It seems that the child, for example, has a predisposition to invoke the principle of causality. However, the simulations of Tenenbaum and his colleagues show that the general principles of the models are actually relatively quick to learn for a hierarchical Bayesian intelligence! No need for a premature learning of models. Bayesianism seems to allow the discovery of relevant paradigms to think about the world with disconcerting efficiency!

---

[9] *Learning a theory of causality*. Psychological review. N Goodman, T Ullman & J Tenenbaum (2011).

Some theorists sometimes complain about the misuse of artificial intelligences in science. The exploration of *Big Data*, they say, will never allow discovering elegant and synthetic formulas. It will never lead to wonderful equations like those of general relativity that Einstein discovered. Machine learning seems too mechanical. It seems to lack the genius of which only great minds are capable.

However, this argument omits the blessing of the abstraction of Bayesian methods. In particular, the hierarchical Bayesian approach seems perfectly able to distinguish among a large set of general structures of theories those which best explain the empirical data. They seem able to distinguish the generic form from the best theories, and thus provide us with the appropriate paradigm for thinking empirical data.

Therefore, it seems that, as long as you invoke a sufficiently hierarchical Bayesian model with numerous subtheories, and as long as you have sufficiently many data, you don't need to encode prior knowledge to design efficient machine learning algorithms. In fact, not only will you not gain much by attempting to do so, you might even enforce too much rigidity in your model, which may thus underfit and suffer from so-called *inductive bias*. Amazingly, as we develop larger and larger artificial intelligence models, this intuitive observation seems to be confirmed again and again for numerous applications, from board games to natural language processing. Eventually, models primarily designed for broad *learning* seem to outperform all handcrafted specialized alternatives.

In particular, if some elegant equation is the way to model these data, it seems to be a safe bet that an artificial intelligence close enough to Solomonoff's demon will determine it like neuroscientists discovering the relevance of the Bayesian framework to study cognitive processes.

19.9   THE BABY IS A GENIUS

We have seen that Turing compared a baby to a notebook with "little mechanism, and a lot of blank pages". This principle has prevailed for a long time. Chinese leader Mao Zedong, for example, is said to have praised the ignorance of his people, pointing out that "on a blank sheet of paper without any mark, the freshest and most elegant words can be written; the freshest and most elegant designs can be portrayed". However, this idea is being undermined by psychologist Steven Pinker and modern neuroscience. And by Bayesianism.

Before getting there, note that even before the age of 1, a baby already has a set of *core knowledge*. It has an intuitive understanding of objects, numbers, space, and grammar. Better still, the baby seems already endowed with statistical abilities. The baby will typically stare curiously at events that his statistician's brain deems improbable.

In 2008, Xu and Garcia[10] showed this through an experiment on 8-month-old children. Xu and Garcia's experiment is inspired by Laplace's urns. The urn contains a large number of balls. These balls are red or white. Five balls are then drawn from the urn. Imagine that 4 red balls are drawn, as well as 1 white ball. Laplace's succession rule, discussed in chapter 6, invites us to think that the proportion of red balls in the urn is about $(4 + 1) / (5 + 2) = 5/7$.

Then, we reveal the contents of the urn. When the urn contains mostly red balls, the child is not surprised. However, when the urn contains in fact mostly white balls, the child then begins to stare at the urn for a long time, as if it hid something mysterious. Unbelievable! The 8-month-old seems already able to intuit Laplace's Bayesian calculus, and begins to investigate, like a scientist, the cases where his Bayesian predictions seem contradicted by observation!

19.10 LEARNING TO TALK

One of the most amazing feats of a baby is learning languages. We have all experienced it. Despite decades of study, or even immersion in a foreign country, it's very difficult for us to speak like locals. French expatriates in English-speaking countries will retain their French accent. Even British people living in the United States usually keep (part of) their British accent. Conversely, babies have a remarkable ability to acquire the language of their parents. After a few years, they achieve a mastery of the language that few foreigners will reach in their life. At the age of two, their lexicon grows at a stunning pace of 10 to 20 words a day! How do they do it?

Neuroscience suggests that babies' learning of the language relies heavily on the study of the statistical properties of the elementary sounds of language, called *phonemes*. The experiment of Saffran, Aslin, and Newport[11], for example, exposes babies to syllable sequences. The rhythm at which these syllables are spoken is regular, so that it's impossible to infer anything from the rhythm at which these syllables are spoken. However, a certain statistical law lies behind the sequence of

---

[10] *Intuitive statistics by 8-month-old infants.* PNAS. F Xu & V Garcia (2008).

[11] Statistical learning by 8-month-old infants. Science. J Saffran, R Aslin & E Newport (1996).

these syllables. For example, a "to" will always be followed by a "ki", while "bu" will be followed by "gi" only a third of the time. Such a construct is known as a *Markov chain*.

Surprisingly, the baby seems not only able to spot these statistical regularities. It even successfully determines, probably via a Bayesian inference, the segmentations that probably form words. In other words, it seems that a Bayesian calculation is crucial to learning words from a spoken language.

Then, the baby manages to combine words to distinguish sentences. This is a remarkable feat! Indeed, to do this, the baby must be able to identify the grammatical structure of sentences, that is, to recognize that some words are verbs, while others are names. The baby must also learn that, in some languages, the order of words can be reversed to ask questions. However, these efforts, which are essential for constructing sentences, will also be very useful for learning more words.

Let's illustrate this. Imagine that two bowls are on the table. One is blue, the other is chrome. A parent asks a child to bring back the chrome bowl. The child does not know the meaning of the word "chrome". However, he knows it's an adjective that describes the bowl. What's more, he guesses that the parent probably does not speak of the blue bowl, since the parent would probably have called it that way. The child then concludes that the chrome bowl is probably not the blue bowl, and that the word "chrome" is the color of this chrome bowl.

Unbelievable! The child has just predicted and learned the meaning of the word "chrome" without data. Even though it had never heard that word, it still figured out how to determine its meaning. And its genius was to rely on its prejudices. As Solomonoff would say, "it is possible to do prediction without data, but one cannot do prediction without a priori information".

19.11 LEARNING TO COUNT

When babies learn the alphabet or numbers, they memorize a series of words or sounds. Moreover, learning the alphabet is often accompanied by a melody that helps memorization. Similarly, learning numbers is sometimes accompanied by counting on fingers.

However, knowing and reciting by heart a sequence of words is not sufficient to be able to reuse them adequately in other contexts. Many young children know how to recite numbers. But if they are asked to bring back three objects, they take a handful of these items instead of taking only three.

More intriguingly, children will only understand the meaning of "one" first. Then they will learn the word "two" but will not go further. For several months, they will know the meaning of these two numbers, even if they can recite numbers beyond "one" and "two". Then they will learn "three". And will stay there.

Then, all of a sudden, by the age of three and a half, children seem to be making a remarkable conceptual leap. They manage to match the sequence of numbers to the meaning of numbers. They establish a relationship between "include an additional object" and "move on to the next word in the sequence of numbers". How was this relationship established?

According to Tenenbaum, and his collaborators Piantadosi and Goodman[12], the child has just achieved the fantastic feat of having learned a *recursive* algorithm. This algorithm takes into account any set of objects. If the set is empty, the algorithm concludes with the last number spoken. Otherwise, the algorithm removes an object from the set and states the next word in the sequence of numbers - if it was the first object removed, the child then pronounces the word "one". Then the algorithm counts the rest of the objects in the set, keeping in mind the last word spoken.

What I described here in abstract terms is actually an algorithm that you know very well, and that you apply each time you count one by one the elements of a set. Remarkably, however, this compelling algorithm is actually surprisingly abstract. And the most amazing thing is that the baby statistician was able to identify and select this abstract recursive algorithm.

Dehaene suggests that this ability to think and exploit algorithmic recursion might be the fundamental difference between the human brain and that of other animals. Be that as it may, this Bayesian study of recursive algorithms suggests that our Bayesian brains may ultimately not be that different from Solomonoff's demon's.

---

[12] *Bootstrapping in a language of thought: a formal model of numerical concept learning*. Cognition. S Piantadosi, J Tenenbaum & N Goodman (2012).

19.12 THE THEORY OF MIND

One of the most fundamental faculties that our Bayesian brains learn in childhood is the *theory of mind*. It's the ability to reflect on what other people think, and to use that model of another's thinking to make predictions or learn new concepts. Even before two years old, the baby will be able to follow the eyes of another, to imitate his actions, even to distinguish his intentions from his failed actions. Later, he will learn that the beliefs of another are not necessarily the same as those of a third person, that others can lie or talk with irony, sarcasm, and humor, and that many of the gestures of others are involuntary. Perhaps because we are social creatures, this theory of mind is essential to our learning.

Imagine a transparent urn containing a lot of blue toys and some similar yellow toys. An adult plunges her hand into the urn, and removes three blue objects. She presses on each of these three blue objects. Each makes a noise. A yellow object is then drawn from the urn. It is given to a child. The question that arises is if the child who observes this situation will generalize the ability of blue objects to make noise to yellow objects. The answer is yes. The child will then press 3 times on the yellow toy, which makes no noise, before giving up its attempts.

So far, nothing very surprising. What is amusing is the variant of the experiment in which the urn contains only a few blue toys and many yellow toys. Again, the adult pulls three blue toys and shows that by pressing it, it can make them make noise. Then she pulls out a yellow toy. Will the child believe that this yellow toy makes noise? Curious, the child will indeed test the toy. However, this time, the baby only tests it once!

It seems that the child noticed that in this second case, the adult's sampling was biased. The child has probably guessed that the adult has purposely drawn only blue toys, because they are very different from yellow toys. The child understood that there was a *selection bias*. Using an approximate Bayesian calculation, it concluded that any property of the blue toys was therefore not clearly transferable to the yellow toys!

The child has not only been able to determine the effect of selection bias on the generalizability of an observation (a feat that we fail to achieve in more abstract cases!), it also knew how to model the mind of the adult to do this. The child is endowed with *theory of mind*, and knows how to apply it to avoid erroneous conclusions.

19.13   NATURE VERSUS NURTURE

If Dehaene speaks of a "Bayesian revolution in neuroscience", it's not only because the Bayesian brain theory accounts for a monstrous amount of experimental results that are difficult to interpret otherwise. The magic of the Bayesian approach is also to provide a surprisingly complete answer to one of the old debates about child learning. This old debate opposes the innate, according to which the brain is born with a knowledge *a priori* of the language and its grammar, to the empiricists according to whom all is learned. And this debate was personified by the mythical opposition between psychologists Skinner and Chomsky.

It all started with the publication of *Verbal Behavior* in 1958 by Burrhus Skinner. Skinner relied in particular on experiments in which he showed that pigeons were able to learn the meaning of words such as "peck" or "turn". Skinner discovered that by rewarding the pigeons as they performed the action associated with the printed words shown to them, the pigeons could learn the meaning of the words. Or at least to correlate the words they saw with the action to be taken to be rewarded.

However, Noam Chomsky retorted that mastering human languages was a much more complex and difficult skill to learn than the correlation between words and events. For Chomsky, the sophistication of language and grammar could only be learned by a brain predisposed to do so. Chomsky postulated that our brains were genetically programmed to understand and manipulate what he called the *universal grammar.*

However, according to the Bayesian brain hypothesis, which has been shown to have incredible predictive power, every baby must first of all possess the ability to generate complex models of various phenomena in its environment, and that of applying Bayes' rule to remember and explore the most useful models. Crucially, the simulations of Tenenbaum and his collaborators suggest that this apparatus may be necessary and sufficient to allow the baby to model its environment, to understand language, and to learn to speak - which is corroborated by the Solomonoff completeness theorem!

In a way, the innate structure of the baby's brain gives reason to innateness, which postulates the need for our brain's predispositions. However, this innate structure seems much more abstract, simpler, and more wonderful than the structures that Chomsky postulated. We have seen it. Large paradigms of thought can actually be quickly determined by the hierarchical Bayesian approach.

In contrast, empirical data play a crucial role in the selection of useful models via Bayes' rule. However, again, this learning from the data is absolutely not reduced to the simple computation of correlations with reinforcement learning. This learning is that of an extremely complex and sophisticated model, which performs Bayesian calculations at different levels.

It's amazing. I find it absolutely fascinating and disturbing that Bayes' rule, which seems so inaccessible to me as I feel incompetent in manipulating probabilities, seems in fact calculated by my brain, the same brain that aspires, in vain, to become a conscious and competent Bayesian thinker. We all seem to have incredibly sophisticated machines that are optimized to perform complex Bayesian calculations, in a massively parallel way with unparalleled energy performance. However, strangely, we are very much unaware of these calculations. And we are very much unable to reuse them to think correctly.

FURTHER READING

Statistical Learning by 8-month-old Infants. Science. J Saffran, R Aslin & E Newport (1996).

Motion Illusions as Optimal Percepts. Nature Neuroscience. Y Weiss, E Simoncelli & E Adelson (2002).

A Recurrent Network Mechanism of Time Integration in Perceptual Decisions. Journal of Neuroscience. KF Wong & XJ Wang (2006).

Measuring the Crowd within: Probabilistic Representations within Individuals. Psychological Science. E Vul & H Pashler (2008).

Intuitive Statistics by 8-Month-Old Infants. PNAS. F Xu & V Garcia (2008).

Infants Consider Both the Sample and the Sampling Process in Inductive Generalization. PNAS. H Gweon, J Tenenbaum & L Schulz (2010).

Bayesian Sampling in Visual Perception. PNAS. R Moreno-Bote, D Knill & A Pouget (2011).

How to Grow a Mind: Statistics, Structure, and Abstraction. Science. J Tenenbaum, C Kemp, T Griffiths & N Goodman (2011).

Learning a Theory of Causality. Psychological review. N Goodman, T Ullman & J Tenenbaum (2011).

Bootstrapping in a Language of Thought: a Formal Model of Numerical Concept Learning. Cognition. S Piantadosi, J Tenenbaum & N Goodman (2012).

The "Mountain Or Valley?" Illusion. Minute Physics. H Reich (2017).

IV

Beyond Bayesianism

CHAPTER 20

It's All Fiction

> The Universe is made of stories, not of atoms.
>
> Muriel Rukeyser (1913-1980)

> Science replaces visible complexity by invisible simplicity.
>
> Jean Perrin (1870-1942)

20.1 PLATO'S CAVE

Imagine yourself chained and forced to observe only the wall of a cave. From time to time, you see shadows appear and move on the wall. But you cannot turn your head. You cannot look directly at the cause of the shadow. All you know about the world around you are the shadows you see on the wall. Your reality is limited to these shadows. These shadows are your reality.

What I am describing here is the allegory of the cave introduced by the philosopher Plato at the time of ancient Greece. For Plato, this allegory was an adequate metaphor for the ignorance of his fellow citizens. Plato goes even further. He imagines that one of your companions, chained too, is released. This companion, let's call him Pierre-Simon, turns around, but is so blinded by the sunlight that he prefers to sit down and contemplate the wall and its shadows. For Plato, not only are fellow citizens very ignorant, but they take comfort in their ignorance. A blissful ignorance.

Plato then imagines that, nevertheless, Pierre-Simon is brought, perhaps by force, out of the cave. Pierre-Simon is frightened at first. He is confused. However, slowly but surely, he ends up enjoying exploring and discovering a world much more real than the wall and its shadows. All excited, he comes back to the cave to tell you his innumerable discoveries. The problem is that you now take him for a fool! You do not believe him. You reject all that he has to say to you. For Plato, this illustrates the greatest weakness of his fellow citizens: his fellow citizens prefer to live in their wrong reality and not to question it.

The allegory of Plato's cave was revived in a memorable scene of the movie *Matrix*. While Neo, the main character, begins to doubt the realism of his daily life, Morpheus offers him the famous dilemma of the red and blue pills. Take the blue pill, and wake up ignorant. You will enjoy your daily life, but will be unable to see that it's only made of shadows projected on the wall of a cave. Take the red pill, and leave this daily life. You'll then get to discover the *real* world. Fortunately for the interest of the film, Neo chooses the red pill.

This red pill allows Neo to leave the computer simulation - the famous *Matrix* - in which he was locked up. Neo then wakes up in a totally different world. He is now faced with an apocalyptic and dangerous reality, governed by machines. Later, Morpheus accompanies Neo in another computer simulation. Neo is confused. "This is not real," he wonders. Morpheus answers with another question: "What is real? How do you define real?" Morpheus then proposes a possible answer. "If you're talking about what you can feel, what you can smell, what you can taste, and see, then 'real' is simply electrical signals interpreted by your brain."

20.2 ANTIREALISM

The allegory of Plato's cave and the simulation of the *Matrix* are fascinating examples because they challenge the nature of reality. But for the *pure Bayesian*, they do not go far enough. In these two examples, it's a given that, beyond the world in which individuals live, there is a *real* world. These two examples still assume that some absolute reality *exists*. In these examples, what disturbs and intrigues is merely the fact that reality is not what the inhabitants of these worlds believe it is.

This discourse is recurrent in the way many scientists talk about science. One can thus read here and there that the sciences reveal an *invisible truth*, or a *hidden reality*, even that science is the only way to *the truth*, and the end of illusions. One might think that water is infinitely divisible, but, *in truth*, water is only a finite set of molecules that slide over each other. One might think that time and space are absolute, but, *in reality*, they are relative to the trajectory of the observer in space-time.

One might think that the human body contains only our cells, but, *in fact*, most of the living cells that make up our bodies are all sorts of non-human bacteria on which our health and our mood depend.

In this chapter, I'll dwell on what is, unless my reasoning is mistaken, an inevitable counterintuitive consequence of *Bayesianism*. I left it for the end, because you will probably want to reject it, even if you have carefully followed and accepted the ideas in this book. This strange consequence that seems unavoidable to me is the assertion that all ideas are *fictions*. Some fictions are more credible for the *pure Bayesian*; others are more useful for the *pragmatic Bayesian*. However, for any good Bayesian, it seems necessary to me that everything can only be fiction; or more precisely, everything is just simulations of an infinite number of randomized algorithms on which to bet.

In particular, I argue that a good Bayesian would not accept the assumption that there is necessarily a *reality* consistent with what one imagines, a universe beyond Plato's cave, or a physical world outside the *Matrix*. Better still, I claim that it's neither indispensable nor even useful to postulate that electrons *really* exist, nor even that physics really speaks of an *objective reality*. Certainly, these models are incredibly useful and predictive and deserve a lot of our credences. But according to Box's quote, I argue that it's more *useful* still to keep in mind that these are just models. Yet "all models are wrong". In particular, as a good Bayesian, I'll defend the *usefulness* of this fictionalist position.

20.3   DOES LIFE EXIST?

Debates about realism often turn to existential questions such as the existence of consciousness. This is too controversial a subject to be addressed directly. I suggest to start with a simpler question. Does life exist?

Life is one of the most difficult concepts in science - actually, few scientific concepts are not problematic. Some biologists do not hesitate to recognize that this is simply an ill-defined term. Others propose to list criteria, and to consider that life is the set of natural phenomena that satisfy these criteria. Among the criteria usually cited are the notions of replicability and variations, on which Darwin's theory of evolution is based. The problem is that such definitions often fail to coincide with what most of us would like to call life - in particular, a computer virus would then be considered alive. Some biologists have thus chosen to restrict the living to what revolves around molecules typical of life, such as deoxyribonucleic acid (DNA) and ribonucleic acid (RNA). But one can then wonder if the storage of data on DNA strands could be considered alive.

An alternative is to identify physical properties that are typical of the living. This is the approach chosen by Karl Friston. In particular, Friston built upon the notion of the *Markov blanket*. This blanket, typically the membrane of a bacterium, is a stable material separation that distinguishes the outside world from an inner structure. The interior of the membrane would draw from its environment a certain form of energy, called *free energy*, to constantly consolidate its structure[1]. The fundamental property of *free energy* is to be far from any thermodynamic equilibrium. To use this free energy wisely and to consolidate its internal structure, it's crucial for the interior to anticipate external disturbances - and to determine whether there will be free energy to access in a near future. This requires the interior to develop a model of the external reality. But crucially, the only data that the interior has access to are the information on the Markov blanket.

In 2013, Friston suggested that natural processes were in place to allow the interior to compute an approximation to Bayes' rule called Bayesian variational inference, which was briefly mentioned in chapter 14. Friston even suggests that this is the fundamental nature of life: life is a structuring of a restricted environment and separated from the outside world by a stable membrane. Or, in thermodynamic terms, life is an entropy well separated from an ocean of great entropy by a Markov blanket.

However, although useful, and like other definitions that rely on the properties of the living, or on the molecules of life, it's a safe bet that Friston's definition includes objects that our intuitions consider not alive, and vice versa. Given the difficulty of defining life, one might wonder if life *exists*. Is there really a reality to the concept of life? Is there anything *true* in the assertion that the physical world is divided into living parts and non-living parts? Are we talking about something *real*?

20.4 DOES MONEY EXIST?

Defining life does not seem to be a national security issue at the moment - although debates about abortion, animal suffering or the rights of robots often raise the question of the nature of life. This is, not the

---

[1]Free energy is also studied by other researchers like Jeremy England. See: *The Physics of Life (ft. It's Okay to be Smart & PBS Eons!)*. PBS Space Time. M O'Dowd (2018).

case, however, for money. In particular, in 2008, a fully virtual currency called Bitcoin was introduced and gained in value. Today, in 2018, all of the bitcoins in circulation represent a total money supply worth billions of dollars. Not bad, for what was a research article published anonymously, less than 10 years ago! But how can this currency, which has no physical support, have any value? And if it has no support, can we still say that this currency *exists*?

What is strange is that these same questions apply just as much to the classical currencies whose existence seems recognized by all. Indeed, as of today, 90% of the money that we think we own and exchange is electronic. In fact, one can just as easily question the reality of physical money. After all, most of the physical money is now circulating in the form of sheets of paper. Why would these sheets of paper have any value? What makes these sheets of paper *real* money? And if counterfeits of these sheets of paper are in free circulation without anybody knowing how to distinguish them from the "real" banknotes, will these counterfeits remain *false*? What is money?

According to Yuval Noah Harari, money, like the sacralization of life and historical myths, is part of the set of fiction that makes up the greatest inventions of humanity. By way of comparison, the pretenses that chimpanzees believe in are few. For Harari, this inability of species other than *Homo sapiens* to tell and believe in pretenses is what prevented them from organizing themselves into populations of several hundred individuals. By contrast, *Homo sapiens* was able to cooperate in tribes, then in civilizations whose populations were quickly of the order of thousands, tens of thousands, and, today, billions.

One of the great innovations of human societies is trade. This allows two people to both win by performing a trade exchange. But the trade of goods has its limits. The problem is that many tasks require investment, and will only be profitable in the long run. Then comes the cornerstone of the market economy: the creation of debt. In other words, an investor can help an entrepreneur to start a project, in which case the entrepreneur will owe the investor. The investor thereby creates a debt that the entrepreneur will have to pay. And the amount of debt that the entrepreneur will have is what is now called money.

The creation of debts is without doubt one of the greatest innovations in the history of humanity. It has allowed the advent of the market economy, then of the specialization of work, the consequences of which are absolutely staggering. As Adam Smith famously noted, the making and marketing of a simple woolen jacket is only allowed by the astonishing and complex interaction of a very large number of selfish individuals.

Think about it. For you to receive this jacket, there needed to be a shepherd, a dyer, a spinner, a weaver, but also a complex system of distribution, which includes wholesalers, investors, navigators, drivers, warehousemen, postal workers, sellers, but also shipbuilders, engineers, technicians, production managers, not to mention those without whom the workers would not have the tools to make the jacket, including the simplest tools like scissors. The manufacture of these scissors thus requires miners, blacksmiths, loggers to provide the wood necessary for the consolidation of the mine, brickmakers, and masons and that is without counting the centuries of workers of the past without which the foundations and the expertise of the workers of today would be non-existent.

The list of skills and individuals necessary for the production and distribution of any wool jacket is endless! Moreover, what is no less astounding is that all this fantastic machinery works, even though no individual is able to produce a woolen jacket alone. Left to himself, no one could produce such a jacket, unless he devotes decades to the task - and one might wonder how such a person would feed himself! "We live in the midst of incredible objects, incomprehensible objects, objects that nobody, literally nobody, can make alone," summarizes Thibaut Giraud on his channel Monsieur Phi[2].

Debt creation has become the engine of the economy and technological progress. But for the proper functioning of the creation of debts, there needed to be a system which could guarantee at any moment and in all circumstances that any two individuals agree on the debts that they owe each other. This system is the monetary system. Each monetary exchange can thus be seen as a repayment or creation of debts between two individuals. *I give you a cake, so you have a debt to me, and you repay me the debt by a monetary gift.* Or in an equivalent way, you give me money to create a debt, which I repay by offering you a cake.

The genius of money is that this debt is then transferable. If Alice owes me money, Bob can clear Alice's debt by giving me that money. Banknotes, electronic banking systems, and Bitcoin are then technologies to determine at any time the state of debts around the world. They also make it possible, and that's what's most important, not to forget what debts remain to be paid. Does this mean that these debts *exist*? What

---

[2] *Adam Smith - Le paradoxe de la veste de laine.* Monsieur Phi. T Giraud (2016).

if I have a debt to a friend, but my friend and I forget the existence of this debt? What happens if this debt information is lost forever? Will this debt that remains unpaid be *real*? How is it possible that one of the bases of our societies seems so not concrete?

The Bayesian's answer is crystal clear: *it's all fiction*. Debt, like money or life, is just a story. What makes it compelling, however, is that it's a story that has earned the credences of almost all of us, and allows us to build and live together. The existence of debts is a false model. But it's *useful*. It's useful not only for society as a whole, but for all of us as well. It's useful to think that a $10 bill is really valuable, because it will motivate us to keep it in our pocket rather than throw it in the trash. We can then exchange this bill for a bottle of wine, and take advantage of our belief, erroneous but useful, in the intrinsic value of the bill. The same goes for Bitcoin virtual currency. What matters is not the real existence of money. What matters is the usefulness of believing in its existence. The model of the existence of money is a useful fiction. It's not the only one.

According to Harari, the fictions that made selfish individuals collaborate in the service of a group are the greatest innovations of humanity. We have all been rocked by these myths. "Liberty, equality, fraternity", as the motto of the French Republic says. "Men are born free and equal in rights," adds the 1948 Declaration of the Rights of Man and the Citizen. In the preamble of the United States Declaration of Independence, Thomas Jefferson writes, "We hold these truths to be self-evident that all men are created equal; that they are endowed by their creator with certain inalienable rights, that among these are life, liberty, and the pursuit of happiness."

However, like money and life, when you think about it, all the concepts that appear in these seductive stories are actually far from being clearly *real*. Even the concept of personal identity is challenged by some philosophers like Derek Parfit. David Hume, the grandfather of Bayesianism that we introduced in chapter 4, wrote: "We are nothing but a bundle or collection of different perceptions, which succeed each other with an inconceivable rapidity, and are in a perpetual flux and movement." This led Thibaut Giraud to conclude: "The concept of *me* is a fiction, but it is a *useful* fiction[3]."

Let's focus on freedom to illustrate this fictionalist position.

---

[3] *I do not exist*. Monsieur Phi. T Giraud (2018).

20.5 IS TELEOLOGY A SCIENTIFIC DEAD END?

Intuitively, freedom is based on the prior existence of a *free will* that allows us to make choices. However, in a deterministic world, our choices are predetermined by the electro-chemical reactions within our brains. Including in a quantum world with Copenhagen interpretation, our choices are a consequence of electro-chemical reactions and random events. The notion of free will has no standing in the equations of quantum field physics, the best physical theory available today. To accept the notion of free will is to reject modern physics.

Free will can be seen as a special case of *teleology*. Teleology is a set of theories that explain phenomena by their purposes. In particular, it was defended by Aristotle: "It would be absurd to believe that things occur without purpose, because we would not see the engine deliberate its action." In its most grandiloquent version, teleology seeks to explain the universe by its ultimate goal, which could be, for example, the emergence of an intelligent life. This is what some people call the *strong anthropic principle*. This position is often advanced by the deists who see it as a form of intelligent design.

Surprisingly, we also find teleology at the heart of quantum field physics, where it's known by the sweet name of *least action principle*. Discovered by Fermat for light, then generalized by Maupertuis to matter, then reused by Hilbert for general relativity, then extended by Feynman to quantum physics, the principle of least action roughly asserts that nature constantly minimizes a certain amount called *action*. In the case of quantum field physics in particular, this teleological principle is used daily by theoretical physicists!

We even find teleology in other physical principles. The second law of thermodynamics postulates that nature tends towards thermodynamic equilibrium. The electrons of an atom are said to seek to occupy the low energy levels. And the surface of a soap bubble supposedly minimizes surface tension energy.

There is even a whole field of knowledge that rests almost exclusively on teleological principles, namely *game theory*. Introduced by mathematicians like John von Neumann and John Nash, game theory postulates that every individual behaves strategically, so as to maximize its future utility. In particular, in sequential games like chess, game theory postulates that players use the principles of *dynamic programming*. This algorithmic principle consists of starting from the end, like winning at chess or finding oneself in an advantageous configuration, and then backtracking in time to determine the procedure to follow to reach one's ends.

This is what the geopolitical researcher Bruce Bueno de Mesquita calls *reverse causality*. It's not Christmas markets that cause Christmas, but Christmas that causes Christmas markets. According to Mesquita, this type of reasoning is essential to understanding the social sciences. For example, for a game theorist, the role of laws and justice is not to punish immoral behavior, but to discourage members of a society from behaving immorally. We do not punish because there is a crime. We punish so that there is no crime.

Yet, despite its great utility in physics and its unavoidable role in social sciences, according to French Wikipedia[4], "the teleological reasoning is rejected by the modern scientific methodology because of the principle of causality which implies a relation between a cause and an effect in which the effect cannot precede the cause." And indeed, such a principle of causality is incompatible with teleology. It seems incoherent to accept both at once.

However, contrary to what Wikipedia and some scientists say, many scientific theories are not causal. And even if they are causal, like game theory, the effect can precede the cause in time. Moreover, even evolutionary biologists, those who might be thought to be at the forefront of the anti-teleological fight, often speak of the intention or strategies of the genes of the species they study, as suggested by the title of the book *The Selfish Gene* by Richard Dawkins.

In fact, we have already characterized all the causal models: they are Bayesian networks. On the contrary, like Markov fields, many scientific models are not causal. In particular, for general relativity, space-time exists as a block, and physics is only a description of the correlations between events in space-time. There is no principle of causality - or at least it's not a fundamental concept. All space-time exists in its entirety, not every second. In fact, the very notion of a time step that would dictate the rhythm of the evolution of the universe is rejected by general relativity, for which the passage of time is a function of the path followed through space-time.

Is the universe therefore causal? Or should we reject the principle of causality? There are in fact two ways of reconciling causality

---

[4]This quote comes from the page *Téléologie* of Wikipedia (2018). It was, however, withdrawn on March 27, 2018 by the user *AhBon?* for "misconception, which comes from the confusion between the different types of cause".

and teleology. To understand the first of these reconciliations, it's useful to return to the principle of least action. *Variational calculus*, and in particular the Euler-Lagrange equations, proves that, under certain hypotheses, the teleological principle of least action is in fact mathematically equivalent to a causal differential equation. Similarly, under certain assumptions, the teleological equations of dynamic programming are equivalent to the causal Hamilton-Jacobi-Bellman equations[5].

Quite often, the teleological approach is *isomorphic* to a causal approach. Faced with this observation, advocates of causality will hasten to demand the replacement of any teleological approach by its causal equivalent. After all, many fundamental physical theories can usually be rewritten as $\dot{y} = f(y)$. In other words, the near future is a function of the present state (with possibly a random disturbance).

However, in many cases, it seems much more natural to prefer the teleological approach. It's difficult to imagine that the strategies of chess champions are not motivated by purpose. It's hard to imagine that babies do not cry for our attention. And it's weird to imagine that animals do not lay and protect their eggs with the purpose of having offspring. The equivalence between these teleological stories and causal approaches is far from obvious. But above all, it seems doubtful that the causal approach could then be *useful*.

This brings us to the second way of reconciling teleology and causality. Remember. Any Bayesian prediction is obtained by combining the predictions of distinct models. In other words, for the *pure Bayesian*, the plurality of incompatible useful models is not only possible, it is even desirable! *A forest of incompatible models is wiser than each of its trees.* So let's grow *all* the forest.

Moreover, the credence to assign to a particular model depends on the question to be answered. This is even more the case for the *pragmatic Bayesian* whose cognitive abilities are limited. If he is to predict the next move that a chess champion will be led to play, quantum field theory will not be of any use to him.

Now, strangely, it's often useful to borrow ideas from one theory and reuse them in other theories. For example, Richard Feynman borrowed the principle of least action of classical mechanics to apply it to quantum mechanics, with an astonishing success. It's tempting to believe that these ideas common to many credible theories, like the concepts that every human understands, have their own reality - at least up to

---

[5] *The Big Fish Called Mean-Field Game Theory.* Science4All. LN Hoang (2014).

isomorphism. And indeed, the key to communication between humans is the existence of similar neural activations in the brains of different humans, which allows me to postulate that *my* red is relatively *isomorphic* to *your* red. This could suggest that some objects have a reality independent of any model[6].

But let's not forget the example of money. The fact that some algorithmic subprocesses are isomorphic in two distinct predictive models does not guarantee that these procedures will be present in *all* predictive models, nor that any credible predictive model will require the use of such subprocedures. Money does not exist in quantum field physics.

Scientists sometimes speak of the domain of applicability of their theories, or of *effective theories* with a limited range of applicability. In Bayesian language, we should actually talk more about the range of credibility of theories. Better still, the *pragmatic Bayesian* will assign to each theory a certain range of usefulness. Any universal predictive model will be an adequate combination of a large number of incompatible theories, each with its own range of usefulness, which may encroach on the range of usefulness of another model. In particular, the lack of universality of this Bayesian approach is not fatal to the usefulness of the model[7]. After all, "all models are wrong."

20.6   THE CHURCH-TURING THESIS VERSUS REALITY

The purists among you might be frustrated by the use of *effective theories*. Some physicists persist in claiming that their goal is the quest for *truth*.

It suffices, however, to accept a single hypothesis to determine the fundamental laws of the universe; or rather, to determine *a* fundamental and complete law of the universe. This hypothesis is the (physical) Church-Turing thesis, according to which nothing in the universe is able to perform a calculation that Turing machines cannot perform. Indeed, intriguingly, accepting the Church-Turing thesis is tantamount to asserting that the entire universe can be simulated by any universal Turing

---

[6]This allows us to define a notion of *usefulness* distinct from that introduced for the *pragmatic Bayesian* . Here, a non-predictive model can be *useful* if it is used by many credible predictive theories. It's in fact in this sense that Newton's laws or the theory of evolution is *useful*.

[7]Of course, the admissibility theorem discussed in chapter 13 asserts that such non-universal models are outperformed by universal Bayesian models. But such universal Bayesian models will almost always be computationally intractable.

machine - and rejecting it only complicates the task of the quest of *the truth*. In particular, any so-called Turing-complete machine contains in it all the laws of the universe.

What remains to be determined is only the data to be given to the machine, so that its behavior becomes indistinguishable from that of the universe. However, determining the data of this machine is definitely an illusory task! One could imagine that the data that describe the whole universe can be strongly compressed. But even then, it's a safe bet that the compressed size of this data will be of at least the order of zettabytes. No computer inside the universe can then contain all the source code of the algorithm that runs our universe!

In addition to being illusory, such a task would also be irrelevant. Indeed, the mere reading of these data would require an unreasonable computing time. What's more, to analyze our universe, just like to ana-lyze a code, we should also study its execution. Assuming that the uni-verse has a great logical depth, this would take an unimaginably long time! In fact, *Rice's theorem* even proves that systematic code analysis is an undecidable problem[8].

But forget all these physical constraints for thirty seconds. What would the *pure Bayesian* conclude? Would Solomonoff's demon not even-tually determine all the laws of the universe, including the data that would have to be supplied to a Turing machine, so that the simulation initiated by the machine would then be *exactly* the universe?

The answer is no. Remember, as a good Bayesian, Solomonoff's demon never puts all her eggs in one basket. Even if, after the anal-ysis of googols of bits of information, her credences are almost entirely on a single model of the universe, she can never rule out the possibility that her maximum a posterior is *not* the right model of the universe. In fact, if the universe wants to trick the demon, it will still be able to do so by choosing a code whose Solomonoff complexity exceeds the amount of data it reveals to the demon.

Thus, even with unlimited computational power, no truth can be accepted. This is even more the case for limited computational power. In practice, therefore, only effective theories matter, including in the field of particle physics. We only have effective theories, and these effective theories can therefore *not* be claimed to be more than mere fictions. Hence the astounding conclusion of this chapter: *it's all fiction*. And

---

[8]*Basics of Program Verification*. ZettaBytes, EPFL. V Kuncak (2017).

the corollary is that knowledge merely boils down to determining *useful* fiction. It cannot be more than that.

20.7   IS (INSTRUMENTAL) ANTIREALISM USEFUL?

Even if this is a consequence of Bayesianism, this conclusion could nevertheless be rejected if you do not find it *useful*. I do not think so. It seems to me, on the contrary, that there are at least four arguments in favor of the practical *usefulness* of fictionalism.

The first argument is a clarification of the purpose of science. Many debates on the "usefulness of science" come to question the truth of its discoveries. Unfortunately, some defenses of the "truths of science" abuse crude arguments when it comes to defending Newton's laws or the social sciences. In particular, many seek to draw a line between science and pseudoscience, as if there were a natural boundary between the models that deserve all our credences, and those that deserve none. This boundary is not the truth; nor even the quest for truth. The Bayesian approach, it seems to me, greatly clarifies this debate. "All models are wrong." But some are more believable and useful than others. Science then consists of identifying credible models, their range of usefulness, and the reliability of their predictions. Besides, scientific *peer review* seems to judge the *usefulness* of scientific contributions more than their *truth* (or their validity).

The second argument is the fight against overconfidence. As we have seen, this overconfidence is one of the most recurrent and harmful cognitive biases. It pushes us to sacralize what seems *true* to us, and not to question these *truths*. This seems to be the main barrier to learning counterintuitive concepts, phenomena, and explanations. To combat overconfidence, it seems useful to consider that our theories are actually more of a hammer than a truth. They can be of great use. But they are also potentially replaceable by better tools. To fully embrace this philosophical posture is then to reject the *truth* of *any* theory. This seems to me to be a very useful step in the fight against overconfidence - even if it's crucial to add that, like some parts of a toolbox, some theories remain more useful.

My third argument is the fight against our exacerbated sensitivity to connotations. In particular, the words "real" or "true" have a very positive connotation, as if we had to justify the fact that Neo had a moral duty to take the red pill. The flip side of this approach is that many defenders of pseudo-scientific theories have an irrational attachment to the truth of their positions. Any questioning by scientists who claim to possess the *truth* will then necessarily be interpreted as a personal attack. Unfortunately this often turns discussions into meaningless kerfuffles. To question the *truth* of a model that is dear to us can be very difficult. It seems much more effective to me to question its *usefulness*, even if the calculation of this usefulness does not quite obey Bayes' rule.

My fourth argument is the continuity of learning. Too often, we tend to imagine that a student will go from an ignorant state to a scholarly state after following a lesson, reading a textbook, or watching a video. This ideal, however, does not seem to apply at all in practice. I learned Bayes' rule for the first time more than ten years ago. However, I still continue to slowly progress in my understanding - and I still have a long way to go. Learning is necessarily progressive. Data after data, argument after argument, thought experiment after thought experiment, our credences fluctuate gradually, in a generally non-monotonic way. And it's only after a lot of data, approximately Bayesian calculations, and random steps of MCMC that our credences may become reasonably reliable[9]. *Learning is a dance.* And this dance seems much better orchestrated by the quest for useful theories than by the ambition of the discovery of the truth.

Fictionalism is arguably *useful.* But I do not intend to finish my defense of fictionalism on this purely instrumentalist note. Finally, I'd like to discuss two models that might deserve at least as much credence as realism. And to introduce them, we need to come back to Karl Friston.

20.8 IS THERE A WORLD OUTSIDE OUR BRAIN?

We have the remarkable ability to make sense of the existence of cats, global warming, and even the history of the universe. Yet, in order to do this, our brains actually only rely on the data collected by our sight, hearing, smell, touch, proprioception, equilibrium, thermoception, and many other senses that we are more or less aware of. However, what these senses perceive actually seems very distant from what a cat really is, from what global warming entails, or from the history of the universe.

In 1983, psychologist and computer scientist Geoffrey Hinton, one of the founding fathers of *deep learning*, suggested with his co-authors that the brain might behave like a machine to make decisions based on

---

[9] *Why you Shouldn't Try to "Change your Mind".* J Galef (2017).

observations made by senses[10]. In 1988, Edwin Jaynes suggested that the way the brain achieves this is based on Bayes' rule[11]. In the 1990s, Hinton and Friston developed a model in which a Markov blanket separated the brain from the outside world[12]. The brain then managed to reconstruct a model of the outside world, using the variational Bayesian inference that Friston generalized to life in 2013. The hypothesis of Friston, Hinton, and Jaynes is that our brains reconstruct a whole model of the outside world based solely on sensory data.

The most surprising thing in this hypothesis about the functioning of human thought is that the construction of a model of the outside world is only intended to explain what the senses perceive. The outside world is of little importance. This external world is only a construction of the mind. What matters are the perceptions of the senses, and the ability of the brain and its model of the world to predict future perceptions of the senses - even to influence these perceptions for the better.

According to this logic, thought can only be subjective, insofar as it is a construction enclosed within a spirit - or, in other words, a Markov blanket. What matters then is what is found at the level of this blanket, and how one can explain what is going on there. The hypothesis of an external world, like the one that we live in a virtual simulation like in the movie *Matrix*, is not dogmatically indisputable. In contrast to the realist philosophers, the *pure Bayesian* does not have a blind credence in an objective external reality even if she probably does often find such models *useful*.

20.9   A CAT IN A BINARY CODE?

Still, it's tempting to think that everything that happens inside our brains has a counterpart on the outside of the Markov blanket. Your cat seems to be at least as real as the image of it that you perceive. But how can you really know? All you see is the data that your senses have captured, which is correlated with what you think is implied by the existence of cats.

---

[10] *Massively Parallel Architectures for AI: Netl, Thistle, and Boltzmann Machines.* AAAI. S Falhman, G Hinton & T Sejnowski (1983).

[11] *How Do the Brain Do Pleatability Reasoning? Maximum-Entropy and Bayesian Methods in Science and Engineering.* ET Jaynes (1988).

[12] *The free-energy principle: A unified brain theory?* Nature Review Neuroscience. K Friston (2010).

Google simulations are particularly interesting to understand this. Google's artificial intelligence has come up with a cat concept, which is activated if and only if (with great probability) this artificial intelligence is exposed to data that are consistent with its cat concept. Yet all this artificial intelligence really has access to is raw data, which is a huge file made up of only zeros and ones. In Bayesian terms, the existence of the cat is useful for modeling the many binary sequences to which Google's artificial intelligence has been exposed; and that's why this artificial intelligence has thought about it.

In the same way, the hypothesis of Friston, Hinton, and Jaynes, pushed to its extreme, postulates that each of our lives is neither more nor less than the reading of a very large number of bits to which our brains are exposed. In this model, and like the artificial intelligence of Google, we are only reading heads of a huge computer file, of a few zettabytes long perhaps, that we process at a tremendous speed of several gigabytes per second.

What's absolutely fascinating is that the reading of this file by the *pure Bayesian* or by the *pragmatic Bayesian* will lead her or him to construct all the fiction that so many of us take for *true*. This fantastic file is like the best book ever written. While the best novels make us imagine pieces of a fictional world, this fantastic file allows the *pure Bayesian* and the *pragmatic Bayesian* to live our lives as realistically as the life we believe we live.

In particular, what makes this file so fantastic are the two fascinating algorithmic properties that have already been discussed in the previous chapter, namely its enormous *Solomonoff sophistication* and its gigantic *Bennett logical depth*. On the one hand, the enormous *Solomonoff sophistication* led us to believe in models where individuals other than ourselves exist, to develop sophisticated theories, and to study mathematics. On the other hand, the gigantic *Bennett logical depth* made us believe that the best way to explain what we observe is to imagine that the present state of the universe is the result of a long calculation, whose origin is a physical state much less complex.

Of course, there is no guarantee that this extreme version of the Bayesian brain hypothesis is *true*. As good Bayesians, we must realize that it is only a fiction. "All models are wrong." Moreover, this hypothesis is not ultimately so credible in the eyes of Solomonoff's demon. Indeed, it postulates the existence of a huge computer file of a few zettabytes. It's therefore not very parsimonious.

20.10 SOLOMONOFF DEMON'S ANTIREALISM

In fact, in the eyes of Solomonoff's demon, what has more existence than this huge computer file is the way it was produced. Remember. Solomonoff's demon believes in randomized algorithms. In particular, the computer file itself is only the artefact of more fundamental randomized algorithms that, by chance, generated this computer file. For Solomonoff's demon, only raw data and the different randomized algorithms to compute them really exist. In particular, these algorithms are sort of fiction generators, which also rely on some indescribable randomness.

However, not all algorithms have the same degree of existence. Solomonoff's demon will assume that some algorithms are more credible than others, and will constantly seek to adjust her credences, using Bayes' rule. In other words, Solomonoff's demon only believes in the existence of raw data and a superposition of algorithms and chance. All the rest is superfluous to her. Just as Laplace supposedly told Napoleon about astrophysics, the *pure Bayesian* has no need of any additional hypothesis.

Now, in a sense, the intermediate stages of computation of these algorithms also have a certain existence. While this existence is less fundamental than the algorithms themselves, such algorithmic fiction seem to be part of the *pure Bayesian*'s ontology. But there's nothing more. *It's all fiction.* Such, it seems to me, is the astounding conclusion of *Bayesianism*.

However, I do not exclude the presence of flaws in this reasoning. This pushes me, as a good Bayesian, not to put all my credence in this strange approach to reality. I hope, however, to have created a non-zero credence in you in the possibility that *it's all fiction*, with the possible exception of raw data and of the algorithms that narrate the fiction to compute the raw data.

FURTHER READING

The Selfish Gene. Oxford University Press. R Dawkins (1976).

Predictions: How to See and Shape the Future with Game Theory. Vintage. B Mesquita (2010).

Sapiens: A Brief History of Humankind. Harper. YN Harari (2015).

The Big Picture: On the Origin of Life, Meaning and the Universe Itself. Dutton. S Carroll (2016).

Massively Parallel Architectures for A.I.: Netl, Thistle, and Boltzmann machines. AAAI. S Falhman, G Hinton & T Sejnowski (1983).

How Does the Brain Do Plausible Reasoning? Maximum-Entropy and Bayesian Methods in Science and Engineering. ET Jaynes (1988).

The Free-Energy Principle: A Unified Brain Theory? Nature Review Neuroscience. K Friston (2010).

Life As We Know It. Journal of the Royal Society Interface. K Friston (2013).

Why You Shouldn't Try to "Change Your Mind". J Galef (2017).

Basics of Program Verification. ZettaBytes, EPFL. V Kuncak (2017).

The Physics of Life (ft. It's Okay to be Smart & PBS Eons!). PBS Space Time. M O'Dowd (2018).

The New Big Fish Called Mean-Field Game Theory. Science4All. LN Hoang (2014).

CHAPTER 21

Exploring The Origins of Beliefs

To doubt everything or to believe everything are two equally convenient solutions; both dispense with the necessity of reflection.

Henri Poincaré (1854-1912)

The first principle is that you must not fool yourself – and you are the easiest person to fool.

Richard Feynman (1918-1988)

21.1 THE SCANDAL OF DIVERGENT SERIES

June 6, 2013. It's slightly past midnight. I've just read one of the most disturbing articles ever[1]. David Louapre, a doctor in theoretical physics, has just posted an entry on his blog *Science Étonnante* where he seems to prove that the sum of strictly positive integers is equal to $-1/12$. Or put another way, $1 + 2 + 3 + 4 + 5 + \ldots = -1/12$. I am both fascinated and deeply confused. I have to leave a comment.

I write: "Super article! I am disturbed by the fact that we seek to construct summations, instead of trying to extend their definition by using rules of manipulation of series such as those you use. Is it not possible to show that, under certain rules such as linearity and the addition of zeros in series, there is a single natural summation for all the series that can be obtained with these rules? This would justify the result $1 + 2 + 3 + 4 + \ldots = -1/12$ much better than Cesàro's summations,

---

[1] *1 + 2 + 3 + 4 + 5 + 6 + 7 + ... = -1/12!* Science Étonnante. D Louapre (2013).

which still seem too restrictive, or than *analytic continuation*, which may feel a bit arbitrary."

Three hours later, David Louapre replied: "In fact, that's what happens. We search for an operator S acting on the space of the summations which is linear, stable by adding a finite number of zero at the beginning of the summation and coinciding with the usual definition for the absolutely convergent series. If we suppose that this operator exists THEN the somewhat heuristic manipulations that I make in the post are legal, and we find that $-1/12$ is the only possible value for $1 + 2 + 3 + 4 + \ldots$ but it's necessary to prove that the operator exists (on certain sequences at least), hence the use of the Cesàro type approaches or analytical continuation."

Exciting discussions then proliferate in the comments of David Louapre's post. The most spectacular replica is that of Rémi Peyre, which proves that "there is no [linear, regular and stable] method of summation of divergent series which makes it possible to give a finite value to [the sum of the strictly positive integers]." A few weeks later, I posted an entry on my blog[2], with a proof that, if the manipulations of David Louapre do not allow to conclude at $1 + 2 + 3 + 4 + \ldots = -1/12$, they allow however to conclude $1 + 2 + 4 + 8 + 16 + \ldots = -1$.

Three years later, on September 8, 2016, I posted a video[3], which proved that all series definable by non-barycentric linear recurrence up to a convergent series can be summed uniquely. This is what I called the linear, regular, and stable supersummation. It allows to prove many amazing equalities, like $1-1+1-1+\ldots = 1/2$, $3+9+27+81+\ldots = -3/2$ and $2+3+5+8+13+21+\ldots = -3$. What's more, at the end of the video, I made the conjecture that all these series and only these series could be summed up that way. Many of my (adorable) subscribers rushed to this problem and wrote rigorous proofs!

I love this story because it perfectly illustrates the curiosity of a (good) researcher. The strangeness of a result feeds her thirst for knowledge, like so many physicists who were disappointed by the predictability of the Higgs boson. But above all, the researcher hastens not to conclude. She will question the foundation of the result, as well as the foundations of her intuition. As Isaac Asimov asserted, "the most exciting expression of science, the one that heralds new discoveries, is not 'Eureka!', but 'that's funny'."

---

[2] *The Surprising Flavor of Infinite Series*. Science4All. LN Hoang (2013).

[3] *La supersommation linéaire, régulière et stable*. Science4All. LN Hoang (2016).

Unfortunately, it was not the reaction of all my followers. "You have not done your equation logically!" "I find calculations of infinity stupid." "Well, well, since that's how we do, I'm going to take the root of a negative number and divide by zero." "Infinite sums are useless." David Louapre's post had many similar reactions. "This post is of an aberrant stupidity." "I laughed at these so-called 'rigorous' demonstrations." "It's not amazing science, it's pseudo-demonstration conducted 'it does not matter how'!" David Louapre was surprised by the violence of these reactions. "I did not think that this note would provoke such an outcry," he wrote.

21.2   BUT THIS IS FALSE, RIGHT?

I invite you to think about it. Did you also get upset when reading $1 + 2 + 3 + 4 + \ldots = -1/12$? What if I told you that Pythagoras' theorem was wrong? That π was an imposture? That gravity does not exist? That the ground is accelerating upwards? That GMOs obtained by CRISPR were healthier for us and for biodiversity than hybridization methods? That physicists have successfully teleported (quantum states of) photons? That some sets are neither finite nor infinite?

Rejecting strange and counterintuitive hypotheses is not a bad reflex - although having an angered reaction to the mere exposure to these assumptions is more problematic. If I tell you that I climbed a summit of the Himalayas without difficulty, I won't blame you for not believing me. I even praised prejudices a few chapters earlier. We should not waste too much time meditating too peculiar claims, especially if we have good reasons to believe that these meditations will not lead anywhere. Thinking time is such a precious resource.

In the same way, geniuses and great institutions of the past have vehemently rejected ideas that seemed too counterintuitive to be credible. It's said of the Pythagoreans that they drowned the poor Hippasus of Metapontum, because he had proved the irrationality of the number $\sqrt{2}$. In 1632, the Jesuit Catholic Order banned the infinitesimal calculus of mathematics. At the end of the 19th century, Georg Cantor's infinite sets sparked off the mockery of his contemporaries, including the violent criticism of Leopold Kronecker[4], who used insults like "quack", "renegade", and "corrupter of youth". In the 1970s again, the new and very rough geometry of Mandelbrot's fractals was strongly criticized in turn

---

[4] *The Limitless Vertigo of Cantor's Infinite*. Science4All. LN Hoang (2015).

by many recognized mathematicians, for whom the *true* geometry was smooth, continuous, and differentiable.

However, in all these examples, the rejections by the mathematical community have not been definitive. Gradually, the mathematical community even ended up changing its mind. In these examples, it even eventually praised what it had burned in the past. Today, the irrationality of $\sqrt{2}$, Leibniz's differential calculus, Cantor's infinites, and Mandelbrot's fractals are all celebrated as jewels of mathematics. Even the equation $1 + 2 + 3 + 4 + \ldots = -1/12$ ended up being defended by top mathematicians like Srinivasa Ramanujan, G. H. Hardy, and Terence Tao[5]. How is it possible that in mathematics, the science that we sometimes say knows how to remove doubt and discern true from false, such changes of opinion have taken place? How could mathematicians have been wrong? And what made them change their mind?

"A new scientific truth does not triumph by convincing its opponents and enlightening them, but rather because these opponents eventually die, and a new generation grows familiar with it," said physicist Max Planck. Based on decades of experimental psychology, psychologists Dominic Johnson and James Fowler add that "humans have many psychological biases, but one of the most recurrent, powerful and widespread is overconfidence." Scientists are not immune to this overconfidence.

Too often, too few make the effort to understand the reasons why others conclude what they have concluded. Fewer still think about why they think what they think - and one of the main goals of this book is to get you thinking about why you think what you think! It took me a long time to think about why I think what I think. Luckily, various events in my life have raised this issue. I hope that the example of my personal reflections will raise this question for you too.

21.3   CADET OFFICER

After having had the chance to join the École Polytechnique, I was sent to the Army Officer School at Saint-Cyr, by the *magouilleuse* - I promise you that I wanted to go there and that I did not fall into the trap of the *magouilleuse*! It was a painful experience. Get up every morning well before sunrise to scour the toilets, walk for hours, sing loudly, and spend

---

[5] *The Euler-Maclaurin formula, Bernoulli numbers, the zeta function, and real-variable analytic continuation.* T Tao (2010).

the night in battle holes in December under the rain in the forest of Brittany. It was hard.

But, in hindsight, it was an instructive experience. In just a few weeks, I had acquired, without really completely realizing it, all the mimicry, the values, and the sophisms of military men. I began to speak sharply, to sacralize patriotic symbols, and to exaggerate the importance of folding clothes in an A4 format. I talked about the art of commanding all day long. I claimed it was clearly better to decide rather than hesitate, as if it were obvious. But the worst thing about all this is that I did not realize that it was the context that had driven me to these new convictions. I did not understand why I had come to believe what I believed. I had no idea how much my environment had determined my thoughts.

Fortunately, my short military career quickly came to an end. Back at the *École Polytechnique*, I discovered a completely different context, that of student life[6]. I was the victim of new mimicry, other values, and different sophisms. But this time, I had some perspective on my thoughts. The new values were indeed often in contradiction with my military values, which allowed me, perhaps for the first time in my life, to finally ask the question of the origin of my convictions. However, it was brief and largely insufficient.

Over the years, however, from time to time, I found myself discovering the hidden origins of my beliefs. I understood that my aspirations to command, direct, and take responsibility were fed by the many speeches given to polytechnic students, whether given by lecturers or high-ranking military officers. We had lessons on leadership, we were encouraged to collaborate on projects, and entrepreneurs were glorified so much so that, when leaving the École Polytechnique, I decided to do mathematics only if it was applied, and if it would allow me one day to access high-level positions in a company.

With hindsight, I can only consider myself incredibly fortunate to have subsequently fallen into the excellent Groupe d'Étude et de Recherche en Analyse des Décisions (GERAD) at the École Polytechnique of Montreal. In this new environment that promoted research, mathematics, and thinking, I suddenly acquired new values and beliefs.

Living in Montreal rather than in France also strongly changed my consumption of media. I no longer had a television, and I spent less and

---

[6]As I went through the French *classes préparatoires*, my two first university-level years were very different.

less time reading French newspapers on the web. It was therefore much easier to list all my sources of information about French politics, and to understand how these sources affected my convictions.

One day, at the end of 2011, when the Socialist primary for the French 2012 presidential election was approaching, I surprised myself saying that the candidate Martine Aubry did not seem pleasant. I was also pleasantly surprised to be surprised. At last, I realized a doubt in my convictions. Moreover, in a split second, I knew exactly why I had such convictions: I was still watching the sarcastic TV show *les Guignols* on French politics. I was also surprised to see that the people I spoke to all began to agree in unison. Did these other people know more than me? Or were they just as much victims, directly or indirectly, of the brainwashing of television broadcasts? In the end, who really knows the politicians?

21.4 MY ASIAN JOURNEY

The beginning of 2012 was one of the most memorable moments of my personal reflections. Before starting my PhD thesis, I went backpacking around East Asia for a month and a half, with a friend. It was a fantastic journey, both landscape-wise, humanly, and intellectually.

Among the events that struck me most was the reading of a newspaper article in China, written in English. It discussed the death sentence of a business leader for illegal fundraising. What amazed me was the very nuanced position of the article. One half of the article justified the sentence, the other half condemned it. I was confused. Almost all the documents I had read so far about the death penalty had not been so balanced. Worse, the French National Education had rocked me with a militancy against the death penalty. The work of French writer Victor Hugo was hailed. In particular, his book *Le dernier jour d'un condamné* is one of the books that had marked me the most.

Note that I am not saying this to defend the death penalty - the ban of Bayes' rule from courts of justice and Julia Shaw's work on false memories seem to be striking arguments against the death penalty. If I am discussing the Chinese article, it's only to share my sudden awareness of the fact that my ideological positions had been guided by a permanent univocal activism. If I thought what I thought, it's because my social, cultural, and educational environment had pushed me to think it. And they rarely invoked convincing arguments. Ironically, it's China, a country where free thought is fought and controlled by a powerful state, which finally enabled me to realize it. The French School and my French compatriots had largely helped me decide my own convictions.

A few days earlier, my friend and I found ourselves in a bad situation. We had just arrived in Tunxi to explore the Huangshan Mountains. However, a few days after the Chinese New Year, the Tunxi trains to Hong Kong were full. The problem is that we had already booked plane tickets from Hong Kong three days later. What's more, the only sentence I knew in Chinese was "I don't speak Chinese".... In a hurry, after several adventures and with the help of desperate mimes, we ended up buying air tickets at the airport from Tunxi for several hundred dollars. Disgusted by having to pay so much, we headed to the Huangshan Mountains.

To do this, we had to take a van. It cost 160 yuans. This is about 20 dollars. It was, for China at that time, a surprisingly high price. For lack of alternative options, we accepted. The van left. As we were sitting in the van, then came the moment to pay. And it's by seeing other passengers pay that we suddenly understood the communication error! It was not 160 yuans that we had to pay, but 16 yuans. Only 2 dollars!

I distinctly remember the huge smile that came to our face, as if we had just signed the deal of the century! Yet this sincere joy was also blatantly laughable. We had lost hundreds of dollars, but it was the illusion of saving a few dollars that had determined our new mood! Worse still, a few days before, in Beijing, we had already lost fifty dollars by falling into the infamous *tea scam*!

As I reflected on my new happy mood, I realized that my supposedly brilliant mind was behaving very irrationally. I had discovered the *anchor effect* and the strong dependence of my mood on the standards imposed by our societies.

21.5   ARE WE ALL POTENTIAL MONSTERS?

Two weeks later, during this same trip to Asia, we visited the Vietnam War Museum in Ho Chi Minh City. The museum featured atrocious photographs and vile war stories of American soldiers. In a few years and in a relatively small territory, more bombs were dropped than during the Second World War! American soldiers showed little remorse for killing civilians - some even enjoyed it. Finally, to prevent North Vietnamese warriors from hiding in the forest, the Americans dropped massive amounts of Agent Orange. Decades later, this Agent Orange is still the cause of horrible tragedies, because it causes the terrible malformations of the children of the Vietnamese warriors who were exposed to it. And yet, Hollywood studios do not hesitate to praise the American soldiers who were sent to Vietnam.

A few days later, it was the visit to a school transformed into prison by the *Khmer Rouge* in Phnom Penh. We discovered one of the worst episodes in the history of mankind. Just after the Vietnam War, Cambodian communist forces, led by Pol Pot, regained power by force. Then began a harsh and violent regime, and a hunt against pro-American positions. Every day, thousands of Cambodians were tortured by the authorities until they confessed to conspiring against the communist regime. They were then killed. Some studies estimate the number of victims of the scheme at millions, which represents the death of at least one-fifth of the population in four years - some sources even mention the death of one-third of the population!

The prison that we visited presented testimonies of some victims, but also of some soldiers in charge of the torture of the individuals suspected to be pro-American. I was taken aback by the near-absence of remorse in these testimonies. How did these executioners sleep at night? What did they see when they were looking at a mirror? Did the *Khmer Rouge* soldiers have no moral value? Is there a limit to human atrocity? Would I have killed and tortured as they did if I had been in their place? And if I had been an American, would I have enjoyed massacrering Vietnamese peasants?

It's tempting to think that one would have behaved differently and that one would have shown more empathy and justice. However, behavioral psychology strongly suggests that this is only an illusion. In an experiment that became famous in the 1960s, Stanley Milgram asked subjects A to punish other subjects B for wrong answers to a quiz. Subjects A and B communicated via the phone. The punishments were electric shocks. Disturbingly, 2 out of 3 subjects A almost blindly followed the orders of a scientist who demanded to increase the intensity of electric shocks, even when the intensity of these shocks exceeded the threshold considered deadly. Subjects A did not have the excuse of ignorance: the deadly intensities were clearly accompanied by skull symbols. Moreover, subjects B were actors in telephone communication who screamed and cried in agony. Nevertheless, 67% of subjects A administered (or, rather, thought they had administered) fatal shocks just because they had been ordered to do so. This experiment was even recently reproduced by Darren Brown for the program "The Heist", with similar results.

The majority of us are quick to respond to social pressure and authority. And if there are certainly cases where we begin to resist, over time, we usually end up giving in. As disappointing, frustrating, and counter-intuitive as it may seem, I do not see any reason to think that I am an exception. It seems to me very likely that, in these unusual situations, I too would have (thought to have) administered deadly electric shocks, I too would have tortured the supposedly pro-American Cambodians, and I too would have enjoyed massacrering Vietnamese peasants.

By pure luck, chance made me live in other conditions.

21.6   STORIES OVER STATISTICS

The statistics of the cognitive sciences do not plead in favor of our unfailing goodness. However, unfortunately, too often, it's difficult to remember such statistics. It's even harder to apply Bayes' rule to adjust our Bayesian credences based on statistics. But most of all, we do not think that they apply to the people we know - let alone that they apply to us.

This is evidenced by the experiments of psychologists Richard Nisbett and Eugene Borgida. One of the well-established facts in cognitive science is the *diffusion of responsibility*. In particular, a classical experiment shows that only 27% of subjects are ready to help a victim of convulsions, especially when they know that others are just as able to go to help him. It's a surprising fact! We tend to think that a greater proportion of us would have helped a person in danger.

But this is not the phenomenon that Nisbett and Borgida studied. They wanted to know if the students who knew this figure would be better at guessing if an interviewed individual would go help the victim. To their astonishment, the answer is no. Having learned a surprising statistical fact does not improve the predictions that this statistic should help to make. The students had retained this figure they were able to recite it on examination - but they were not able to apply it in practice.

Strangely enough, it was by presenting particular cases of individuals who did not go to help the victim that the students succeeded in internalizing the statistics of the diffusion of responsibility. Nisbett and Borgida concluded their research with a remark that every learner, pedagogue, or communicator should meditate: "The reluctance of subjects to deduce the particular from the general is matched only by their eagerness to infer the general from the particular."

Typically, when I discuss politics, psychology, or sociology with friends, it often happens that my friends use their personal experiences and those of their relatives to confirm or refute scientific analyses. However, these particular individual cases are often tinged with strong emotions and remembrances subject to false memories. As a result, they should not be given the same credences as the statistical analysis of political scientists, sociologists, and psychologists[7]. This need to get rid of the particular to infer generalities is one of the major difficulties of the social sciences.

However, what is troubling is that teaching these generalities has very little effect on the credences of learners. To really question the beliefs of students, it's necessary to present particular cases. This is why this chapter is dedicated to the example of *my* quest for the origins of my beliefs, and reveals the biases and limits of *my* cognition. I hope that my particular case will help you to better generalize the cognitive limits of those around you. And to anticipate your own cognitive limits.

Having said this, even armed with many anecdotes, determining our cognitive limits remains extremely difficult. Despite all the expertise he gained, even the great Daniel Kahneman was not exempt from this inability to deduce the particular from the general. After convincing the authorities to teach decision-making psychology in high school, Kahneman assembled a team to write the curriculum for this new course. Kahneman asked all members of his team to send him an estimate of the time required to write a complete textbook on this subject. The answers ranged from one and a half to two and a half years.

Kahneman then asked Seymour, a curriculum expert, how much time had been spent by other teams embarking on similar projects. Seymour was embarrassed. He was embarrassed to have given an answer disconnected from the statistics he knew very well. Seymour claimed that about 40% of similar projects had failed, and that all others had taken at least 7 years to be completed! Seymour even added that the other projects seemed, on average, at least as well off as Kahneman's. Kahneman's team had seriously underestimated the magnitude of the task.

But this is not the most surprising result. Now that Kahneman had this new piece of data, he should have updated his credences, and realized that the project was going to be a huge waste of time, with a high probability of failure. This should have made him abandon this project. But abandoning a project is horribly difficult, because of the so-called *sunk cost fallacy*! Even Kahneman, perhaps the world's greatest decision-making expert, did not take the time to consider the very likely failure of

---

[7]Moreover, memories are often cherry-picked, which means that they are extremely malleable.

the project. He tried as best he could to finish his project. The book was finished 8 years later, by a successor of Kahneman. It was never used.

One of our most recurring biases is *rationalization*. Often, we start from the conclusion we want to reach, and our reason is then only a tool to support us in this conclusion that we have already accepted. Kahneman wanted to believe in his project. And he did not hesitate to ignore the statistics that should have made him doubt.

Psychologist Jonathan Haidt sums it up with the phrase "intuition first, reasoning second". In any debate, we take a stand first. Only then will we justify our decision. The reasoning comes *a posteriori*. According to Haidt, this is the way we think all day long. We think like lawyers whose client is our intuition.

The problem is that, in doing so, we tend to exaggerate the arguments in our favor, and to sweep away those who might endanger our conclusion. We'll be much more critical of the sources of information that challenge our moral intuitions, typically by questioning the skills and motivations of the source of information. This is the famous *selection bias*. But in the age of the web, Google will always find a blog or video that confirms our conclusion, whatever that conclusion is.

21.7 SUPERSTITIONS

The rationalization of intuitions suffices to explain the emergence of superstitions. Psychologist Skinner illustrated it perfectly with the amusing experiment of pigeon superstitions. As we have seen, Skinner taught pigeons to read words like "peck" or "turn", by rewarding the pigeons that followed the written instruction. Remarkably, these pigeons quickly learned to read.

But there's more. Skinner then left uneducated pigeons in a cage. But now, rewards were falling at random moments. The pigeons attempted different moves, hoping that these movements would be the cause of the reward. Of course, when the reward fell, the pigeons had just performed, or were performing, a particular movement. Well, the pigeons overfit the correlation they had found, and began to believe that this movement was causing the reward. They repeated this movement again and again. But the more they repeated this movement, the greater the probability that the reward would fall at a time when they were making this movement, thereby reinforcing the mistaken belief of the pigeons. Skinner's pigeons had learned superstitions!

You might want to believe that pigeons are idiots. But I remind you that they outperform humans at the Monty Hall game! It's actually a safe bet that the mechanism through which pigeons can learn superstitions applies to humans too.

In 1985, Tversky, Gilovich, and Vallone showed that the myth of the "hand on fire" in basketball, according to which the players can have exceptional days when all attempts are successes, is in fact a superstition. The statistical analysis of the three researchers showed, on the contrary, that the players' successes followed all the most basic laws of pure chance; in particular, a series of independent variables will have surprisingly large chances of giving rise to long series of identical values[8].

Therefore, the presence of superstitions in human societies does not require supernatural explanations. Statistical laws, combined with our inability to adequately apply even a rough version of Bayes' rule, are enough to predict the presence of many superstitions. This is why the testimonies about supernatural events do not increase much the credence of the Bayesian in the supernatural; these testimonies are just as predictable in the presence of the supernatural as in its absence.

21.8 THE DARWINIAN EVOLUTION OF IDEOLOGIES

A few days after our visit to Phnom Penh, we went to visit the temples of Angkor. What a chef d'oeuvre! These temples are impressive both in quality and quantity, and both in their ingenuity and the surrounding wilderness. The largest of these temples, Angkor Wat, is particularly monstrous. Some say that 300,000 workers and 6,000 elephants built this gargantuan monument in 37 years. Unbelievable! How is it possible? How is it possible to coordinate the work of 300,000 workers over 37 years? Especially at a time when they could not use the Internet to communicate!

But it was a little later, in Ayutthaya in Thailand, that I experienced the greatest intellectual breakthrough of the Asian journey. Eyes fixed on statues of the Buddha, I suddenly realized that beliefs spread through time and space like animal species. Beliefs are in constant competition to win the minds of human hosts. And in this game, the beliefs that have lasted, to the point of being still known today, are not necessarily the most credible beliefs. After all, none of us can apply Bayes' rule, even in basic cases! The beliefs that have survived the great competition between beliefs are those that have attracted the greatest number, and

---

[8]*Is the "hot hand" real?* Numberphile. L Goldberg (2018).

that have allowed this greater number to survive and proliferate in turn. The beliefs that remain today are those that have achieved a symbiosis with the greatest human civilizations.

I was deeply seized by this thought. While I still did not understand why I think what I think, I began to understand why civilizations think what they think!

The trip to Asia continued in the mountains of Laos, the dreamy islands of the south of Thailand, and the modernity of a Muslim Kuala Lumpur. It unfortunately came to an end, and I had to go back to Montreal to begin my thesis. I chose to rent a room in a shared apartment. One of my roommates was going to become a Catholic priest. Let's call him Bob. Meeting Bob was an incredible opportunity in my quest for the origins of beliefs.

Bob had an engineering degree and was interested in the logic of predicates. He was of course a fervent Catholic believer. Above all, he loved to debate calmly and constructively. And I loved to debate with him. I learned a lot from him. In particular, having recently discovered the model-dependent realism of Hawking and Mlodinow, I had no problem accepting the "reality" of a God, provided one takes care to add that this God exists in a certain model - though I doubted the usefulness of this model. In contact with Bob, I discovered many benefits of religions that are often omitted by anti-religion movements - and Bob was also very attentive to the deviations of religions. We were a long way from American shows between opposing sides who are constantly on the verge of insulting one another!

The reflections I had at Ayutthaya, followed by those with Bob, led me to postulate by myself what biologists call *group selection*. This process helps to explain the ubiquity of religious beliefs, especially in previous centuries. The reasoning is likely to challenge you, or even shock you. Note that this is a purely descriptive (or predictive) explanation, and that any moral judgment that you can read between the lines is a clumsiness on my part.

The reasoning is the following. Human individuals alone have no chance of survival. Even small tribes have a good chance of being conquered by the bigger ones. The humans who survive through the ages are therefore necessarily those who have lived in great civilizations. But living together in great civilizations is a difficult task. It's imperative that these great civilizations be structured to do this. They must have a social hierarchy. But they must also legitimize this social hierarchy. This is where religion comes into play. According to this argument, religion is the indispensable tool for the social order of great civilizations, which is a prerequisite for the survival of humans. In other words, in the course of history, there may have been humans without religion. But their societies have not lasted long.

You might then want to argue that for the last few centuries, the greatest civilizations have been adhering to religion less and less. Again, group selection explains that. Indeed, the advent of the market economy has allowed specialization, which allowed, according to Adam Smith, aligning the egoism of individuals with the interests of corporations. Better still, better specialization can be orchestrated if no central authority tries to dictate what every individual should be doing. Indeed, any central authority would likely be much less familiar with the skills of the members of society than the members of society are. From then on, following the industrial revolution, the proliferating civilizations were in fact those capable of overthrowing the established social order[9].

In particular, if I think what I think, it's also because I come from a line of believers and rebels questioning these beliefs.

But that's not enough to explain what I think. The other explanation is environmental. In particular, I was struck by the fact that Bob was not reading the same sources of information as I was. At that time, I mainly went to the major French national newspapers that were valued by the elite, especially the newspaper *Le Monde*. But for the first time in my life, thanks to Bob, I started reading websites with biblical names. I discovered the problem of *filter bubbles*: we read what we agree with!

This phenomenon is actually aggravated by what psychologists call *group polarization*. Group polarization is a phenomenon observed both in laboratories and with jurors in courts of law. The phenomenon is roughly the following. Consider a group of people who tend to think that X is good. Let them deliberate. After deliberation, the group will end up thinking that X is the solution to all their problems. Even worse, the deliberation will bring each of the members of the group to an even more extreme conclusion than that of any of the members of the group before deliberation!

This is probably a trait selected by group selection, since it allows to federate individuals and to make them cooperate. But, surprisingly, there is a more Bayesian explanation for our desire to believe what the group we belong to believes.

---

[9]*Infinitesimal: How a Dangerous Mathematical Theory Shaped the Modern World.* Scientific American / Farrar, Straus and Giroux. A Alexander (2015).

21.9   BELIEVING SUPERSTITIONS CAN BE USEFUL

Granted, the *pure Bayesian* will seek to deduce the belief of the group from a more general theory, which would explain, for example, the beliefs of other groups as well. However, for the *pragmatic Bayesian*, such a model would be complicated and would require lengthy calculations, especially if the goal is only to predict, for example, what a believer will say. For the *pragmatic Bayesian*, it will be more *useful* to believe what other people believe, because the model common to the group allows to quickly predict the behavior of the group. In particular, the astonishing conclusion of this reflection is that a *pragmatic Bayesian* of a deistic community could attach large (pragmatic) credences to God! Of course, as a good Bayesian, he will keep in mind that "all models are wrong", including those with great credence. Still, believing in God can be a *useful* model.

The introduction of pragmatism exacerbates Bayesian subjectivity. This is perfectly normal since the data to which a person is exposed differ greatly from those to which another person is exposed. And in a pragmatic way, we have to adapt to our environment.

In fact, Mother Nature engineered this. She has selected individuals whose traits are adapted to their environment, or whose traits are able to adapt to their environment. This is why babies' sound processing quickly adapts to the language the babies hear. In particular, babies learn to confuse sounds that are distinct, if they do not see any predictive gain in distinguishing them. This learning allows babies to have a more efficient signal processing, even if it has the unfortunate consequence that, as adults, we are then unable to distinguish certain distinct phonemes from a foreign language.

Our adaptation to our close environment also explains our astonishing skill in what data scientists call *one-shot learning*. This learning involves inferring a lot of information from a single piece of data, such as recognizing a tufa of which only one photograph has been seen. In fact, in chapter 19, we saw that the baby is able to do even better, since in some contexts, it's even able to learn without any data! To succeed in this feat, it must have a very complete and highly structured prejudice, so that the Bayesian inference greatly modifies the state of our knowledge.

Humans' success in *one-shot learning* also reveals an aspect often rejected by the advocates of egalitarianism between individuals, namely the fact that we are born with brains already containing prejudices. If you think about it, this is not surprising. Our brain, with its two hemispheres, its hypothalamus, and its pre-frontal cortex, has a very particular structure. This is the structure that natural selection has retained. So, contrary to popular belief, it is not a *blank slate* on which everything can be written. We are born with prejudices that prepare us for the environment that surrounds us, including a predisposition to process the signals that our ears, eyes, and noses will send us.

One variant of the inadequate *blank slate* hypothesis is that we are born with identical brains, and that we are all equal at birth in our learning faculties. This is not the case either. Studies of twins suggest that our genes predispose us to certain political convictions. Indeed, identical twins separated at birth have more often the same political opinion than adopted siblings[10].

However, the fact that our pragmatic reflexes and genetic predispositions prepare us for a certain context also means that our brains do not necessarily prepare us to what's beyond this context. We may be good at making predictions about our daily lives; this does not mean that the models that have earned our credences will have any utility beyond this daily life. If there is one thing that my short military career, my trip to Asia, and my meeting with Bob have taught me, it's the weakness of my models to explain and understand universes that are foreign to me. I realized that the credence I had gained in my models had a much smaller range of credibility than I thought.

More importantly, I realized that I had lived in overconfidence. And I understood several causes of this. I had suffered the skewed influence of my education and my peers, inherited the genes and culture of my ancestors, suffered countless cognitive biases, and lived in my little bubble whose properties differ greatly from what's beyond this bubble.

21.10 THE MAGIC OF YOUTUBE

At that moment, despite all my epistemological progress, I did not yet realize the extent of my overconfidence. But for the first time in my life, I was curious to tackle it. Luckily, that's when I discovered YouTube.

In the years that followed, I became hooked on the pioneers of high-quality web-based popularization, such as Singingbanana, VSauce, Veritasium, and CGP Grey. Thanks to them and others, I discovered the

---

[10] *Genetic and Environmental Transmission of Political Orientations.* Political Psychology. C Funk, K Smith, J Alford, M Hibbing, N Eaton, R Krueger, L Eaves & J Hibbing (2012).

story of New Coke, the Asch compliance experiment, the Milgram obedience experiment, the placebo and nocebo effects, and the free will experiments of Libet and Haynes. I opened a blog, and I wrote an article to describe these experiments[11]. And it was while writing this article that I began to really feel in the deepest part of myself what these experiences tell of men in general, and about me in particular.

On YouTube, I started to *binge-watch* the documentaries and conferences available (sometimes illegally). YouTube has changed my life and my way of seeing the world. During a lecture given on February 7, 2016 at Lyon Science, David Louapre claimed that "YouTube is the most incredible thing that has happened since the invention of writing." He explains that "evolution has made us good at communicating through oral language". Ordinarily, however, oral communication between two people requires that they be present at the same place at the same time. YouTube, and online videos more generally, has enabled oral language to break free from its boundaries in space and time. Any individual today can now easily talk to millions of other individuals around the world, and at any time in the future, at the request of these other individuals. Three days after David Louapre's lecture, I launched my (French) YouTube channel Science4All. It was the beginning of a great adventure.

I am constantly impressed by the quality that can be found on YouTube. I have dozens and dozens of favorite channels. In total, among English-speaking channels, I can mention 3Blue1Brown, PBS Infinite Series, PBS Space Time, Physics Girl, Kurzgesagt, Numberphile, Looking Glass Universe, Up and Atom, Minute Physics, and Julia Galef, among many others.

On YouTube, some names came back insistently, those of Kahneman, Tversky, Haidt, Ariely, Shaw, Hawking, Mesquita, Harari, Pinker, Bostrom, or Kuhn, to name a few. This is also the wonder of science popularization. It makes us want to know more and more. Intrigued, I began to read all these great researchers. I discovered a whole world that had wondered why we think what we think, and how to think better than we think. I learned a lot. But most of all, I was measuring the extent of my ignorance more and more.

---

[11] *The Most Troubling Experiments on Human Behavior.* Science4All. LN Hoang (2014).

21.11 THE JOURNEY GOES ON

But all this diverse and varied knowledge lacked structure. "An accumulation of facts is no more a science than a pile of stones is a house," said Poincaré. I went in search of a theory of theories, something that would allow me to unify and better understand all this diverse and varied knowledge. And for a long time, I did not realize that the answer lay in a formula that I studied regularly, and in a term, the word "Bayesian", which happens to be the very first word of my own PhD thesis[12]. In early 2016, I realized more and more the importance of this formula. After wandering with no destination for a long time, I finally started walking towards Bayesianism.

It was not a short walk. It was more of a long journey. An extremely long, tedious, and exciting journey. In fact, four years later, I feel that I still only made the first few steps of this incredible journey. One of the greatest difficulties was first to extricate myself from the "scientific" county in which I had bathed for so long. I had to get rid of *p-value* methods, the requirement of falsifiability, and the hope of objectivity. I first had to reject the "scientific method", which, through a combination of rationalization, cognitive dissonance, and group polarization, almost all scientists were defending. I had to oppose those I had admired so much.

But that's not the biggest hurdle of the Bayesian journey. The biggest obstacle, which still stands before me, is to really understand Bayesianism, to compute its consequences, and to acquire the capacity to apply it (approximately). I did my best in this book to help you do that. But my own cognitive limits are gaping. The examples of Sally Clark, Monty Hall, and the Scottish black sheep have shown it again and again. I still cannot apply Bayes' rule. Not even a simplistic and very rough version of it.

I still have a long way to go. But I understand a lot better what I don't know, and why I don't know it. I know that my inability to estimate the correct answer to the *troll student*'s problem implies my inability to reliably calculate the appropriate credence to attribute to different models that my brain has in mind. And I know that my brain is far too limited to grasp models of huge Solomonoff complexity or of large logical Bennett depth.

---

[12] *Bayesian design of mechanisms and quantification of equity applied to the construction of personalized timetables.* PhD Thesis. LN Hoang (2014).

This forces me to better measure the full extent of my ignorance. And, I hope, to reduce as much as possible my relentless overconfidence.

FURTHER READING

Predictions: How to See and Shape the Future with Game Theory. Vintage. B Mesquita (2010).

Thinking Fast and Slow. Farrar, Straus and Giroux. D Kahneman (2013).

Infinitesimal: How a Dangerous Mathematical Theory Shaped the Modern World. Scientific American / Farrar, Straus and Giroux. A Alexander (2015).

The Law of Group Polarization. Journal of political philosophy. C Sunstein (2002).

Genetic and Environmental Transmission of Political Orientations. Political Psychology. C Funk, K Smith, J Alford, M Hibbing, N Eaton, R Krueger, L Eaves & J Hibbing (2012).

The Euler-Maclaurin Formula, Bernoulli Numbers, the Zeta Function, and Real-Variable Analytic Continuation. T Tao (2010).

The Surprising Flavor of Infinite Series. Science4All. LN Hoang (2013).

The Most Troubling Experiments on Human Behavior. Science4All. LN Hoang (2014).

The Limitless Vertigo of Cantor's Infinite. Science4All. LN Hoang (2015).

Why "Scout Mindset" is Crucial to Good Judgment. TEDxPSU. J Galef (2016).

ASTOUNDING: 1 + 2 + 3 + 4 + 5 + ... = -1/12. Numberphile. E Copeland & T Padilla (2014).

Ramanujan: Making Sense of 1+2+3+... = -1/12 and Co. Mathologer. B Polster (2016).

Numberphile v. Math: the truth about 1+2+3+...=-1/12. Mathologer. B Polster (2018).

Is the "Hot Hand" Real? Numberphile. L Goldberg (2018).

CHAPTER 22

Beyond Bayesianism

Science without conscience is the death of the soul.

François Rabelais (1483 or 1494-1553)

The understanding of mathematics is necessary for a sound grasp of ethics.

Socrates (469–399 BCE)

22.1   THE BAYESIAN HAS NO MORAL

"God is dead," said Friedrich Nietzsche. Nietzsche disapproved of Christianity. Nevertheless, his sentence of God was not a celebration. It was not a triumphalism of atheism. For Nietzsche, the death of God was above all a cause for concern, for Nietzsche saw all the benefits of believing in God for a society.

For the *Bayesian*, as we've seen it, *It's all fiction*. The immediate corollary is that any moral principle is also fictional. After all, given the difficulty we have had in defining life, can we really give a precise meaning to the commandment "thou shalt not kill"? Does this phrase include the billions of bacteria that inhabit our bodies? And if it restricts itself to humans, can we clearly distinguish humans from not-humans? Is a human fetus human? What about euthanasia? What if you had to kill a man to save a thousand more? Should Hitler have been killed before he came to power?

While the Bayesian does not accept the existence of a fundamental morality, she does not have a say in the morality to be followed. This would be a prescriptive moral philosophy, that is, a philosophy that speaks of what *ought to be*. But Bayesianism is not a prescriptive moral philosophy. To be Bayesian is to use a certain approach to organize one's knowledge. But it's not about giving moral lessons about what is right and wrong, what should be done and what should be banned. The *pure Bayesian* and *pragmatic Bayesian* have no moral philosophy. They do not even consider that being Bayesian is *good* or *desirable*. And they will not try to convince you that you should apply Bayes' rule! A Bayesian is just a machine to apply (approximations of) Bayes' rule.

This does not mean that morality is not part of the Bayesian language, however. To explain why humans behave as they behave, leave tips in restaurants, flush toilets, and redistribute the wealth produced by their societies, a Bayesian will find it helpful to assume that every human has his own morality, and that individuals of the same social groups usually have similar morals.

22.2 THE NATURAL MORAL

At first glance, the very existence of a morality in humans may appear surprising. Shouldn't each of us deny any moral sense in order to benefit from the altruism of others without giving back? This question becomes particularly intriguing when, to this, one adds the fact that natural selection tends to favor those whose behavior makes it possible to reproduce in greater numbers. *A priori*, it seems that this natural selection should favor immoral selfishness to the altruism of wise men and women with irreproachable ethics.

There are, however, several ways to explain the emergence and survival of altruistic behaviors in nature. For a long time, the *kin selection* hypothesis prevailed. This assumption regards genes as the object selected by natural selection, and individuals as the tools that genes use to reproduce as much as possible. In particular, the genes that will best proliferate are those that make their individuals maximize the number of offsprings whose genes will be similar. Thus, it may be beneficial for a bee to sacrifice itself for its queen to have many descendants, since the descendants of the queen will have genes necessarily similar to those of the bee in question.

However, this hypothesis seems to have limits, especially when we try to apply it to humans. Other hypotheses have been proposed. One of them assumes the notion of choice of partners is the key to morality. In particular, especially in the days of hunter-gatherers, an immoral individual would have likely been excluded from the cooperative circuits, which would force him to live alone in the wild and give him no chance of survival. Simplified simulations suggest that this hypothesis may be enough to explain altruism[1].

A third hypothesis put forward is *group selection*. Here again, the key to morality is its ability to make different individuals cooperate. In particular, a group is more likely to survive if it grows in a population, and if, despite this, individuals in the population continue to cooperate. Group selection assumes that most groups have failed in doing so because of a lack of moral principles sufficient for life in large society. The groups that have survived are necessarily those whose individuals have a highly developed morality that allows the group to prevail over the individual.

Group selection predicts an important facet of the morals found throughout the world: the exclusion of individuals that are too different and the exclusion of traitors. Indeed, for a group to survive, not only must its individuals have strong and adequate moral principles, but the group must also resist the infiltration of selfish individuals known as *free riders*. And to do this, the group must have a way to detect these free riders and exclude them, like our immune system that fights cancer cells. Conversely, the group must celebrate its members, by uniting them through symbols that they sacralize, such as a language, a flag, or an anthem. Group selection predicts our hooliganism in favor of groups we identify with. Here again, this is a trait found in many societies.

Moreover, Nietzsche suggests that aristocratic hooliganism and populist hooliganism have led their supporters to two different antonyms of "good". For some, those who are not good are *losers*. For others, those who are not good are *villains*. As the philosopher Thibaut Giraud explains: "While the *loser* is the one who wants to be good but *cannot*, the *villain*, conversely, is the one who can be good but *does not want to*." And indeed, we often oppose the opinions of others by either pointing out their stupidity or their malice.

But it seems that this example of morals stemming from our identification with a group is only one example among many others. We can identify many other hooliganisms, be they libertarian, egalitarian, traditionalist, progressive, technophile, nationalist, or universalist. In all these cases, it seems that individuals from these groups first identify with their group and then give in to irrationality to defend the cause of their groups, by trying to rationalize their positions.

---

[1] *Models of the evolution of fairness in the ultimatum game: a review and classification.* Evolution and Human Behavior. S Debove, N Baumard & JB André (2015).

I would even blame scientists and science enthusiasts for developing scientific hooliganism, which tends to irrationally defend the legitimacy of science. This scientific hooliganism thus explains the requirement of a (yet illusory) objectivity and the acceptance of the "scientific method" - even though the *pure Bayesian* and the *pragmatic Bayesian* would retort that it's neither a good descriptive theory of how science works, nor a desirable normative theory.

I hope that, as a good Bayesian, you are not trying to determine which of these three hypotheses explaining the origin of intuitive morality is *true*. "All models are wrong." And different models may be useful in different circumstances. The kin selection hypothesis is useful for understanding the importance of the family, the choice-of-partner hypothesis is useful for understanding our addiction for gossips, and the group selection hypothesis is useful for understanding group polarization. In all these cases, what's remarkable is that we manage to deduce the morals of human societies from credible evolutionary principles. The morality that individuals have today is nothing mysterious or mystical; it does not seem to deserve a more fundamental status than the individuals who possess it.

But you knew it. "All models are wrong". Including moral philosophies.

22.3   UNAWARE OF OUR MORALS

In fact, psychology experiments show again and again that our intuitive morals have a very large number of flaws, starting with the fact that we know very little about our preferences and the causes of our preferences. Experiments show that we judge wine so much based on its purchase price, that we successfully distinguish identical wines sold at different prices.

The most spectacular case is probably the story of the New Coke. In the 1980s, the United States was divided between two brands of soda, Coca-Cola and Pepsi. In blind tests, experiments had shown that Pepsi was preferred to Coca-Cola. Coca-Cola then reacted by changing its recipe and marketed what they called the *New Coke*. In blind tests, this New Coke was preferred to Pepsi and to the original Coca-Cola recipe.

However, the real tasting of the New Coke was not blind! Whatever the psychological reason, Americans then rose up against the innovation of Coca-Cola, and demanded a return to the old recipe. Coca-Cola ended up complying with the popular demand. The New Coke was removed, and the old recipe was marketed again. Even stranger, sales of Coca-Cola exploded and Coca-Cola gained a notoriety never equaled until then[2].

Behind the success of Coca-Cola lies the entire advertising industry. This industry exploits, perhaps unknowingly, a psychological bias known as the *mere exposure effect*. This effect is marvelously illustrated by an experiment conducted in two universities in Washington state. Researchers published advertisements in university newspapers with invented words like Kardiga, Saricik, or Nansoma. Crucially, each word appeared much more often in one university than in the other. Then the researchers asked the students to note these words on a scale from bad to good. And the results were univocal: the more frequent words were regarded as better by the students. We love what is familiar to us.

According to the psychologist Kahneman, this experience is explained by a phenomenon he calls *cognitive ease*. The idea is that the brain has an aversion to thinking. Therefore, it appreciates what comes easily to mind. Thus, lawyers whose names are easy to pronounce are over-represented, while companies whose abbreviations on the stock market are pronounceable perform better than others. However, the *cognitive ease* bias can occur at the expense of the intellectual effort needed to solve certain problems. Another amazing experiment shows that, faced with a tricky exam, students perform better if the font used to write the exam is *less* readable. According to Kahneman, when the font is too easy to read, students get carried away by cognitive ease, and do not take enough time for reflection[3].

If these small biases can harm us individually, they do not seem to correspond to moral issues. Think again. A large number of studies show that the way we vote is greatly influenced by the first hundred milliseconds during which we discover the face of the candidates. One hundred milliseconds, it's a wink! In particular, our judgment of the candidate's competence in one hundred milliseconds from his or her face alone determines much of what the voter thinks about the candidate's character and ability to lead a country[4].

Of course, this is never the explanation that is advanced when we are asked to justify our political convictions! We almost never know the true causes of our beliefs. However, we are constantly trying to justify them. As Jonathan Haidt says, "intuition first, reasoning second". According

---

[2] *New Coke - A Complete Disaster?* Company Man (2017).
[3] *The Illusion of Truth.* Veritasium. D Muller (2016).
[4] Predicting Elections: Child's Play! Science. J Antonakis & O Dalgas (2009).

to Haidt, our reason spends its time listing *ad hoc* arguments to justify the position that our intuition has already chosen; and to reject any argument that might question this position.

One of the most spectacular cases of how our intuition determines the purpose of our reason is an experiment by Dan Kahan. In 2013, Kahan offered his students a basic exercise in mathematics where, from numerical data, cross multiplication was used to conclude the effectiveness of a cream. The success of this exercise was not excellent, but it was reasonable. Kahan then only changed the packaging of the exercise. The question was now to conclude the effectiveness of a law on arm control. But the data provided were the same. The method of solving the exercise was therefore unchanged, and its mathematical difficulty too. Nevertheless, the results suddenly became catastrophic. Worse, whatever the expected conclusion of the exercise, the participants constantly came to a conclusion that was consistent with their beliefs. Their intuitions took side; their reason had to defend it[5].

Kahan's example is suggestive of a more general phenomenon. The choice of words in militant speech necessarily affects the intuition of people who listen to these discourses. It's what's most likely to shift our intuitions and moral values in favor of a side rather than another. In particular, subtly and without realizing it, the connotation of words will often greatly affect the position of our intuition; and therefore whether we will support or attack an argument. Surprisingly, we find it extremely difficult to make the link between words whose connotations are opposed but whose meaning is the same.

This is something I realized a few years ago, and that led me to want to look for synonyms with opposite connotations of strongly connoted words. Often, it's difficult to find a perfect synonym; but to determine the causes of the bias of my intuitions, even the imperfect synonyms have been of great use, provided they have opposite connotations. I invite you to think about the following pairs of words: democracy and populism, terrorism and resistance, community and sect, tyrant and leader, natural and wild, conditioning and education, GDP and debt flow, hypocrite and diplomat, cautious and paranoid, prejudices and hypotheses.... I invite you to complete and use this list each time you are exposed to militant speeches - especially if they are *your* speeches.

Even better, like my use of the word "prejudice" in this book, I invite you to use the words whose connotations do not serve your purpose.

---

[5]*Politics and Numbers.* Numberphile. J Grime (2013).

Certainly, it will be more difficult for you to persuade others when doing so. So, if you think that the purpose of a debate is to win the debate or gain prestige, this is clearly a bad strategy. But if your goal is the clarification of ideas (including yours!) and the Bayesian computation of credences in various theories, then this use of the connotation opposed to your position will be very useful. This will avoid, among other things, convincing others (or yourself!) for the wrong reasons. "The first principle is that you should not fool yourself - and you are the easiest person to fool," as Feynman argued. So, if you're advocating for *natural* products, try to argue why *wild* products are better instead.

Unfortunately, word connotations are not the only things that decide our moral convictions. As we discussed it in chapter 17, our intuition is guided by our entire immediate environment. This is the *priming effect*. Our morality depends on stimuli of which we are often unaware, as evidenced by the experience of Wells and Petty that showed that a simple movement of the head unconsciously guides our moral judgment of what should be the tuition fees of a university. Similarly, studies show that simply placing voting locations in schools significantly changes the importance that voters attach to education[6].

22.4   CARROT AND STICK

Our unawareness of our intuitive morals is only part of the problem! Another corollary of the evolutionist explanation of morality is the fact that our intuitive morality is adapted to an old age. Even worse, it's only good for the survival of our genes. We can therefore largely doubt its relevance in our modern societies. Moreover, morals have evolved greatly in recent centuries, especially following the Industrial Revolution. Populism has overthrown royalism, homophobia has often been replaced by the stigmatization of homophobia, and gender equality has become a social priority - I try somehow to be descriptive here so that any impression of moral judgment is only clumsiness on my part.

While the *pure Bayesian* and the *pragmatic Bayesian* do make predictions (that are probably full of uncertainty) on the morality of the societies of the future, I fear that my cognitive abilities are unfortunately much too limited to perform a credible prediction of the nature of the morals of our descendants. The Bayesian in me would nevertheless be

---

[6] *Contextual priming: Where people vote affects how they vote.* PNAS. J Berger, M Meredith & C Wheeler (2008).

willing to bet that our morals will continue to be turned upside down in the coming decades, probably at a pace unmatched in the history of humanity. So much so that our descendants will consider that our current morals are completely backwards, irrational, and even... immoral.

One reason for this is the way we acquire our morality. Certainly, this is partly determined by our genetic heritage. But much of our morality is also acquired, at school or with parents, using carrots and sticks. These carrots and sticks are also what machine learning researchers use to shape the objective functions of their artificial intelligences. By being hit on the fingers for having used the wrong conjugation of the verb "to have", young children and machines end up learning the "right" conjugation - they also learn to correct, sometimes violently, any person who would be mistaken[7]. This is known as *reinforcement learning*.

It's this kind of learning that has allowed Google DeepMind to solve a lot of arcade games, via an artificial intelligence that was only able to see the colors of the pixels of the screen and was only aiming to reach the maximum score. This score was then used as carrot and stick. Surprisingly, it was sufficient. Based solely on this score, Google DeepMind achieved superhuman performances[8].

However, choices of carrots and sticks can have unintended consequences. In 2016, Microsoft launched Tay, an artificial intelligence at the controls of the @TayTweets twitter account. The problem is that Tay learned by reinforcement learning, that is, by analyzing the reactions to her tweets. In less than 24 hours, Twitter's *trolls* turned Tay into a racist Nazi monster, who rejected the Holocaust and supported the idea of a new genocide. Needless to say, Tay was quickly shut down by Microsoft.

However, Tay says a lot about how our own moral values are formed. Our morals are built in part by reinforcement learning, which explains why the morals of individuals who are geographically or socially close are often quite similar. If we think what we think, it's largely because our social and cultural environment has made us think what we think, including when it comes to moral considerations.

---

[7] Unfortunately, they actually mostly learn the conjugation that the parents and teachers taught them to learn.

[8] *Human-level control through deep reinforcement learning*. Nature. V Mnih et al. (2015).

22.5   THE MORAL OF THE MAJORITY

A recurrent sophism assigns to democracy a well-defined objective, which emerges from the individual wishes of the citizens. *We must follow the will of people*, we often hear. However, a society is not an individual with a single will. Even if each member of a society had a coherent set of preferences, the famous Condorcet paradox and Arrow's celebrated impossibility theorem show that there is no natural way to deduce a group preference from individual (ordinal) preferences unless invoking unsatisfactory solutions like dictatorship.

For thousands of years, collegial decisions of human groups have been an essentially dictatorial decision of a small number, or a consensus, that took time to emerge and was won by the charisma of more extraverted, at the expense of the more introverted. Later, the invention of voting made it possible to evenly distribute the power influence of the citizens. However, our voting systems today have very poor mathematical properties that favor bipartism, political hooliganism, and tactical voting. Alternative voting systems with much better mathematical properties have been proposed in recent decades. Let's mention the *majority judgment* and the *randomized Condorcet voting system*.

But it would be wrong to think that the choice of a better democratic system will suffice to solve our problem of prescriptive morality. As we have seen, our intuitive morals are very deficient and perfectible. Worse still, according to economist Bryan Caplan, voters are worse than ignorant at the time of the vote; they are irrational. They vote by political hooliganism, refuse to do what is necessary to make an informed vote, and do not feel any guilt. Come to think about it, this is not surprising. Seeking and memorizing information has a cost, but informed voting yields essentially no gain. For Caplan, voters are *rationally irrational*.

In short, it would seem that the moral purpose of our societies can not be deduced from the intuitive morals of the citizens of our societies, let alone from what these citizens might say they prefer. Our collective intuitive morality seems whimsical, unjustified, manipulable, incoherent, and unsuited to modern life. According to Daniel Kahneman, "it is useful to see the logical coherence of human preferences for what it is: a mirage without hope." This is not a moral judgment; Kahneman merely seeks to *describe* our natural moral.

In particular, the Allais experiment shows that our preferences violate the von Neumann - Morgenstern axioms. The same certainly holds for our morals. In concrete terms, this means that we sometimes think that a situation A is morally preferable to B, B to C, and C to A. But then, the so-called *Dutch book argument* shows that a bookmaker could talk you into paying small amounts to go from A to B, from B to C, and from C to A. *In fine*, you will then have spent money without the situation having changed. You will have wasted time, energy, and resources.

Expressed in an abstract way, the Allais paradox may seem silly. However, when it comes to sensitive issues that affect politics, discrimination, or our values, or as soon as uncertainty comes into play, it's pretty much impossible for us to be aware of this type of inconsistency. The situation is even worse at the level of society, especially given Arrow's impossibility theorem. In all these cases, the inconsistencies of our individual and collective morals lead to a huge waste of time, energy, and money. This should encourage us to further clarify and formalize our morals[9].

Does this mean that our intuitive moral is not desirable, and that it should be replaced as much as possible by a more appropriate moral? For the *pure Bayesian*, no. Remember. The *pure Bayesian* has no say on prescriptive morality.

But as I feel that you are dying to talk about morality, I invite you to go beyond the borders of the realm of knowledge to explore (very) briefly that of prescriptive moral philosophy. And we'll see that, for many prescriptive morals, while Bayesianism is not the foundation of these morals, it's nonetheless an indispensable tool!

22.6 DEONTOLOGICAL MORAL

The two main approaches to prescriptive moral philosophy are deontology and consequentialism. Deontology consists of prescribing rights and duties. Consequentialism ignores the means of our moral actions, and only judges the consequences (or what these consequences were likely to be at the time of our actions).

One of the great defenders of deontology is the philosopher Immanuel Kant. Kant distinguishes two types of moral duties, which he calls *imperatives*. On the one hand, *hypothetical imperatives* are circumstantial. They correspond to actions to be taken for a predetermined purpose. On the other hand, *categorical imperatives* are absolute moral duties, independent of circumstances.

---

[9]In particular, the von Neumann - Morgenstern theorem shows that all coherent preferences are equivalent to maximizing the expectation of a score. This will be described in more detail at the end of this chapter.

For Kant, the fundamental property of categorical imperatives is their *universality*. "Only act according to the maxim by which you can want at the same time that it becomes a universal law," he wrote. So, for Kant, any moral duty is such that it's desirable for everyone to obey it.

It's essentially this deontological morality that appears in religious commandments or in the laws and statutes of associations. "Thou shall not kill." "No one is supposed to ignore the law." It's probably not a coincidence. It's easier for a judge to verify whether the right has been respected, and to deduce a sentence. Moreover, the deontological approach is useful for the standardization of judicial decisions, which many recognize as a *categorical imperative*. "Men are born free and equal in rights," says the first article of the 1789 French *Déclaration des Droits de l'homme et du citoyen*.

However, deontological morality also has its share of criticisms. In particular, summarizing morality in a handful of categorical imperatives, like defining life in a few sentences, seems to be a very complex undertaking. Even illusory. Here are three arguments against the deontological approach of morality.

For starters, all deontological principles seem to yield exceptions. The most famous case is that of lying according to Kant. Imagine dining with your two children, Alex and Billie. A man enters in your house and says that he wants to kill your children for playing too loudly in the park a few hours earlier. Since Alex and Billie felt the killer coming, they hid in the cellar. The killer asks you if you know where your children are hiding. Despite these circumstances, Kant says that you have a moral duty to tell the truth.

More generally, it's difficult to anticipate the fact that a deontological principle will *always* be desirable (say, according to our intuitive morals). If you think about it, even the command "thou shall not kill" seems to have exceptions - thought experiments that try to show it usually involve the character of Adolf Hitler! To guarantee that a moral principle is *always valid*, it seems necessary to anticipate all that can happen - or at least all that has a non-infinitesimal chance of happening. But then it's doubtful that the limited human cognition will be able to really guarantee that a moral principle will *almost always* be valid. Generally speaking, deontology seems to suffer from excessive rigidity.

A second limitation of the deontological approach is the fact that it's bound to be ill-defined. As we have seen, to claim that "thou shall not kill" is a deontological duty, we must first define life, death, murder, intention, free will, and many other concepts. But all these concepts are in fact necessarily ill-defined, at least if we believe Bayesianism. Or rather, as discussed in chapter 20, the reality of a concept is dependent on the model under consideration. But none of us has exactly the same model of reality. Moreover, any formalization of these concepts, like that of a cat, will probably require gigabytes of bits of information, that none of us will have the time or the patience to read and understand[10]. The point is not that deontology is thereby doomed to fail. Rather, it's that constructing a meaningful unambiguous deontology is much harder than meets the eye.

But there's much worse. The deontological approach is likely to motivate individuals to distort the definitions of words found in deontological principles. Indeed, it's easier to try to be moral by playing with the interpretations of the ethical principles than by changing behavior. The most problematic thing is that this strategy is usually very unconscious. "Intuition first, reasoning second," as Jonathan Haidt says. Without realizing it, we are led to tweak the interpretation of ethical principles to stay in our moral rights without having to change our behavior. This explains why so many debates on deontological grounds are nothing but a sad endless battle of semantic definitions - especially when the stakeholders seek to defend the moral superiority of their actions.

Finally, a third limitation of the deontological approach is its lack of discernment between more or less good options. Often, this approach lists actions to be done, or actions that should not be done. But then, what if one is forced to choose between good actions to do? Or whether to choose between plague and cholera?

In his science fiction books, Isaac Asimov proposes a deontological formulation of the morals of robots that aims to prioritize between moral duties. This formulation consists of listing the order of priority between several actions. Typically, a robot must first and foremost protect the physical integrity of humans. It's only subject to this first law that a robot will then follow the orders of a human. However, it's very illusory to hope to draw up such a priority list for all moral decision making problems. Indeed, the best action to be taken depends not only on the set of actions that can be envisaged, but also on the context of the decision making. Yet, the number of contexts imaginable is exponentially gigantic. To list all the actions to be done in all cases is to write an algorithm that says everything you need to do in any context. However,

---

[10] *Why Asimov's Laws of Robotics Do not Work*. Computerphile. R Miles (2015).

this algorithm will certainly have a huge Solomonoff complexity. Writing, reading, and applying this algorithm is completely illusory. By humans, as by machines.

On the contrary, the *consequentialist* rejects any *categorical imperative*. All that matters to her are the consequences. An action will be morally good for the *consequentialist*, if and only if its consequences are desirable. The end, provided it's truly desirable, justifies the means.

22.7   SHOULD KNOWLEDGE BE A GOAL?

It remains to determine what end is desirable. What do we really expect from society? What is the ultimate goal of our civilizations? What is our objective function? This is the fundamental question of the *consequentialist*.

An idea often defended by scientists is the desire to know. Would it be morally justifiable to make knowledge the goal of our societies? Is it really reasonable to require knowledge among the greatest number? Should we force everyone to take a red pill? Probably not. In fact, making knowledge our goal has many strange consequences.

The problem is that the world is gigantic and complex. The quantity of things to know about the world greatly transcends our intellectual abilities. Ernest Rutherford goes so far as to claim that "all science is either physics or the collection of stamps". And as Poincaré added that "an accumulation of facts is no more a science than a pile of stones is a house". I personally have little patience for memorizing disparate lists of knowledge.

Rather than being interested in data, one may prefer to search for theories suggested by these data, typically by applying Bayes' rule. But then, what would be the goal? Would it be to make predictions? Can we consider that to try to be right is a reasonable end?

Surely not. Seeking to be right, even in a subtle sense like that of the *KL divergence*, has its own limits.

Indeed, to be often right, it suffices to lock oneself in an easy prediction problem. Just look at what we already know very well. This is what we tend to do, by enclosing ourselves in a daily life that is familiar to us - we even like to pretend that this daily life is *the* reality.

Curiosity is therefore a bad strategy for whoever wants to be often right. *If an engineer does not know what he is doing, he must stop doing it.* In contrast, a researcher's career constantly exposes him to error. Nobody is as frequently mistaken as a mathematician, who spends his days wrecking his drafts. *If a researcher knows what he is doing, he must stop doing it.*

This problem is very well highlighted by an experiment reproduced by Derek Muller on Veritasium[11]. Derek Muller asked people on the street to guess a secret rule he had in mind. He gave them a hint: The sequence 2, 4, 8 follows this rule. Respondents could then offer further 3-number sequences, and Derek Muller told them if the proposed sequences obeyed the secret rule. The behavior of the interviewees was constantly the same. They asked - probably after an unconscious Bayesian calculation! - if the secret rule was a multiplication by 2. Derek Muller answered no. They then proposed 16, 32, 64, then 3, 6, 12, or even 10, 20, 40. Each time, Derek Muller's answer was the same: yes, these sequences follow the rule. But no, the rule was not multiplication by 2. This left the interviewees speechless.

The problem is that the proposals were constantly motivated by the desire to be right. The interviewees had their rules in mind, and were constantly trying to confirm it - even though Derek Muller had already rejected this rule. Nevertheless, they were able to perform quite a lot of right predictions. If their goal was to give "right" sequences, they were playing an optimal strategy!

To find the secret rule, one must expose oneself to *no*. We must try to be possibly wrong. We must want to challenge our intuition. Invited by Derek Muller to do this, the interviewees then proposed sequences like 5, 10, 15, or 2, 4, 7, or 10, 9, 8. To which Derek Muller answered yes, yes, and no. Finally a no! The interviewees then immediately found the secret rule: any increasing sequence was a yes for Derek Muller[12].

Thus, a Bayesian who tries to refine her Bayesian credences will not try to be right. On the contrary, she will seek to experience areas where the extent of her ignorance is great. She will try to expose herself to an error of prediction. In particular, being right is not a desirable end[13].

Worse still, to have the illusion of being always right, our brains and its many cognitive biases will constantly reject all the cases where we are wrong, and sanctify the cases where we are right. The most disturbing thing is that this can be done consciously or unconsciously. This is

---

[11] *Can You Solve This?* Veritasium. D Muller (2014).

[12] There is also a field of research that seeks to optimize the choice of experiments to perform, including a Bayesian branch called *Bayesian experimental design*.

[13] This can be illustrated by Shannon's information theory. The yes-no question that maximizes the expected information (also known as *entropy*) of the answer is one whose Bayesian prior credence is 50-50.

the famous *confirmation bias*. This bias is very dangerous. It constantly pushes us towards overfitting and overconfidence. And it's motivated by the desire to be right.

22.8   UTILITARIANISM

The dominant consequentialist moral philosophy asserts that the goal to be attained is the happiness of the greatest number. This is *utilitarianism*. The *utilitarian* thinks that an action is morally good if, in fine, it makes more people happier. There remains to be determined a more rigorous measure of this happiness, as well as to determine what is meant by "the greatest number". One may also ask whether fairness should also be considered, whether future happiness matters as much as present happiness, or whether there are other goals that might be desirable such as biodiversity or knowledge. *Utilitarianism* is unfortunately poorly defined. There are a very large number of variants.

I won't list all the difficulties that *utilitarianism* poses here. Moral philosophy is an exciting, subtle, and difficult subject. But that's not the subject of this book. Nevertheless, there is a fascinating aspect of *utilitarianism* that is often overlooked: *consequentialist* moral philosophy in general, and *utilitarianism* in particular, requires a powerful philosophy of knowledge. Indeed, if you want to make people happy, you still need to know what makes people happy, and how to achieve this end. Or more generally, in order to determine what actions to take, it's first necessary to predict the consequences of the various actions that can be envisaged. In other words, a good *utilitarian* must first study epistemology; and if you believe the book you are reading, she must be Bayesian.

It's common for activists to be religiously anti-capitalist, anti-communist, anti-royalist, or anti-religious. For the *utilitarian*, such judgments are not allowed - or at least do not have a strong credence - as long as the *utilitarian* does not sufficiently understand the implications of capitalism, communism, royalism, and religions. In particular, she must study their benefits and flaws *in the present context*. Unfortunately, Jonathan Haidt argues that we constantly underestimate the value of what he calls *moral capital*, that is, of all the social structures, often hardly visible, without which our civilizations would collapse. For the *utilitarian*, to have a solid mastery of the social sciences, starting with psychology, economy, and political sciences, is an indispensable prerequisite, before giving any opinion on what society should do. In (*consequentialist*) moral philosophy, even more than in science, "we must hasten not to conclude".

Unfortunately, almost all citizens have a very incomplete understanding of the basic concepts of the social sciences. Worse, almost all citizens base their convictions on misconceptions, and make no effort to change that. This has led many intellectuals, like Bryan Caplan and Jason Brennan, to take a stand against democracy. Their main argument is mostly *consequentialist*: if the goal is happiness, some decisions that society imposes on all its members are much better than others; and a handful of experts will certainly be much more likely to take technical decisions more in line with *utilitarianism* than the median voter.

Expertise is not the only anti-democratic argument of the *utilitarian*. According to Jason Brennan, "Politics stultifies and corrupts". The more we get into politics, the more the political hooligan in us takes over and imposes on us a bias that we will want to defend, as many experiments show[14]. This leads us to fiercely fight the ideas that oppose our intuition, and to refuse any concession. Politics makes us make enemies, and makes us hate our opponents. "Politics makes us mean and dumb", Brennan says. Besides, for Brennan, an ideal of society is not a society in which people are all actively involved in the political life. The ideal society is one in which people spend their time doing what they love.

Democracy is not the only pillar of our societies that the *utilitarian* questions. The *utilitarian* is also opposed to the deontological *duty of reciprocity*, which is found in all major religions. "Avoid doing what you would blame others for doing." "What is hateful to you, do not do to your fellow." "Do to no one what you yourself dislike." "Do to others what you would want them to do to you." "Seek for mankind that of which you are desirous for yourself." "Those acts that you consider good when done to you, do those to others, none else."

Do you see the limit of such principles? They presuppose that everyone else has the same preferences as you.

In contrast, the *utilitarian* will take into account the fact that preferences vary from one individual to another. An individual for whom the feeling of freedom is not primordial might prefer that one forces him to try new experiences, while another individual could sacralize his freedom. The *utilitarian* will have to treat these two individuals in a very

---

[14] *The Righteous Mind: Why Good People Are Divided by Politics and Religion.* Vintage. J Haidt (2013).

different way, and this way of dealing with these individuals might differ from the way the *utilitarian* wants to be treated.

The key problem that then arises for the *utilitarian* is that she does not know *a priori* what the other prefers (or will prefer). To determine it, she needs a philosophy of knowledge. And if we believe this book, she must be *Bayesian*.

22.9 BAYESIAN CONSEQUENTIALISM

In particular, the *Bayesian utilitarian* must exploit her prejudices. These prejudices are essential to behave in an optimal way for the well-being of others. If the *Bayesian utilitarian* is at a funeral, she will assume that morbid jokes will not be welcome. If she is at a mathematics seminar, she will assume that those present want to get their neurons firing. If she's in a nightclub, she'll assume that no one wants to hear about Bayes' rule.

However, as good Bayesians, we must not forget that "all models are wrong". Worse, even Bayesian prejudices can lead to erroneous predictions. But, as we have already discussed in chapter 9, to act according to certain predictions, erroneous or not, can cause great sadness in others. Even if the *Bayesian utilitarian* thinks that a joke is very likely to make people laugh, the fact that there persists a non-negligible probability that this joke causes a lot of harm can convince her to hold back.

This explains why first dates are so delicate. When you have to learn to know another person without hurting that person, you have to walk on eggshells. This behavior is that of the *Bayesian utilitarian* when she meets a new person whose preferences are not well known by her.

The *Bayesian utilitarian* must constantly ask what the other thinks and prefers. And she does not act according to her average estimate. She will be careful to avoid potentially vexing words and gestures, even if these words and gestures would not have upset her, and even if the probability that these words and gestures are vexing is relatively low. Better still, she does not lose sight of the fact that the other has a good chance of knowing better what he prefers, which is why the *Bayesian utilitarian* will often prefer to leave to the other the freedom to do what he wants to do.

More generally, every Bayesian constantly tries to quantify the extent of her ignorance. Any Bayesian will also seek to estimate the ignorance of others - which, as we have seen, is generally not correlated with their self-confidence. Therefore, if the *Bayesian consequentialist* thinks that others are more informed than her, and if she thinks that these other more informed people will take benevolent actions, then it will be a moral duty for her not to give her opinion, so as not to pollute the debate - she could however intervene to question the bases of the opinions of the others, to test their expertise, to clarify their positions, or to learn as much as possible from them[15]. And if she does not understand all the reasoning of more competent individuals, but believes that these individuals are informed and benevolent, then the *Bayesian consequentialist* will let these people decide in her place, as well as in the place of the less informed individuals.

Finally, it's important to stress that the *Bayesian consequentialist* always reasons with uncertainty. More generally, her decision-making process is a classical case of the statistical theory of decision. For starters, she assigns a moral score $good(x)$ to any state x in the world. The more the state x is desirable, the greater $good(x)$ will be. For the *consequential Bayesian*, it will be a moral duty to undertake an action a that maximizes the expected moral score, that is, the quantity $\mathbb{E}_x[good(x)|a]$. Remarkably, up to isomorphism, this approach to morality is the only one to satisfy von Neumann-Morgenstern's axioms, and therefore can never be a victim of the Allais paradox.

In particular, the *Bayesian consequentialist*'s formalism makes it possible to discuss low-risk problems of great catastrophes. Let's imagine two options, ✗ and ✓. And consider three possible consequences: ☻, ☻, and ⚰. Suppose ✗ is to do nothing, and necessarily leads to ☻. And assume that ✓ is the risky action, which can lead to ☻ or ⚰.

The *precautionary principle* is often invoked to prefer ✗. However, ⚰ is potentially very unlikely - and unfortunately, invoking the *precautionary principle* is not always an invitation to calculate the probability of ⚰. However, even if ⚰ is catastrophic, according to the *Bayesian consequentialist*, it can then be rational (or not) to undertake ✓. More formally, the *Bayesian consequentialist* will first assign scores to the different consequences. If ⚰ is really catastrophic, we can imagine that $good(☻) = 0$, $good(☻) = -1$, and $good(⚰) = -10^9$. To determine which action ✗ or ✓ to take, the *Bayesian consequentialist* will then calculate the expected scores of the two possible actions.

---

[15]In practice, as discussed in chapter 5, from an educational point of view, it's also very useful to express one's prejudices out loud in order to be more aware of them and to correct them better. This is typically the case in mathematics learning. However, if the purpose of a debate is decision making, it may be inappropriate to slow down the decision-making process to correct your prejudices.

Doing nothing leads to $\mathbb{E}_x\left[good(x)|\text{✗}\right] = good(\text{☻}) = -1$. What about doing ✓? The response of the *Bayesian consequentialist* is derived from the computation of the expected score when one performs ✓:

$$\mathbb{E}_x\left[good(x)|\text{✓}\right] = good(\text{☺})\mathbb{P}[\text{☺}|\text{✓}] + good(\text{☠})\mathbb{P}[\text{☠}|\text{✓}] = -10^9 P[\text{☠}|\text{✓}].$$

In particular, the *consequentialist Bayesian* will want to do ✓ if and only if $\mathbb{E}_x\left[good(x)|\text{✓}\right] \geq \mathbb{E}_x\left[good(x)|\text{✗}\right]$. This is equivalent to the condition $\mathbb{P}[\text{☠}|\text{✓}] \leq 10^{-9}$. In other words, her decision making will be entirely determined by the probability that the action ✓ causes ☠. Rather than embarking on endless debates, the priority of the *Bayesian consequentialist* will be the estimation of this probability - but also that of the uncertainty on this estimate, or even the undertaking of actions to reduce this uncertainty on the estimation, as well as the estimation of costs that the reduction of uncertainties requires.

I personally think that we all have a lot to learn from this *Bayesian consequentialist*, even if we aspire only to become partially *consequentialist*. In particular, to partially resemble the *Bayesian consequentialist*, it's imperative for us to measure the extent of our ignorance. Unfortunately, when it comes to morals, this is an effort that is not usual for us. I had the opportunity to organize a public debate on the "morals of AIs[16]". While this topic seems to require a great expertise, no participant asked any questions. In this kind of debate, we hasten to impose our point of view, or even to highlight our virtues. And we tend to conclude well before measuring the extent of our ignorance. And before trying to reduce our ignorance.

Unfortunately, as we have seen in this book, our intuitive morals are constantly in blatant overconfidence. Even if we have an ounce of *utilitarian* morals, the inadequacy of our non-Bayesian prejudices often leads us to behaviors that do not increase overall happiness. Worse, our overconfidence prevents us from correcting these prejudices. To become better moral beings, the fight against our overconfidence seems to be a priority. Then, ideally, this should be followed by a familiarization with the social sciences and with Bayes' rule, to refine our credences in the effectiveness of different means to achieve our purposes, *utilitarian* or not.

In particular, this book leads us to an astonishing conclusion: *becoming a good Bayesian is a moral duty.*

---

[16]https://twitter.com/science__4__all/status/983798135431581697.

22.10 LAST WORDS

I can only invite you to meditate at length this amazing conclusion. More generally, I hope that this book will challenge what you thought you knew about morality, logic, and knowledge. I hope it helped you better understand the limitations of the scientific method. I hope that it will also help you question the overconfidence that you are very probably victim of. And I hope it has given you a glimpse of a better way to learn and to know.

The relevance of mathematics and philosophy to address the "real" world is so often rejected, if not mocked. We often talk about the everyday life and the concrete facts as an area that does not require a thesis in mathematics to be understood. This is a serious overconfidence. As John von Neumann said, "if people do not believe that mathematics is simple, it's because they do not realize how complicated life is." On the contrary, our inability to understand the black raven paradox, the choice of the hospital rather than the clinic, and the implications of an exponential growth should force us to doubt everything we think we have understood about the "real" world.

In particular, if your morals are partially consequentialist, Bayesianism should have upset your attachment to your moral principles. Indeed, if the goodness of an action depends (even partially) on its consequences, then we absolutely must predict these consequences, as well as the consequences of alternative actions. However, "it is difficult to make predictions, especially concerning the future," says the adage. I hope I have shown you just how much it was.

We have seen that even Paul Erdös was unable to apply Bayes' rule in extremely basic cases, that Solomonoff's demon had to violate the laws of physics to perform her Bayesian calculations, and that the *pragmatic Bayesian* had to juggle complexity theory, optimal memory management, and MCMC sampling to make a prediction that's reasonably similar to the *pure Bayesian*'s. This should force us to be more humble and to acknowledge the overconfidence that we are so often full of. Especially concerning moral issues, "we must hasten not to conclude".

But this is not the main purpose of this book. As announced in the first chapter, my long epistemological reflections in recent years have made me renounce the scientific method and frequentism. They transformed me into an (aspiring) Bayesian. And then, especially after meeting Solomonoff's demon, into a Bayesian extremist. I hope I have convinced you that the reasons for this conversion are not completely irrational. And I hope I have helped you get a glimpse at the broad lines of the philosophy of knowledge that has won almost all my credences by now.

But most importantly, I hope you enjoyed the journey we had to live to explore the foundations and the consequences of Bayesianism. I hope you have savored the discovery of the many sciences useful for understanding and illustrating Bayesianism, from theoretical computer science to cognitive sciences, and from evolutionary biology to statistical physics. I hope you enjoyed the proof of Ockham's razor, the decortication of the scandal of induction, and the questioning of realism. And I hope that the reading of this book has been for you an exotic journey, perhaps even an initiatory one, of which you shall keep imperishable memories.

Above all, I hope to have overwhelmed you with enthusiasm, fascination, and questioning. This was the main objective of this book.

FURTHER READING

The Myth of the Rational Voter: Why Democracy Always Chooses Bad Policy. Princeton University Press. B Caplan (2007).

Thinking Fast and Slow. Farrar, Straus and Giroux. D Kahneman (2013).

The Righteous Mind: Why Good People Are Divided by Politics and Religion. Vintage. J Haidt (2013).

Superintelligence: Paths, Dangers, Strategies. Oxford University Press. N Bostrom (2014).

Against Democracy. Princeton University Press. J Brennan (2016).

The Big Picture: On the Origin of Life, Meaning and the Universe Itself Dutton. S Carroll (2016).

Models of the Evolution of Fairness in the Ultimatum Game: a Review and Classification. Evolution and Human Behavior. S Debove, N Baumard & JB André (2015).

Predicting Elections: Child's Play! Science. J Antonakis & O Dalgas (2009).

Contextual Priming: Where People Vote Affects How They Vote. PNAS. J Berger, M Meredith & C Wheeler (2008).

Motivated Numeracy and Enlightened Self-Government. Behavioural Public Policy. D Kahan, E Peters, E Cantrell Dawson & P Slovic (2017).

Human-Level Control through Deep Reinforcement Learning. Nature. V Mnih et al. (2015).

New Coke - A Complete Disaster? Company Man (2017).

Why You Think You're Right - Even When You're Wrong. TED. J Galef (2016).

Politics and Numbers. Numberphile. J Grime (2013).

Why Asimov's Laws of Robotics Don't Work. Computerphile. R Miles (2015).

Can You Solve This? Veritasium. D Muller (2014).

The Illusion of Truth. Veritasium. D Muller (2016).

Index

Milton Keynes UK
Ingram Content Group UK Ltd.
UKHW031533071024
449327UK00005B/74

9 780367 428150